Monitoring and Control of Electrical Power Systems Using Machine Learning Techniques

Monitoring and Control of Electrical Power Systems Using Machine Learning Techniques

Edited by

Emilio Barocio Espejo
University of Guadalajara
Guadalajara, Mexico

Felix Rafael Segundo Sevilla
Zurich University of Applied Science
Zurich, Switzerland

Petr Korba
Zurich University of Applied Science
Zurich, Switzerland

ELSEVIER

Elsevier
Radarweg 29, PO Box 211, 1000 AE Amsterdam, Netherlands
The Boulevard, Langford Lane, Kidlington, Oxford OX5 1GB, United Kingdom
50 Hampshire Street, 5th Floor, Cambridge, MA 02139, United States

ISBN: 978-0-323-99904-5

For information on all Elsevier publications
visit our website at https://www.elsevier.com/books-and-journals

Publisher: Joseph Hayton
Acquisitions Editor: Rachel Pomery
Editorial Project Manager: Maria Elaine Desamero
Production Project Manager: Maria Bernard
Cover Designer: Greg Harris

Typeset by VTeX

Contents

Contributors

César Angeles-Camacho
National Autonomous University of Mexico, Mexico City, Mexico

Mario Roberto Arrieta-Paternina
National Autonomous University of Mexico, Mexico City, Mexico

Alexandre Balbinot
Federal University of Rio Grande do Sul, Porto Alegre, Brazil

Emilio Barocio Espejo
University of Guadalajara, Guadalajara, Mexico

Ernesto Beltran
University of Guadalajara, Guadalajara, Mexico

Ramón J. Betancourt
University of Colima, Colima, Mexico

Christoph Brosinsky
Technische Universität Ilmenau, Ilmenau, Germany

Daniele Carta
Forschungszentrum Jülich GmbH, Julich, Germany

Gabriela Castillo-García
National Autonomous University of Mexico, Mexico City, Mexico

Oswaldo Isaac Cortes Robles
University of Guadalajara, Guadalajara, Mexico

Jochen L. Cremer
Delft University of Technology, Delft, Netherlands

José Antonio de la O Serna
Universidad Autónoma de Nuevo León, Monterrey, Mexico

Dunstano del Puerto Flores
University of Guadalajara, Guadalajara, Mexico

José Guadalupe Fuentes-Velázquez
National Autonomous University of Mexico, Mexico City, Mexico

Luis Alonso Trujillo Guajardo
Autonomous University of Nuevo León, San Nicolás de los Garza, Mexico

Daniel Guillen
Tecnológico de Monterrey, Monterrey, Mexico

Julio Cesar Hernández Ramírez
Autonomous University of San Luis Potosí, San Luis Potosí, Mexico

Mert Karaçelebi
Delft University of Technology, Delft, Netherlands

Eleftherios O. Kontis
Aristotle University of Thessaloniki, Thessaloniki, Greece

Petr Korba
Zurich University of Applied Sciences, Winterthur, Switzerland

Roberto Leborgne
Federal University of Rio Grande do Sul, Porto Alegre, Brazil

Gabriel Mejía-Ruiz
National Autonomous University of Mexico, Mexico City, Mexico

Theofilos A. Papadopoulos
Democritus University of Thrace, Xanthi, Greece

Grigoris K. Papagiannis
Aristotle University of Thessaloniki, Thessaloniki, Greece

Paolo Attilio Pegoraro
University of Cagliari, Cagliari, Italy

Ana Karen Apolo Peñaloza
Federal University of Rio Grande do Sul, Porto Alegre, Brazil

Juan M. Ramirez
Center for Research and Advanced Studies of the National Polytechnic Institute, Mexico City, Mexico

Juan Segundo Ramírez
Autonomous University of San Luis Potosí, San Luis Potosí, Mexico

Juan Sebastián Rocha-Doria
National Autonomous University of Mexico, Mexico City, Mexico

Ramón Daniel Rodríguez-Soto
University of Guadalajara, Guadalajara, Mexico

Fernando Salinas Salinas
Autonomous University of Nuevo León, San Nicolás de los Garza, Mexico

Antonio Concha Sánchez
University of Colima, Colima, Mexico

Betsy Sandoval Guzmán
University of Guadalajara, Guadalajara, Mexico

Felix Rafael Segundo Sevilla
Zurich University of Applied Sciences, Winterthur, Switzerland

Antonio Vincenzo Solinas
University of Cagliari, Cagliari, Italy

Sara Sulis
University of Cagliari, Cagliari, Italy

Nitin Sundriyal
Center for Research and Advanced Studies of the National Polytechnic Institute, Mexico City, Mexico

Mazheruddin H. Syed
University of Strathclyde, Glasgow, United Kingdom

Vicente Torres-García
National Autonomous University of Mexico, Mexico City, Mexico

Alejandro Zamora-Méndez
Michoacan University of Saint Nicholas of Hidalgo, Morelia, Mexico

Preface

Machine learning techniques have found their way into applications in the field of electrical power systems in the twenty-first century. The main reason is the strong pressure from politics in all developed countries around the globe to reach carbon neutrality in the near future. In many countries, there is a strong push to be independent of fossil fuels and, at the same time, there is a nuclear phase-out. This leads to a massive integration of volatile renewable energy sources into the existing grids. This has a strong impact on the stable operation of the electric power system, which is becoming a challenging problem. It has already become evident that we may not always be able to handle this challenge yet. To be more specific, we remind the reader of the recent incidents, such as the separation of the Continental European power system in January 2021 (which was successfully mitigated in the end), the great Texas blackout of 2021, and other similar incidents.

Ensuring security and quality of power supply is essential for our modern society. Thanks to fast power electronics, information, and communication technologies developed and available recently. New insights into all grid levels across the entire power system can be obtained by means of properly analyzed fast sampled and often time-synchronized measurements. More specifically, we experience a massive roll out of smart meters in low-voltage grids and installations of phasor measurement units in every substation of the high-voltage grid worldwide. These technologies yield a big amount of data, which can be used to monitor, analyze, and to detect the presence and the source of faults. Automatic control loops can be designed to speed up the appropriate action required to secure the power supply; this can hardly be done fast enough manually by a human operator.

This book gives an overview of applications for monitoring and control of electric power systems using machine learning techniques. It has been our intension to bridge the gap between theory and practice. We trust that we have done so in this text by presenting guiding principles developed to solve practical problems encountered in distribution and transmission grids in our research projects and to demonstrate the achieved functionality on concrete examples.

This book is intended for students and researchers interested in advanced methods for monitoring, control, and optimization of power quality using machine learning techniques and engaging engineers from the utilities and the power system industry.

Experts from all over the globe have been recruited to contribute to this book. It would not have been possible to come up with this book without the diligent work and support of several international research teams from Switzerland, Mexico, Italy, Greece, Germany, Brazil, and the UK; many thanks to all of them for their valuable contributions to the monitoring and control of power systems and to this volume. The book covers advanced applications and solutions for monitoring and control of electrical power systems in transmission and distribution, provides deep insight into power system dynamics, power quality disturbance detection and classification through machine learning, deep learning, and spatio-temporal algorithms. Supplementary online components and datasets for machine learning training processes and open-source grid models are included on GitHub.

December 2022

Emilio Barocio
Guadalajara, Mexico
Rafael Segundo
Winterthur, Switzerland
Petr Korba
Zürich, Switzerland

1 Derivation of generic equivalent models for distribution network analysis using artificial intelligence techniques

Eleftherios O. Kontis[a], Theofilos A. Papadopoulos[b], Mazheruddin H. Syed[c], and Grigoris K. Papagiannis[a]
[a]Aristotle University of Thessaloniki, Thessaloniki, Greece, [b]Democritus University of Thrace, Xanthi, Greece, [c]University of Strathclyde, Glasgow, United Kingdom

1.1 Introduction

Traditionally, power systems analysis is performed by using detailed models [1], [2]. However, the penetration of distributed energy resources (DERs), the use of new types of loads, e.g., non-linear power electronic loads, electric vehicles, etc., the need for more operational flexibility and the application of advanced voltage and frequency control strategies prevent system operators to develop and maintain accurate and up-to-date system models [3]. Moreover, detailed information of the network structure, network assets, and control parameters of DERs is rarely available to system operators [4]. Thus specific power system buses, e.g., at key substations and feeders, are represented by equivalent models, simulating the aggregated behavior of the downstream network components (lines, transformers, etc.), the actual load, and the possibly existing DERs [5].

Equivalent modeling approaches are generally categorized into component- and measurement-based [5]. The component-based or knowledge-based approach is a bottom-up methodology; thus implies the detailed knowledge of the various network components, hindering its practical application [6], [7]. On the other hand, the measurement-based or behavior-based modeling approach, as a top-down methodology, is used to derive equivalent models with/without prior knowledge of the actual system, i.e., gray- /black-box equivalent models, respectively [7]. This method does not require detailed network knowledge; instead relies on field measurements to derive the model parameters by applying identification techniques. During the last decades with the maturing of smart grid technologies and the improvement on data capture and storage capabilities, this approach has become more popular and has been adopted for the modeling of active distribution network (ADNs) and microgrids (MGs) par excellence.

Monitoring and Control of Electrical Power Systems Using Machine Learning Techniques
https://doi.org/10.1016/B978-0-32-399904-5.00007-7

Irrespective of the approach used for the model development, aggregated models can be grouped into static and dynamic [8], [9], [10]. Static equivalents express the real and reactive power at the connecting bus of interest at any time instant as a function of the voltage magnitude and/or frequency at that specific time instant [6]. The most known static models are the exponential, polynomial, linear, comprehensive, static induction motor, and power electronic-interfaced models [5]. Such models can be used for cases exhibiting near instantaneous time changes in power, following a voltage and/or frequency deviation, i.e., steady-state studies. They can be also used if the interest is on the new steady-state, rather than on the initial transient, e.g., long-term voltage stability studies. On the other hand, for dynamic studies involving small-signal and transient stability [4], frequency analysis [11], and short-term voltage stability [12], dynamic equivalents exhibiting time-dependent responses are preferred. They are formulated in terms of difference or differential equations to express the real and reactive power at any time instant with respect to voltage and/or frequency [6]. Although static models are not appropriate substitutes for dynamic studies, about 70% of power system operators worldwide still use static load models to conduct contingency analysis and power system stability studies [13]. This issue becomes more critical particularly in cases of high penetration of induction motors, e.g., residential air conditioners, or DERs [5].

A common drawback of the equivalent modeling methods [7], [8], [9], [10], [11] lies in the fact that their parameters are valid only for the operating condition from which they have been originally derived. This is because of their high dependency on the distribution network (DN) operating conditions (voltage level, network configuration, etc.), the load stochastic behavior, and the intermittent nature of DERs [4], [14]. To determine generic model parameters for measurement-based dynamic equivalent models, several approaches have been proposed in literature. In [14], [15], and [16], statistical analysis parameter identification is applied, by assigning the generic model parameters with the corresponding median or mean values obtained under different operating conditions. Other multisignal processing analysis techniques have been also introduced in [4], [17], [18], [19], [20], where the generic model parameters are derived in the least-square sense [17–20], or by adopting linear approximation functions [21], [22]. Although, these approaches extend the applicability of the equivalent models, their robustness is limited within the range of the available training data set, thus cannot predict new conditions beyond the original observation range [23]. Another significant issue is that human intervention is required to process and classify the available data into groups of similar characteristics, e.g., data groups obtained under similar network conditions [16], [17], or determine representative records.

The emergence of artificial intelligence (AI) has introduced new flexible, self-learning and automated techniques to develop generic black-box models. In these models, the DN is represented by a single artificial neural network (ANN) utilizing a large amount of data [24], [25]. However, such black-box models have no physical meaning and mathematical formulation background; thus they do not provide any insight into the system dynamics [23] and are practically infeasible, as they cannot be integrated into simulation environments (commercial software, etc.) and interact with other modules. In this context, other AI-based approaches have been proposed by deploying static or dynamic equivalent model formulations. In particular, in [15] generic models for conventional equivalent models, i.e., ZIP and exponential recovery (first- and second-order) are developed by using ANNs. Similarly, in [11] support vector clustering is adopted to derive generic parameters for a transfer function-based equivalent model. In [23], clustering and ANN are combined to derive robust sets of parameters for a variable-order equivalent model [26]. First, clustering is used to automatically divide the available set of recorded data into groups with similar characteristics. At the next stage, for each cluster the model parameters are estimated by applying the vector-fitting technique and are extended to a wide range of operating conditions via ANN modeling. Clustering is also used in [27] to classify data into clusters presenting similar dynamic behavior; at a second stage, representative model parameters of a discrete transfer function model are calculated in terms of nonlinear least squares (NLS) optimization. Similarly to [23] and [27], in [7] clustering is used to determine the optimal number of representative operating conditions and a long short-term memory (LSTM) recurrent neural network (RNN) is used to estimate the parameters of composite dynamic equivalent model. Finally, in [28], a combination of optimization-based estimation algorithms and AI techniques are adopted to obtain adaptable generic models of exponential recovery, composite, and first-order functions-based. In this work, two decision trees are used to relate a given disturbance with one of the examined models (classification) and with a set of parameters that best fit the selected model with the disturbance characteristics (prediction), respectively.

Based on the above analysis, it can be deduced that several conventional and AI-based approaches have been proposed in the literature for the derivation of generic sets of equivalent model parameters. However, their accuracy and robustness has not been systematically investigated and evaluated. Therefore this chapter formulates in the MATLAB® environment [29] six generalization approaches (based on the findings and recommendations of [15], [16], [23] and [27]) that can be used to derive robust/generic equivalent models for DN analysis. To provide didactic examples to the reader, the examined generalization approaches (GAs) are used to derive robust/generic sets of parameters for a static as well as a dynamic equivalent model. The former can be used for the steady-state analysis of DNs and the latter to represent DNs in stability studies. The performance of all the examined GAs is thoroughly tested using measurements obtained from a laboratory scale DN, consisting of static loads, induction motors, synchronous generators, and converter-interfaced DERs. This way, a comparative assessment of all examined GAs under practical conditions is conducted. The presented analysis provides a comprehensive assessment and insights concerning the distinct advantages and drawbacks of all methods. Finally, it is worth noting that all MATLAB codes used to implement the examined equivalent models and the examined GAs and all laboratory measurements used to evaluate the performance of the examined methods, are freely available at [30].

The rest of the chapter is organized as follows: In Section 1.2, the structure of the equivalent models, adopted to provide didactic examples to the reader, is presented alongside with the parameter estimation procedure. In Section 1.3, the examined GAs are described in detail. In Section 1.4, the performance of the examined GAs is evaluated using laboratory measurements. Section 1.5 briefly presents the MATLAB codes used for the implementation of the examined approaches. Finally, Section 1.6 discusses the main findings of the research and concludes the chapter.

1.2 Examined equivalent models

In this section, the structure of the adopted equivalent models is described and the parameter estimation procedure is briefly explained.

1.2.1 Structure of the examined equivalent models

Two representative equivalent models for DN analysis are examined. The first one is a static equivalent model, and the second one is a first-order dynamic counterpart. The steady-state model can be used to represent DNs in steady-state studies, and the dynamic model can be used to simulate DNs in stability studies. Concerning the static model, the well-established polynomial (ZIP) model is considered [5], [10]. The examined dynamic equivalent is a modified version of the conventional exponential recovery model (ERM) [31]. As discussed in [26], the conventional ERM fails to simulate and analyze reverse power flow phenomena that may occur in ADNs. Thus the modified ERM (MERM) of [26] is used to facilitate the analysis of these phenomena. The mathematical formulation as well as further details for both equivalent models are provided in the following subsections.

1.2.1.1 The ZIP model

The polynomial model is also referred as the ZIP model, since it consists of constant impedance (Z), constant current (I), and constant power (P) components. The ZIP model is described by the following equations:

$$P_{ZIP} = P_0\left[p_1\left(\frac{V}{V_0}\right)^2 + p_2\left(\frac{V}{V_0}\right) + p_3\right], \tag{1.1}$$

$$Q_{ZIP} = Q_0\left[q_1\left(\frac{V}{V_0}\right)^2 + q_2\left(\frac{V}{V_0}\right) + q_3\right] \tag{1.2}$$

with

$$\sum_{i=1}^{3} p_i = 1, \tag{1.3}$$

$$\sum_{i=1}^{3} q_i = 1. \tag{1.4}$$

Here, P_{ZIP} and Q_{ZIP} denote the real and reactive power estimates, provided by the ZIP model, respectively. V is the network voltage and V_0 is the pre-disturbance voltage magnitude. P_0 and Q_0 are the predisturbance total real and reactive power consumption, respectively. Parameters p_1 and q_1 represent the relative participation of constant impedance component; parameters p_2, q_2 and p_3, q_3 represent the relative participation of constant current and constant power components, respectively [5]. The relative participation of each component (Z,I,P) in the total mix lies in the range between 0 and 1 p.u., giving rise to their overall sum being equal to 1 p.u. This variant of the model is called the constrained ZIP model [5]. However, it may fail to provide accurate results, especially in networks hosting DERs. In these cases, the so-called accurate ZIP model can be used. In this variant, individual p_i and q_i parameters can be less than 0 p.u. and/or larger than 1 p.u. Once again, the sum of all p_i parameters (and the sum of all q_i parameters) should be equal to 1. Although the parameters of this variant do not have any physical meaning, the model can be more accurate compared to the constrained ZIP counterpart [5]. In this work, only the accurate ZIP model is considered.

1.2.1.2 The modified ERM

Real and reactive power is expressed by the MERM using the following equations [26]:

$$P_{MERM} = P_s - (P_s - P_t) \cdot e^{-\frac{t}{T_p}} \tag{1.5}$$

$$Q_{MERM} = Q_s - (Q_s - Q_t) \cdot e^{-\frac{t}{T_q}}, \tag{1.6}$$

where

$$P_s = P_0\left[\alpha_{p1}\left(\frac{V}{V_0}\right) + \alpha_{p2}\right], \quad \alpha_{p1} + \alpha_{p2} = 1 \tag{1.7}$$

$$P_t = P_0\left[\beta_{p1}\left(\frac{V}{V_0}\right) + \beta_{p2}\right], \quad \beta_{p1} + \beta_{p2} = 1. \tag{1.8}$$

$$Q_s = Q_0\left[\alpha_{q1}\left(\frac{V}{V_0}\right) + \alpha_{q2}\right], \quad \alpha_{q1} + \alpha_{q2} = 1 \tag{1.9}$$

$$Q_t = Q_0\left[\beta_{q1}\left(\frac{V}{V_0}\right) + \beta_{q2}\right], \quad \beta_{q1} + \beta_{q2} = 1. \tag{1.10}$$

Here, P_{MERM} and Q_{MERM} denote the real and reactive power estimates provided by the MERM. P_s and Q_s are two first-order polynomial functions used to describe the steady-state behavior of real and reactive power, respectively. Parameters α_{p1}, α_{p2} are the polynomial coefficients of P_s, whereas α_{q1}, and α_{q2} are the polynomial coefficients of Q_s. Additionally, P_t and Q_t are two first-order polynomial functions used to describe the transient behavior of real and reactive power, respectively. P_t is simulated using parameters β_{p1} and β_{p2}, whereas Q_t is approximated using coefficients β_{q1} and β_{q2}. Finally, T_p and T_q are the recovery time constants of real and reactive power responses, respectively [26].

1.2.2 Parameter estimation

The parameters of the examined equivalent models are estimated in terms of NLS optimization [32], [33], using the trust-region-reflective algorithm [29]. The aim of the optimization is to minimize via successive iterations the objective function of (1.11):

$$J = \sum_{n=1}^{N} \left(y[n] - \hat{y}[n]\right)^2. \tag{1.11}$$

Here, N is the total number of the available samples. $y[n]$ is the actual (measured) real or reactive power response of the examined DN at the n-th sample. $\hat{y}[n]$ is the estimated (by the adopted equivalent model) response at the n-th sample.

For the ZIP model, $\hat{y}[n]$ is approximated by using (1.1) and (1.2). In this case, constraints of (1.3) and (1.4) should be considered. Concerning the MERM, $\hat{y}[n]$ is approximated by using (1.5) and (1.6), taking into account (1.7)–(1.10). For both equivalents, no upper and lower bounds for the model parameters are considered.

1.3 Derivation of generic equivalent models

To develop generic equivalents, the methodology of Fig. 1.1 is adopted [23]. More specifically, a database is initially developed, containing M data sets. Each data set $i = \{1, .., M\}$ is comprised of three equally sized vectors $\boldsymbol{V}_i$, $\boldsymbol{P}_i$, and $\boldsymbol{Q}_i$ and of two auxiliary variables denoted as LM_i and GM_i. $\boldsymbol{V}_i$ corresponds to the line-to-line RMS time domain response of voltage at the point of interconnection (POI) with the external power system. Moreover, $\boldsymbol{P}_i$ and $\boldsymbol{Q}_i$ denote the total (three-phase) RMS time domain responses of real and reactive power at the POI. $\boldsymbol{V}_i$, $\boldsymbol{P}_i$, and $\boldsymbol{Q}_i$ practically reflect the dynamic behavior of the DN during the i-th disturbance; they are recorded during voltage disturbances/events. Variables LM_i and GM_i are used to reflect the load composition and the type of the DERs connected to the DN during the i-th disturbance. As discussed in [23], the values of LM_i and GM_i can be defined either using forecasts or ideally using smart meter recordings.

Once the database has been created, preprocessing is applied, aiming to determine predisturbance steady-state values of voltage ($V_{0,i}$), real ($P_{0,i}$), and reactive power ($Q_{0,i}$) for the M available data sets. These values are used to provide further insight concerning the DN operating conditions prior to the i-th disturbance. Subsequently, the M data sets are randomly divided into two groups. The first group contains M_D data sets and is used to derive generic parameters for the examined equivalents, i.e., model parameters that can be used to analyze efficiently a wide range of network conditions. Generic parameters are derived by adopting the two-step procedure, described below:

- **Step-1**: *Parameter Estimation*. The available dynamic responses of voltage, real, and reactive power are used to identify parameters for the examined equivalent models. In particular, for each one of the M_D data sets and for each equivalent

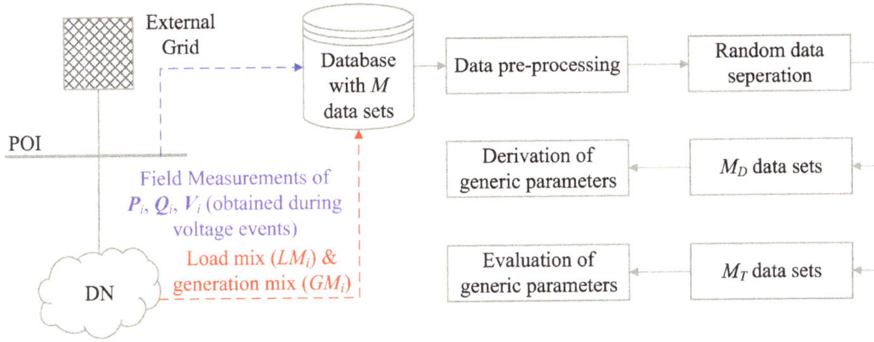

Figure 1.1 Concept for the derivation of generic equivalent models for DN analysis.

model, a dedicated set of parameters is identified. More specifically, for each data set, the NLS optimization of (1.11) is performed to estimate the equivalent model parameters. Concerning the ZIP model, sets $\boldsymbol{\theta}_{p,j}^{ZIP} = [p_1, p_2, p_3]$ and $\boldsymbol{\theta}_{q,j}^{ZIP} = [q_1, q_2, q_3]$ are identified for the j-th data set ($j = \{1, .., M_D\}$). $\boldsymbol{\theta}_{p,j}^{ZIP}$ is used to simulate the real power behavior, and $\boldsymbol{\theta}_{q,j}^{ZIP}$ is utilized to analyze the reactive power behavior. Concerning the MERM, the modeling of real and reactive power responses is achieved using sets $\boldsymbol{\theta}_{p,j}^{MEMR} = [\alpha_{p1}, \alpha_{p2}, \beta_{p1}, \beta_{p2}, T_p]$ and $\boldsymbol{\theta}_{q,j}^{MEMR} = [\alpha_{q1}, \alpha_{q2}, \beta_{q1}, \beta_{q2}, T_q]$, respectively.

- **Step-2**: *Application of a Generalization Approach.* A generalization approach is applied, aiming to derive generic sets of parameters for the examined equivalent models. Six GAs are considered and compared. All approaches are described in detail in the next subsection.

1.3.1 Examined GAs

The examined GAs include the following: i) methods based on statistical analysis, ii) methods based on clustering and statistical analysis, iii) methods based on ANNs, and iv) methods based on clustering and ANNs. A more in depth analysis follows.

1.3.1.1 *Generalization via statistical analysis*

In this approach, generic model parameters are identified using statistical analysis. In particular, the generic parameters are determined as the mean values of the identified model parameters. It is worth noting that this approach is proposed by CIGRE for the derivation of robust/generic equivalents [5].

1.3.1.2 *Generalization via clustering techniques and statistical analysis*

In this approach, clustering is applied to automatically divide the available training data into a small number of groups, which present similar characteristics. Clustering can be performed either based on the predisturbance operating conditions [23], i.e., based on $V_{0,j}$, $P_{0,j}$, and $Q_{0,j}$, or based on the real and reactive power dynamic responses after a disturbance [27], i.e., based on $\boldsymbol{P}_j$, and $\boldsymbol{Q}_j$.

To apply the clustering based on the predisturbance operating conditions, the following procedure is adopted [23]: $V_{0,j}$, $P_{0,j}$, and $Q_{0,j}$ are combined for each one of the available M_D training data sets into a single vector $\boldsymbol{X}_j = [V_{0,j}, P_{0,j}, Q_{0,j}]$; thus the set $\boldsymbol{X} = \{\boldsymbol{X}_1, \boldsymbol{X}_2, ..., \boldsymbol{X}_{M_D}\}$ is created. Subsequently, $\boldsymbol{X}$ is forwarded as input to a clustering algorithm, i.e., the k-means++. k-means++ categorizes the M_D available training data into L ($\leq M_D$) clusters $\boldsymbol{OC} = \{OC_1, OC_2, ..., OC_L\}$, aiming to minimize the within cluster sum of squares [23], as shown in (1.12).

$$\underset{\boldsymbol{OC}}{\arg\min} \sum_{l=1}^{L} \sum_{X_j \in OC} ||\boldsymbol{X}_j - \boldsymbol{\mu}_l||^2. \tag{1.12}$$

Here $\boldsymbol{\mu}_l$ is the mean value of operating conditions contained in the l-th cluster.

Alternatively, to apply clustering in terms of the available dynamic responses, the following procedure is proposed: $\boldsymbol{P}_j$, and $\boldsymbol{Q}_j$ are combined for each one of the available M_D training data sets into a single vector $\boldsymbol{S}_j = [\boldsymbol{P}_j, \boldsymbol{Q}_j]$; thus the set $\boldsymbol{S} = \{\boldsymbol{S}_1, \boldsymbol{S}_2, ..., \boldsymbol{S}_{M_D}\}$ is created. Subsequently, $\boldsymbol{S}$ is forwarded as input to the k-means++ algorithm, which categorizes the M_D dynamic responses into D ($\leq M_D$) clusters $\boldsymbol{DR} = \{DR_1, DR_2, ..., DR_D\}$ aiming to minimize the within cluster sum of squares [23], as shown in (1.13).

$$\underset{\boldsymbol{DR}}{\arg\min} \sum_{d=1}^{D} \sum_{\boldsymbol{S}_j \in DR} ||\boldsymbol{S}_j - \boldsymbol{\mu}_d||^2. \tag{1.13}$$

Here $\boldsymbol{\mu}_d$ is the average dynamic behavior, i.e., average real and reactive power dynamic response of the data contained in the d-th cluster.

To derive the optimal number of clusters (i.e., L or D), the knee-point criterion for the curve of the within cluster sum of squares to between cluster sum variation ratio (WCBCR) is used. Following this criterion, the optimal number of clusters is defined by the knee of the WCBCR curve [34].

Once the clusters have been determined, generic parameters for the examined equivalents are derived as follows: For each cluster, a statistical analysis of models parameters is performed. Eventually, for each cluster and for each one of the examined models, the mean values of all parameters are identified and used as generic sets of parameters.

1.3.1.3 Generalization using ANNs

In this generalization approach, the available M_D training data sets are used for the development of two-layer feed-forward ANNs. The aim of the ANNs is to capture general relationships between the parameters of the examined equivalent models and the operating conditions of the DN. Using this information, robust sets of parameters can be derived [23]. For each one of the examined equivalent models, two discrete

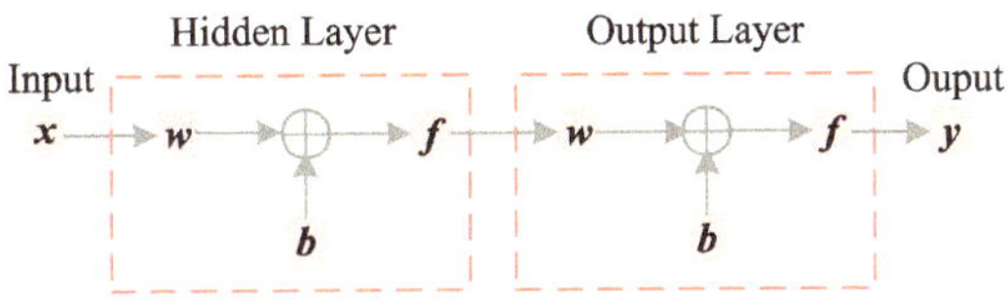

Figure 1.2 Structure of the proposed ANNs.

ANNs are developed. The first one is dedicated to the modeling of real power and the second one to the modeling of reactive power. The structure of the proposed ANNs is presented in Fig. 1.2. As shown, the ANN structure consists of a hidden and an output layer. The input/output relationship in both the output and the hidden layer is defined as

$$y = f(\boldsymbol{W}^T \boldsymbol{x} + \boldsymbol{b}). \tag{1.14}$$

Here, $\boldsymbol{y}$ is the output matrix, and $\boldsymbol{w}$ and $\boldsymbol{b}$ are the weight and bias matrices, respectively. f is a transfer function (TF) that produces the final outcome of the layer [35]. The TF for the output layer should be the purelin (linear) function to allow the outputs of the ANN to acquire any finite value [23], [35]. Concerning the hidden layer, the TF can be either the *tan-sigmoid* (*tansig*) or the *log-sigmoid* (*logsig*) function [23], [35]. The input matrix $\boldsymbol{x}$ of the proposed ANNs include the predisturbance steady-state values of voltage, real and reactive power at the POI, i.e., $V_{0,j}$, $P_{0,j}$, and $Q_{0,j}$, as well as the load and the generation mix of the examined DN. The load mix ($\boldsymbol{LM}_j$) is expressed using an 1X2 vector. The first element of this vector is used to denote the total power consumed by dynamic loads, e.g., motors; the second element is used to express the total power consumed by static loads. Additionally, the generation mix ($\boldsymbol{GM}_j$) is expressed using an 1X3 vector. The first, second, and third element of this vector denote the power generated by synchronous machines, asynchronous generators, and converter interfaced units, respectively. The input matrix for both the ZIP and the MERM is written as

$$\boldsymbol{x} = \begin{bmatrix} P_{0,1} & \cdots & P_{0,j} & \cdots & P_{0,M_{DT}} \\ Q_{0,1} & \cdots & Q_{0,j} & \cdots & Q_{0,M_{DT}} \\ V_{0,1} & \cdots & V_{0,j} & \cdots & V_{0,M_{DT}} \\ \boldsymbol{LM}_1 & \cdots & \boldsymbol{LM}_j & \cdots & \boldsymbol{LM}_{M_{DT}} \\ \boldsymbol{GM}_1 & \cdots & \boldsymbol{GM}_j & \cdots & \boldsymbol{GM}_{M_{DT}} \end{bmatrix}. \tag{1.15}$$

Here, M_{DT} denotes the number of data sets used for the training of the ANNs. Further information concerning how the M_{DT} data sets are defined are provided in the paragraphs that follow.

The targets ($\boldsymbol{t}$) of the proposed ANNs differ based on the examined equivalent model. The targets for the ANNs that are considered for the generalization of the ZIP model parameters are summarized in (1.16) and (1.17). In particular, (1.16) describes the target matrix for the modeling of real power by the ZIP model and (1.17) describes the target matrix for the modeling of reactive power.

$$t_p^{ZIP} = \begin{bmatrix} p_{1,1} & \cdots & p_{1,j} & \cdots & p_{1,M_{DT}} \\ p_{2,1} & \cdots & p_{2,j} & \cdots & p_{2,M_{DT}} \end{bmatrix} \tag{1.16}$$

$$t_q^{ZIP} = \begin{bmatrix} q_{1,1} & \cdots & q_{1,j} & \cdots & q_{1,M_{DT}} \\ q_{2,1} & \cdots & q_{2,j} & \cdots & q_{2,M_{DT}} \end{bmatrix}. \tag{1.17}$$

Note that the following sets of parameters: $\boldsymbol{\theta_{p3}} = [\theta_{p3,1}, ..., \theta_{p3,j}, ...\theta_{p3,M_{DT}}]$ and $\boldsymbol{\theta_{q3}} = [\theta_{q3,1}, ..., \theta_{q3,j}, ...\theta_{q3,M_{DT}}]$, are not included in target matrices $\boldsymbol{t_p^{ZIP}}$ and $\boldsymbol{t_q^{ZIP}}$, respectively. This approach is adopted since $\boldsymbol{\theta_{p3}}$ and $\boldsymbol{\theta_{q3}}$ can be directly computed from (1.3) and (1.4), respectively, using sets of parameters $\boldsymbol{\theta_{p1}} = [\theta_{p1,1}, ..., \theta_{p1,j}, ...\theta_{p1,M_{DT}}]$, $\boldsymbol{\theta_{p2}} = [\theta_{p2,1}, ..., \theta_{p2,j}, ...\theta_{p2,M_{DT}}]$, $\boldsymbol{\theta_{q1}} = [\theta_{q1,1}, ..., \theta_{q1,j}, ...\theta_{q1,M_{DT}}]$, and $\boldsymbol{\theta_{q2}} = [\theta_{q2,1}, ..., \theta_{q2,j}, ...\theta_{q2,M_{DT}}]$.

Targets for the ANNs referring to the generalization of the MERM parameters are summarized in (1.18) and (1.19). Specifically, (1.18) describes the target matrix for the modeling of real power and (1.19) describes the target matrix for the modeling of reactive power.

$$t_p^{MERM} = \begin{bmatrix} \alpha_{p1,1} & \cdots & \alpha_{p1,j} & \cdots & \alpha_{p1,M_{DT}} \\ \beta_{p1,1} & \cdots & \beta_{p1,j} & \cdots & \beta_{p1,M_{DT}} \\ T_{p,1} & \cdots & T_{p,j} & \cdots & T_{p,M_{DT}} \end{bmatrix} \tag{1.18}$$

$$t_p^{MERM} = \begin{bmatrix} \alpha_{q1,1} & \cdots & \alpha_{q1,j} & \cdots & \alpha_{q1,M_{DT}} \\ \beta_{q1,1} & \cdots & \beta_{q1,j} & \cdots & \beta_{q1,M_{DT}} \\ T_{q,1} & \cdots & T_{q,j} & \cdots & T_{q,M_{DT}} \end{bmatrix}. \tag{1.19}$$

Sets $\boldsymbol{\theta_{\alpha_{p2}}} = [\alpha_{p2,1}, ..., \alpha_{p2,j}, ...\alpha_{p2,M_{DT}}]$ and $\boldsymbol{\theta_{\beta_{p2}}} = [\beta_{p2,1}, ..., \beta_{p2,j}, ...\beta_{p2,M_{DT}}]$ are omitted from target matrix $\boldsymbol{t_p^{MERM}}$, since they can be directly computed from (1.7) and (1.8), respectively, using $\boldsymbol{\theta_{\alpha_{p1}}} = [\alpha_{p1,1}, ..., \alpha_{p1,j}, ...\alpha_{p1,M_{DT}}]$ and $\boldsymbol{\theta_{\beta_{p1}}} = [\beta_{p1,1}, ..., \beta_{p1,j}, ...\beta_{p1,M_{DT}}]$. Similarly, $\boldsymbol{\theta_{\alpha_{q2}}} = [\alpha_{q2,1}, ..., \alpha_{q2,j}, ...\alpha_{q2,M_{DT}}]$ and $\boldsymbol{\theta_{\beta_{q2}}} = [\beta_{q2,1}, ..., \beta_{q2,j}, ...\beta_{q2,M_{DT}}]$ are not included in target matrix $\boldsymbol{t_q^{MERM}}$.

During the training phase, ANNs iteratively adjust their weights and biases to minimize the following mean square error (*mse*):

$$mse = \frac{1}{E}\sum_{e=1}^{E}\left(t(e) - y(e)\right)^2. \tag{1.20}$$

Here, E denotes the total number of elements included in the corresponding target ($\boldsymbol{t}$) and output ($\boldsymbol{y}$) matrices.

To avoid overfitting problems, the early stopping technique is used [23], [35]. To implement this technique, the M_D available data sets are randomly divided into two discrete groups. The first group contains M_{DT} data, and the second group contains M_{DV} data. The first group is the training data set used to update weights and biases of the ANNs. The second one constitutes the validation data set, and it is used to assess the accuracy of the ANN in new conditions, i.e., conditions not used during the training phase. The mse is computed for both the validation and the training data sets. The corresponding mse is denoted as $vmse$ and $tmse$, respectively. During the initial stage of the training procedure, both $tmse$ and $vmse$ decrease. However, in case the ANN starts overfitting the training data, the $vmse$ increases. In cases where the $vmse$ increases for six consecutive iterations, then the training is terminated to avoid overfitting issues. Weights and biases of the ANNs are automatically set to the values corresponding to the minimum achieved validation error [35].

1.3.1.4 Generalization using clustering techniques and ANNs

In this generalization approach, the available M_D data sets are initially classified into different clusters. The clustering can be performed either based on the predisturbance operating conditions or based on the available dynamic responses of real and reactive power. Afterwards, for each cluster, dedicated ANNs are developed to generalize the parameters of the examined equivalents. All ANNs are developed based on the principles discussed in subsection 1.3.1.3.

1.3.1.5 Summary of the considered GAs

On the basis on the previous analysis, six GAs (GA) are considered and compared; a summary of them is provided below. The implementation of the examined GAs is also depicted in Fig. 1.3 by means of flowcharts.

- GA#1: In this approach, generic sets of parameters are identified by means of the corresponding mean values, i.e., mean values of the identified parameters. GA#1 constitutes the most simple generalization approach.
- GA#2: To derive generic sets of parameters, the following procedure is applied: Initially, the available data are classified into clusters. The clustering is performed using (1.12), i.e., it is performed based on the pre-disturbance operating conditions. Finally, for each cluster representative (generic) parameters are identified by means of the corresponding mean values, i.e., mean values of parameters that belong to each cluster.
- GA#3: This approach is similar to GA#2. However, in GA#3 the clustering is performed based on real and reactive power responses. Thus (1.13) is used to categorize the available data into clusters.
- GA#4: In this approach, generic sets of parameters are derived using ANNs. The structure of the ANNs is presented in subsection 1.3.1.3.
- GA#5: In this approach, generic sets of parameters are derived as follows: Initially, available data are classified into several clusters. The clustering is performed based on the predisturbance operating conditions. The clustering is performed via (1.12). Subsequently, for each cluster dedicated ANNs are developed aiming to capture general relationships between model parameters and the operating conditions of the DN.
- GA#6: This approach is similar to GA#5. However, in this case, the clustering of the available data is performed via (1.13).

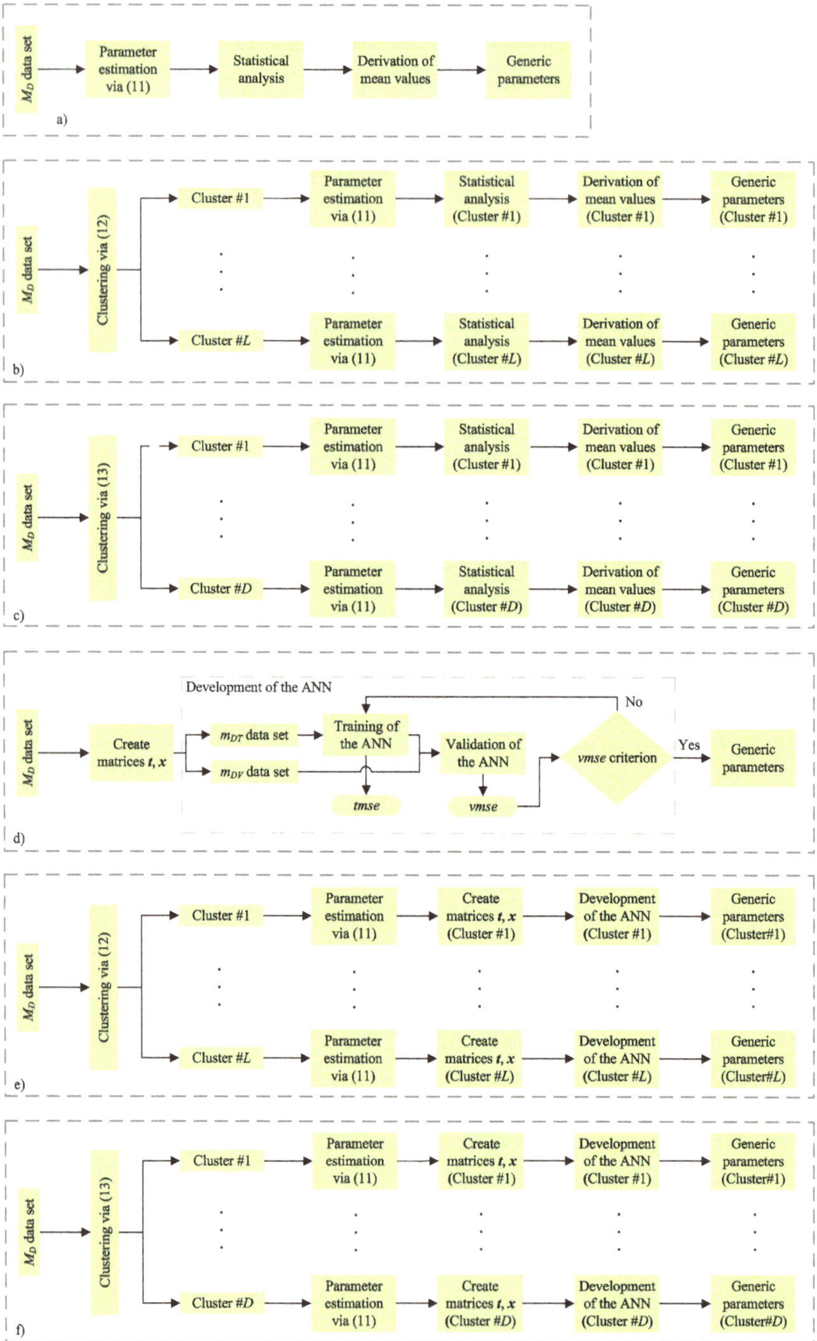

Figure 1.3 Overview of the examined GAs: a) GA#1, b) GA#2, c) GA#3, d) GA#4, e) GA#5, and f) GA#6.

1.3.2 Evaluation metrics

The performance of the examined approaches is evaluated (cross-validated) using the remaining M_T data sets. For this purpose, the generic parameters, identified by each one of the examined GAs, are used to regenerate (estimate) active and reactive power responses of the M_T data sets. ZIP estimations for real and reactive power are computed via (1.1) and (1.2), respectively, using as input the corresponding voltage response. Concerning MERM, real and reactive power responses are estimated using (1.5) and (1.6), respectively.

To evaluate the performance of the derived generic models, two indexes (metrics) are used, namely the root mean square error ($RMSE$) [23] and the steady-state relative error (SSE) [32]. $RMSE$ is used to assess the accuracy of the estimated dynamic responses; SSE is used to quantify the percentage error between the estimated and the actual new steady-state. $RMSE$ and SSE are formally defined as

$$RMSE = \sqrt{\frac{1}{N}\sum_{n=1}^{N}\Big(y(n) - \hat{y}(n)\Big)^2}, \tag{1.21}$$

$$SSE(\%) = \left|\frac{y_{ss} - \hat{y}_{ss}}{y_{ss}}\right| \cdot 100\%, \tag{1.22}$$

where y_{ss} and $\hat{y}_{ss}$ are the actual and the estimated new steady-state powers after the examined disturbance, respectively.

1.4 Methods evaluation using laboratory measurements

The performance of the examined GAs is evaluated in this section using laboratory measurements.

1.4.1 Description of the lab setup

A number of experiments was carried out within the Dynamic power systems laboratory[1,2] (DPSL) at the University of Strathclyde to generate the data required for the validation of the examined GAs. The DPSL microgrid offers a flexible 400 V, 50 Hz reconfigurable three-phase network. A single line diagram of the network, utilized for the experiments, is presented in Fig. 1.4. The network is divided into two sub-grids energized through a three-phase bidirectional power converter operating as a programmable voltage source (PVS).

Sub-grid #1 comprises a 2 kVA synchronous generator (SG), a 5.5 kVA asynchronous motor (AM1), and a 10 kW/7.5 kVAR static load bank (SLB). The SG is

[1] https://www.ulabequipment.com/facility/dynamicpowersystems.
[2] https://pureportal.strath.ac.uk/en/equipments/dynamic-power-systems-laboratory.

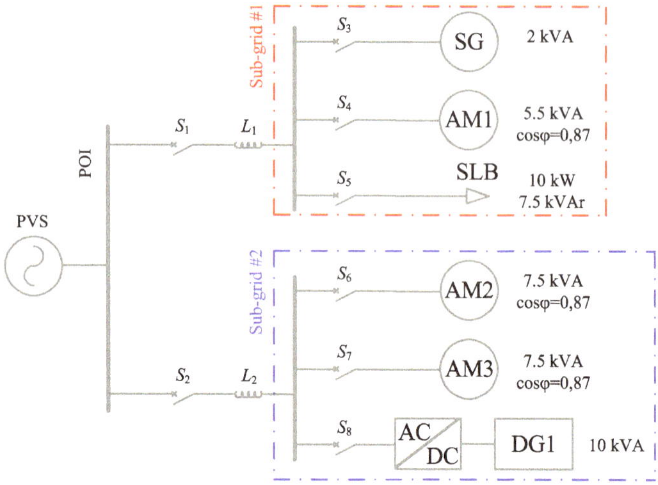

Figure 1.4 Examined lab configuration.

representative of synchronous generation within DN operating on active power (P) vs. frequency (f) and reactive power (Q) vs. voltage (V) droops. The SLB has 64-steps to control its active power, offering control of reactive power through power factor set point (0.5 to 1). The AM1 is operated as a load, representative of rotating loads within the DN, with full control over its torque.

Sub-grid #2 comprises two asynchronous motors and a 10 kVA inverter. The two asynchronous machines, AM2 and AM3, are both rated at 7.5 kVA with a lagging power factor of 0.87. The two machines are operated as generators, with control of their output from 0 to rated power. The 10 kVA inverter is representative of power electronic interfaced DERs and is operated as a current source controlling its active and reactive power through a set point.

Switches S_1 (for sub-grid #1) and S_2 (for subgrid #2) offer the capability to switch in/out the entire subgrids, whereas switches S_3–S_8 enable the connection/disconnection of each component individually, allowing the analysis of a variety of passive DN and ADN configurations. To investigate different operational conditions, the voltage at the POI is varied between 360 V (0.9 pu) to 440 V (1.1 pu) using the PVS. Voltage disturbances, varying between −0.1 p.u. and 0.1 p.u., are introduced to investigate DN dynamics. Voltages and currents at the POI are recorded with a sampling rate of 500 samples per second (sps), and they are used to calculate the dynamic responses of active and reactive power at the POI.

Using the above-mentioned lab setup, a total number of 500 scenarios, representing discrete DN configurations, operating conditions, and voltage disturbances was generated. The recorded data were used to evaluate the performance of the examined generalization approaches.

1.4.2 Parameter identification and derivation of generic models

Initially, available data are used to estimate parameters for the examined equivalents. An instance of the parameter estimation procedure is presented in Fig. 1.5. In this specific case, a voltage step-down disturbance with magnitude equal to −4% is considered. As shown, the ZIP model fails to analyze the dynamic behavior of real and reactive power. However, the ZIP model can efficiently estimate the new steady-states. Indeed, the corresponding SSE for the modeling of real and reactive power is merely 1.02% and 3.34%, respectively. This reveals that the ZIP model can be used for the steady-state analysis of DNs. The MERM efficiently analyzes the dynamic behavior of both real and reactive powers. Thus it can be used to analyze the dynamic properties of the examined lab-scale DN.

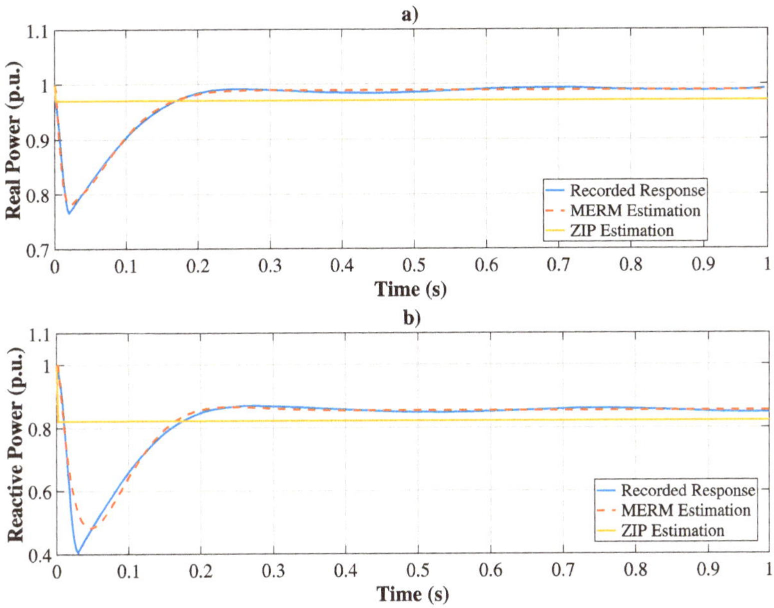

Figure 1.5 Instance of the parameter estimation procedure. a) Real and b) reactive power.

The variability of all the identified model parameters across the available training data set is demonstrated in Fig. 1.6. Results indicate that parameters of the ZIP model and, especially, parameters of the MERM vary significantly across the examined scenarios. Therefore the examined GAs are used to derive generic/robust sets of parameters for the considered equivalents.

To apply clustering-based GAs, the optimal number of clusters should be initially defined. For this purpose, the WCBCR index is calculated for both clustering approaches (clustering based on the predisturbance operating conditions and clustering based on the resulting dynamic responses of real and reactive power). The computed WCBCR indexes are presented in Fig. 1.7. As shown according to the knee-point cri-

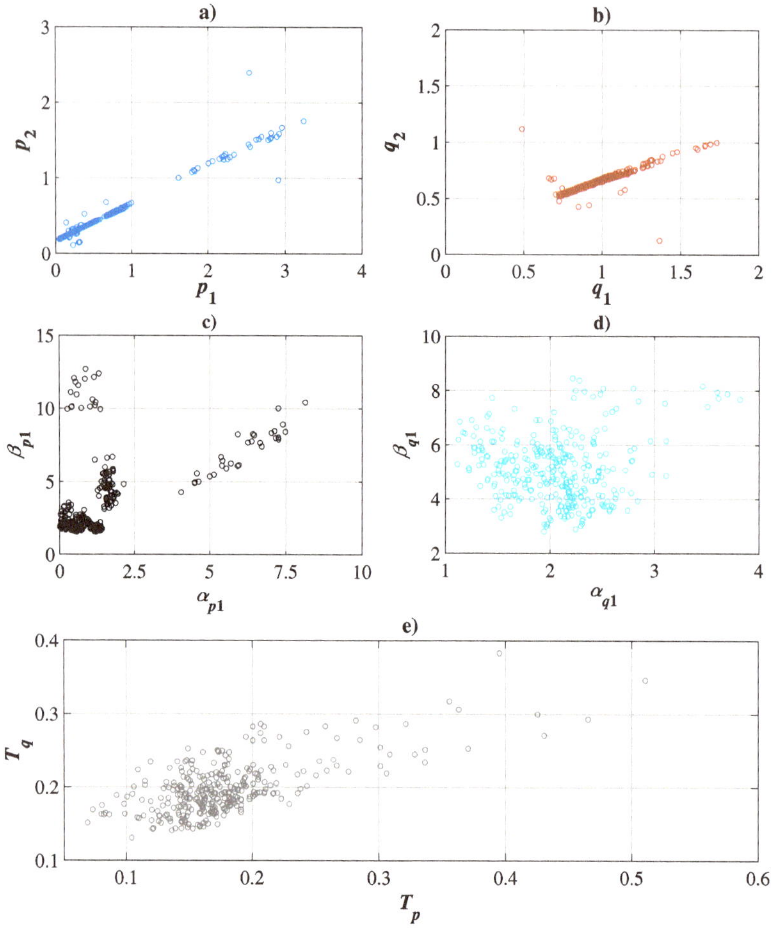

Figure 1.6 Variability of model parameters. ZIP: a) Parameters for the modeling of real power, and b) parameters for the modeling of reactive power. MERM: c) Polynomial coefficients for the modeling of real power; d) polynomial coefficients for the modeling of reactive power; e) time constants for the modeling of real and reactive power.

terion, a number of four clusters, i.e., $L=D=4$, is required for both the examined approaches.

To apply ANN-based GAs, the following parameters should be determined: i) the TF for the hidden layer. As already discussed, the available options are the *tansig* and the *logsig* functions. ii) The optimal number of neurons per layer. iii) The algorithm for the training of ANNs. Three training algorithms, which are compatible with the early stopping technique, are examined, namely the Levenberg–Marquardt (LM), the conjugate gradient back propagation (SCG), and the resilient back propagation (RP).

To define the abovementioned parameters, a parametric analysis is performed, assuming all possible combinations of TFs and training methods. Moreover, the number

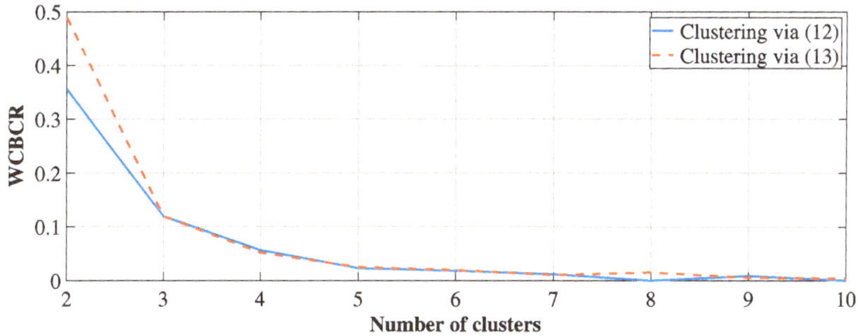

Figure 1.7 WCBCR index for different number of clusters and for discrete clustering approaches.

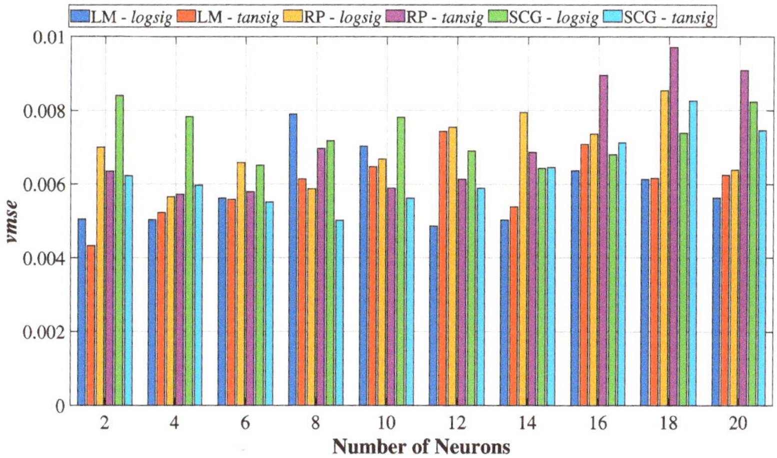

Figure 1.8 Modeling of real power via the ZIP model. Impact of several parameters on *vmse*.

of neurons is considered variable from 2 up to 20. For each possible combination of number of neurons, TF, and training algorithm, a set of 100 discrete initial conditions for the weights and bias matrices is randomly generated using the Monte Carlo (MC) method. The mean values of the *vmse*, as resulted from the MC simulations, are plotted in Figs. 1.8 – 1.11.

The corresponding results reveal that the LM is the most appropriate algorithm for the training of the ANNs, and the *tansig* is the most appropriate TF for the hidden layers of the developed ANNs. The number of neurons depends on the structure of the examined equivalent model. Indeed, as shown in Figs. 1.8 and 1.9, a number of 2 neurons is adequate for the generalization of the ZIP model parameters. Results of Figs. 1.10 and 1.11 indicate that a number of at least 8 neurons is required for the generalization of the MERM parameters.

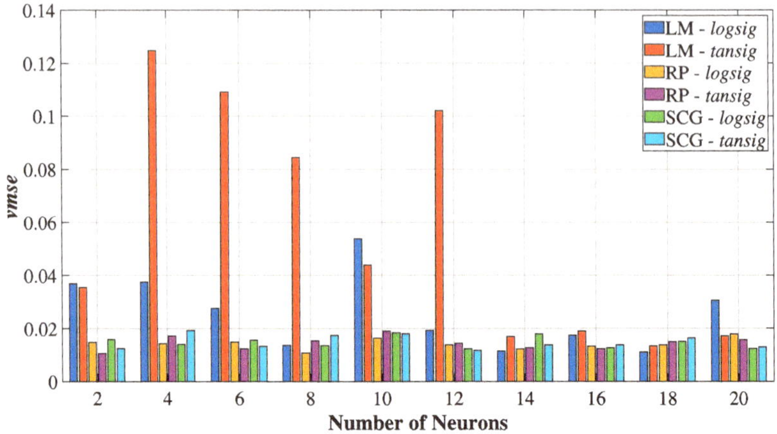

Figure 1.9 Modeling of reactive power via the ZIP model. Impact of several parameters on $vmse$.

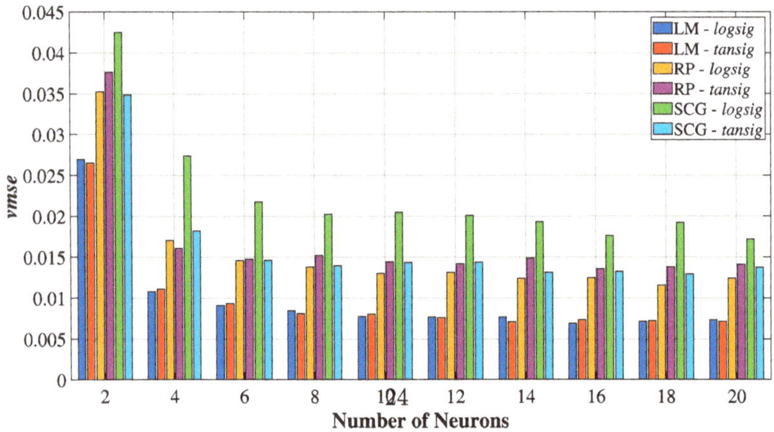

Figure 1.10 Modeling of real power via the MERM. Impact of several parameters on $vmse$.

1.4.3 Evaluation results

To thoroughly evaluate the performance of the examined GAs, the MC method is applied. In particular, the following procedure is adopted: The available laboratory measurements are randomly separated into M_D and M_T data sets. This random separation is performed 100 times. This way, 100 discrete MC simulations are preformed. In all MC simulations, M_D= 425 data sets (with M_{DT}=350 and M_{DV}=75), whereas M_T=75 data sets. In each MC simulation, the M_D data sets are used to derive generic sets of parameters, and the M_T data sets are used to cross-validate the performance of the examined GAs. For each MC simulation, $RMSE$ and SSE metrics are computed for each one of the available M_T data sets. Eventually, the mean value of the adopted indexes is computed for each case, i.e., mean values across the M_T data. As already

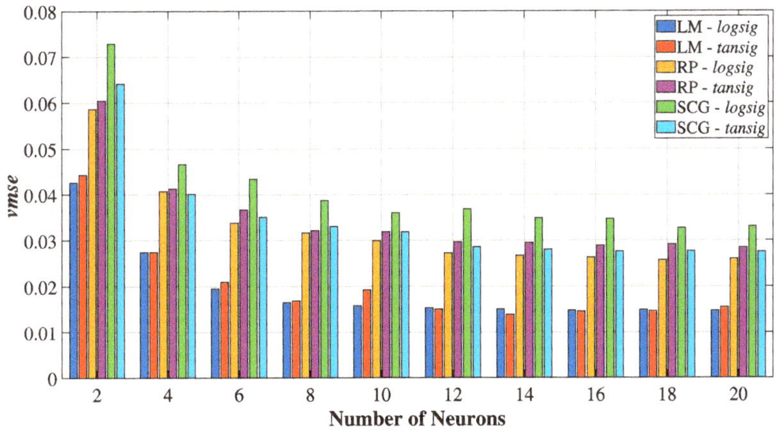

Figure 1.11 Modeling of reactive power via the MERM. Impact of several parameters on $vmse$.

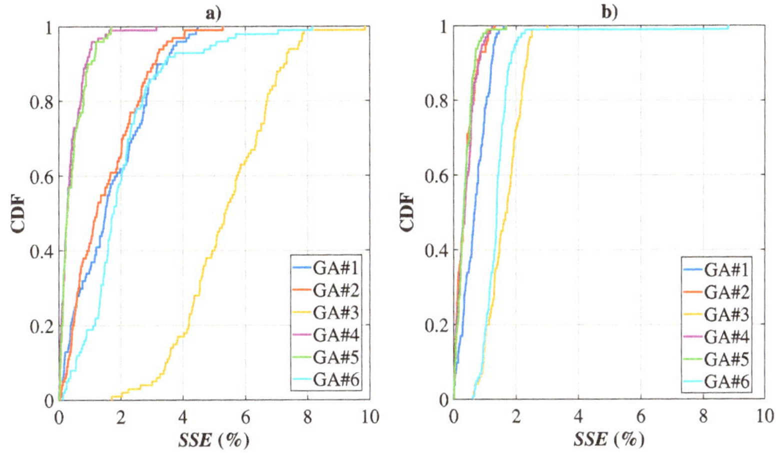

Figure 1.12 Modeling of the steady-state behavior of the DN using the ZIP model. SSE for a) real and b) reactive power.

discussed, the ZIP model can be used for the steady-state analysis of DNs and the MERM to analyze the dynamic behavior of DNs. Therefore the performance of the ZIP model is evaluated by using the SSE metric, whereas the accuracy of the MERM is quantified in terms of the $RMSE$.

In Fig. 1.12, the ability of the ZIP model to analyze the new steady-state of the examined DN after a disturbance is evaluated by means of cumulative distribution functions (CDFs). As shown, GAs that utilize clustering based on dynamic responses, i.e., GA#3 and GA#6, provide the worst SSE. This is evident for both real and reactive power modeling. Indeed, ZIP model is not able to capture DN dynamics, thus clustering based on dynamic responses cannot provide accurate and reliable generic

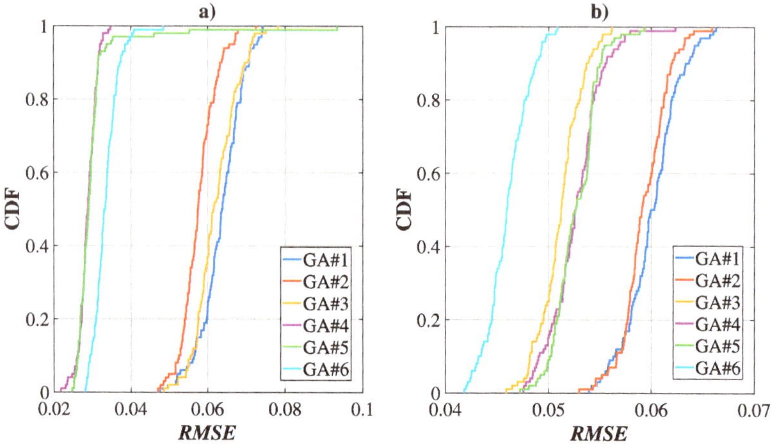

Figure 1.13 Modeling of the dynamic behavior of the DN using the MERM model. $RMSE$ for a) real and b) reactive power.

parameters. On the other hand, GAs that use clustering techniques based on predisturbance operating conditions and ANNs provide the most accurate results, i.e., the lowest SSE. Indeed, for those methods, the SSE for the modeling of both real and reactive powers is in all cases lower than 2%.

In Fig. 1.13, the ability of the MERM to analyze the dynamic behavior of the examined DN is evaluated by means of the $RMSE$. Considering both real and reactive power responses, results indicate that generalizations approaches based on clustering techniques and ANNs provide the most accurate estimates, resulting in the lowest $RMSE$ values. In this case, clustering based on dynamic responses results in more accurate estimates compared to all other examined approaches. However, it is worth noting that these methods are not suitable for online (real-time) applications. Indeed, GA#3 and GA#6 require the dynamic responses of real and reactive power to identify the most appropriate cluster for the derivation of generic parameters for the analysis of the DN. Nevertheless, these responses are available only after the occurrence of system disturbances. Thus the system operator is not able to assess a priori the dynamic behavior of the DN, i.e., prior to the occurrence of a disturbance. On the other hand, clustering methods that determine generic parameters via pre-disturbance operating conditions can support real-time applications. Indeed, in these approaches system operator can assess a priori the dynamic behavior of the DN by analyzing the steady-state operating conditions of the DN. This analysis can be performed using real-time measurements.

1.5 Implementation of the examined GAs using Matlab

All codes, used to implement the examined GAs have been developed using the MATLAB software [29], and they are freely available to the interested reader through the repository of [30]. The codes can be used as didactic examples for the development of new AI-based generic equivalent modeling approaches. In the same repository, the reader can also find all laboratory measurements used to evaluate the performance of the examined GAs. These laboratory measurements can serve as benchmark responses for the development of new generalized equivalent models in the future.

1.5.1 Matlab files for the derivation of generic equivalent models

The MATLAB files, included in the repository of [30] and required for the development of the examined GAs, are summarized below:

- MATLAB function "Apply_ANN": used for the development of the examined ANNs.
- MATLAB file named as "Clustering": used for the implementation of the examined clustering methods.
- MATLAB function "errperf": used to compute the adopted evaluation metrics.
- MATLAB function "eval_fun": used to compute the WCBCR index.
- MATLAB function "MERM": used to identify parameters for the MERM.
- MATLAB function "MERM_response": used to estimate the response of the MERM under a known voltage disturbance. Appropriate, robust/generic MERM parameters should be used.
- MATLAB function "ZIP": used to identify parameters for the ZIP model.
- MATLAB function "ZIP_response": used to estimate the response of the ZIP model under a known voltage disturbance. Appropriate, robust/generic ZIP parameters should be used.

Using the abovementioned MATLAB files, the reader can develop all the GAs examined in this chapter. It is worth noting that further instructions for the use of all the MATLAB functions are provided within the corresponding code in the form of comments.

1.5.2 Lab measurements

All lab measurements used for the training and the validation of the examined GAs are stored in the .mat file named "Recorded_responses." Instructions for the processing and the interpretation of the measurements can be found in the "Notes.txt" file.

1.5.3 Didactic example

In [30], a didactic example is also provided. More specifically, the reader shall download the zipped folder (called "codes") and extract its content. Subsequently, the reader can launch the example by typing "Implementation_example" at the MATLAB command window.

In the provided implementation example, GA#1 and GA#4 are considered and compared. Both GAs are used to derive generic sets of parameters for the ZIP model. For both GAs, a set of 100 MC simulations is performed to statistically analyze their performance. In each MC simulation, the available data/responses are randomly separated into two discrete groups. The first group is used for the training of the GAs, and the second group is used to cross-validate their performance. The main outputs of this didactic example are the ZIP model parameters across the examined test cases (lab experiments) and the evaluation metrics across the examined MC simulations. Additionally, CDFs of the SSE for the modeling of real and reactive power across the 100 MC simulations are automatically plotted.

1.6 Conclusions

In this chapter, six GAs for the derivation of generic equivalent models for DN analysis are formulated and compared. The examined GAs include the following: i) methods based on statistical analysis, ii) methods based on clustering techniques and statistical analysis, iii) methods based on ANNs, and iv) methods that use both clustering techniques and ANNs. The MERM and the ZIP models are used as typical examples of equivalent models to demonstrate the use of the examined GAs. The performance of all the examined methods is assessed using measurements acquired from a laboratory-scale DN.

The conducted analysis reveals that the ZIP model can be used only for the steady-state analysis of DNs, whereas the MERM can be used to simulate DN dynamic behavior. Concerning the examined GAs, the corresponding validation results indicate that the conventional method, which utilizes statistical analysis to derive generic parameters, exhibits the poorest performance. On the other hand, methods that utilize clustering techniques and two-layer feed-forward ANNs provide the most generic equivalent models, resulting in the lowest modeling errors. Clustering can be performed either based on the DN dynamic responses or based on the DN predisturbance operating conditions. The latter approach can support online applications, i.e., it can be used to assess the DN behavior during and after a disturbance by simply using steady-state measurements that are always available to the system operator. Thus it is proposed as the most promising option for the online derivation of generic equivalent models.

Finally, it is worth noting that all MATLAB codes used to implement the examined GAs and the laboratory measurements used to evaluate the performance of the examined methods are freely available at [30]. This way, the interested reader can reproduce the results of the presented analysis. These codes can also serve as didactic examples for the development of new, more sophisticated, AI-based modeling approaches for the analysis of DNs.

Acknowledgment

This research is co-financed by Greece and the European Union (European Social Fund – ESF) through the Operational Programme "Human Resources Development, Education and Lifelong Learning" in the context of the project "Reinforcement of Postdoctoral Researchers – 2nd Cycle" (MIS-5033021), implemented by the State Scholarships Foundation (IKY). This work was also part funded by the European Union's Horizon 2020 research and innovation programme under grant agreement No 870620 – ERIGrid 2.0 project.

References

[1] Task Force on Identification of Electromechanical Modes. Identification of Electromechanical Modes in Power Systems, Technical Report PES-TR15, IEEE Power & Energy Society, Piscataway, NJ, USA, 2019.
[2] V.V. Feng Ma, Right-sized power system dynamic equivalents for power system operation, IEEE Transactions on Power Systems 26 (4) (2011) 1998–2005.
[3] W.K.A. Ishchenko, J.M.A. Myrzik, Dynamic equivalencing of distribution networks with dispersed generation using Hankel norm approximation, IET Generation, Transmission & Distribution 1 (2007) 818–825, (7).
[4] E.O. Kontis, G.A. Barzegkar-Ntovom, K.A. Staios, T.A. Papadopoulos, G.K. Papagiannis, A toolbox for analyzing and testing mode identification techniques and network equivalent models, Energies 12 (13) (2019).
[5] Cigre Working Group C4.605, Modelling and Aggregation of Loads in Flexible Power Networks, Technical Report, 2014.
[6] E.O. Kontis, T.A. Papadopoulos, A.I. Chrysochos, G.K. Papagiannis, Measurement-based dynamic load modeling using the vector fitting technique, IEEE Transactions on Power Systems 33 (1) (2018) 338–351.
[7] P. Wang, Z. Zhanga, Q. Huang, X. Tang, W.-J. Lee, Robustness-improved method for measurement-based equivalent modeling of active distribution network, IEEE Transactions on Industry Applications 57 (3) (2021) 2146–2155.
[8] A. Arif, Z. Wang, J. Wang, B. Mather, H. Bashualdo, D. Zhao, Load modeling—a review, IEEE Transactions on Smart Grid 9 (6) (2018) 5986–5999.
[9] Load representation for dynamic performance analysis (of power systems), IEEE Transactions on Power Systems 8 (2) (1993) 472–482.
[10] Bibliography on load models for power flow and dynamic performance simulation, IEEE Transactions on Power Systems 10 (1) (1995) 523–538.
[11] H. Golpîra, H. Seifi, M.R. Haghifam, Dynamic equivalencing of an active distribution network for large-scale power system frequency stability studies, IET Generation, Transmission & Distribution 9 (2015) 2245–2254, (9).
[12] L. Robitzky, D. Mayorga-Gonzalez, C. Kittl, C. Strunck, J. Zwartscholten, S. Muller, U. Hager, J. Myrzik, C. Rehtanz, Impact of active distribution networks on voltage stability of electric power systems, in: Proceedings of the IREP Bulk Power Systems Dynamics and Control Symposium, 2017, pp. 1–6.
[13] J.V. Milanović, K. Yamashita, S. Martínez Villanueva, S.Z. Djokic, L.M. Korunović, International industry practice on power system load modeling, IEEE Transactions on Power Systems 28 (3) (2013) 3038–3046.
[14] S. Mat Zali, J.V. Milanović, Generic model of active distribution network for large power system stability studies, IEEE Transactions on Power Systems 28 (3) (2013) 3126–3133.

[15] K.S. Metallinos, T.A. Papadopoulos, C.A. Charalambous, Derivation and evaluation of generic measurement-based dynamic load models, Electric Power Systems Research 140 (2016) 193–200.
[16] D.P. Stojanović, L.M. Korunović, J. Milanović, Dynamic load modelling based on measurements in medium voltage distribution network, Electric Power Systems Research 78 (2) (2008) 228–238.
[17] B.-K. Choi, H.-D. Chiang, Y. Li, H. Li, Y.-T. Chen, D.-H. Huang, M. Lauby, Measurement-based dynamic load models: derivation, comparison, and validation, IEEE Transactions on Power Systems 21 (3) (2006) 1276–1283.
[18] H. Renmu, M. Jin, D. Hill, Composite load modeling via measurement approach, IEEE Transactions on Power Systems 21 (2) (2006) 663–672.
[19] I.F. Visconti, D.A. Lima, J.M.C.d.S. Costa, N.R.d.B.C. Sobrinho, Measurement-based load modeling using transfer functions for dynamic simulations, IEEE Transactions on Power Systems 29 (1) (2014) 111–120.
[20] G. Chaspierre, G. Denis, P. Panciatici, T. Van Cutsem, An active distribution network equivalent derived from large-disturbance simulations with uncertainty, IEEE Transactions on Smart Grid 11 (6) (2020) 4749–4759.
[21] P.N. Papadopoulos, T.A. Papadopoulos, P. Crolla, A.J. Roscoe, G.K. Papagiannis, G.M. Burt, Black-box dynamic equivalent model for microgrids using measurement data, IET Generation, Transmission & Distribution 8 (5) (2014) 851–861.
[22] E.O. Kontis, A.I. Chrysochos, G.K. Papagiannis, T.A. Papadopoulos, Development of measurement-based generic load models for dynamic simulations, in: 2015 IEEE Eindhoven PowerTech, 2015, pp. 1–6.
[23] E.O. Kontis, T.A. Papadopoulos, M.H. Syed, E. Guillo-Sansano, G.M. Burt, G.K. Papagiannis, Artificial-intelligence method for the derivation of generic aggregated dynamic equivalent models, IEEE Transactions on Power Systems 34 (4) (2019) 2947–2956.
[24] J. Denecke, I. Erlich, Dynamic equivalents of active distribution networks, in: 2017 IEEE Power Energy Society General Meeting, 2017, pp. 1–5.
[25] C. Zheng, S. Wang, Y. Liu, C. Liu, W. Xie, C. Fang, S. Liu, A novel equivalent model of active distribution networks based on lstm, IEEE Transactions on Neural Networks and Learning Systems 30 (9) (2019) 2611–2624.
[26] E.O. Kontis, T.A. Papadopoulos, A.I. Chrysochos, G.K. Papagiannis, On the applicability of exponential recovery models for the simulation of active distribution networks, IEEE Transactions on Power Delivery 33 (6) (2018) 3220–3222.
[27] G. Mitrentsis, H. Lens, Data-driven dynamic models of active distribution networks using unsupervised learning techniques on field measurements, IEEE Transactions on Smart Grid 12 (4) (2021) 2952–2965.
[28] L. Chávarro-Barrera, S. Pérez-Londoño, J. Mora-Flórez, An adaptive approach for dynamic load modeling in microgrids, IEEE Transactions on Smart Grid 12 (4) (2021) 2834–2843.
[29] The Mathworks, Inc., Natick, Massachusetts, Matlab version 9.5.0.944444 (R2018b), 2018.
[30] https://power.ee.auth.gr/en/downloads/.
[31] D. Karlsson, D.J. Hill, Modelling and identification of nonlinear dynamic loads in power systems, IEEE Transactions on Power Systems 9 (1) (1994) 157–166.
[32] G. Barzegkar-Ntovom, T. Papadopoulos, E.O. Kontis, Robust framework for online parameter estimation of dynamic equivalent models using measurements, IEEE Transactions on Power Systems 36 (3) (2021) 2380–2389.

[33] Y. Ge, A.J. Flueck, D.-K. Kim, J.-B. Ahn, J.-D. Lee, D.-Y. Kwon, An event-oriented method for online load modeling based on synchrophasor data, IEEE Transactions on Smart Grid 6 (4) (2015) 2060–2068.
[34] G.J. Tsekouras, N.D. Hatziargyriou, E.N. Dialynas, Two-stage pattern recognition of load curves for classification of electricity customers, IEEE Transactions on Power Systems 22 (3) (2007) 1120–1128.
[35] H. Demuth, M. Beale, Matlab Neural Network Toolbox User's Guide, Version 4.

Disturbance dataset development for machine-learning-based power quality monitoring in distributed generation systems: a practical guide

2

Oswaldo Isaac Cortes Robles[a], Emilio Barocio Espejo[a], Juan Segundo Ramírez[b], and Julio Cesar Hernández Ramírez[b]
[a]University of Guadalajara, Guadalajara, Mexico, [b]Autonomous University of San Luis Potosí, San Luis Potosí, Mexico

2.1 Relevance of the disturbance dataset in the training/assessing of machine-learning-based models for power quality monitoring

Nowadays, because of the constant electrical power systems (EPS) evolution, and the great technology growth, data plays a major role in the analysis of the EPS. In this regard, machine-learning-based-models have proved to be effective in pattern recognition to predict, classify, cluster, relevant information about the system status. However, the model training is a crucial task that directly depends on the training dataset. Moreover, system data have been affected by the decentralization of the traditional EPS produced by the penetration of distributed generation sources, bringing challenges in the development of datasets for the training and testing of the machine-learning-based models.

2.1.1 Power quality monitoring in distributed generation systems

In recent years, the electrical power systems (EPSs) have undergone a significant change in their structure because of the increasing penetration of distributed generation sources, producing distributed generation systems (DGSs) (see Fig. 2.1). Thus notwithstanding the great advantages of the DGSs, such as the system power reliability, the located generation, transmission loss reduction, among others, the distributed generation scheme also has brought new challenges to the analysis and control.

Among the challenges related to the DGSs is the operating settings of generators, schedule of the batteries, and controllable loads to offset the changing loads, and

Monitoring and Control of Electrical Power Systems Using Machine Learning Techniques
https://doi.org/10.1016/B978-0-32-399904-5.00008-9

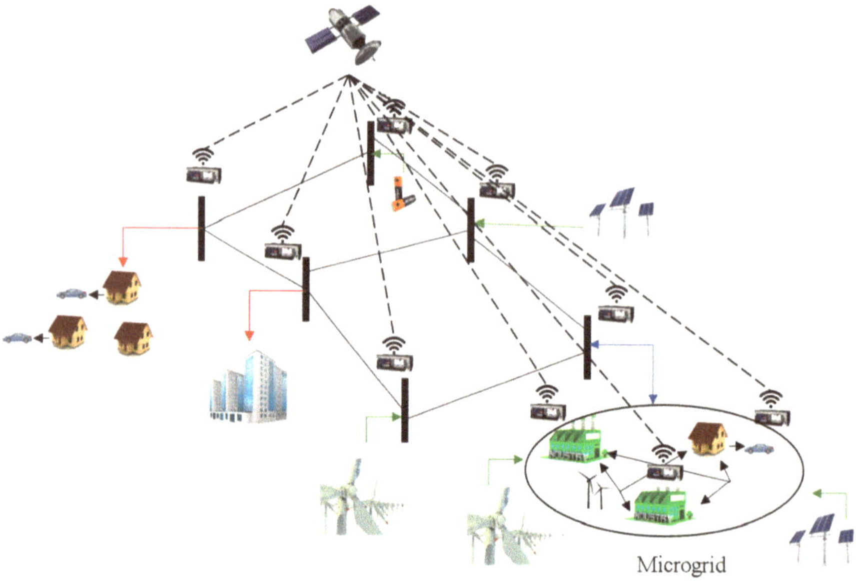

Figure 2.1 Distributed generation systems scheme.

consequently, maintaining the power balance. Besides, the DGSs schemes often incorporate electronic control devices, producing undesirable variations in the power, current, and voltage waveforms, leading to power quality (PQ) issues, which affect the electrical protection settings, produces overheat and failure of the electrical equipment, and increases the economic losses [1]. A critical requirement is then to provide information about the generated power and of the PQ of the DGS. Therefore PQ monitoring is a key activity for the analysis and control of the DGSs [2], which require dedicated metering equipment to measure and process the electrical signals through advanced mathematical models and signal processing techniques to facilitate the interpretation of the system operator, and therefore, to carry out effective control actions. Hence, PQ monitoring should be fast, robust, and resilient against multiscale disturbances in the measured signals.

Recently, monitoring techniques as point-on-wave (POW) has received wide attention because of its GPS synchronized sampling and waveform points identification [3], [4], [5]. POW allows to monitor and record waveforms with accurate time stamp, and carry out an analysis based on the measured synchrophasor data. In this regard, a crucial activity of PQ monitoring is the detection and classification of disturbances that affect the electric waveforms. For this aim, in the last decades, researchers have directed their effort to the development of PQ disturbance classification methods based on machine learning because of their proven highly effective classification performance [6]. Reported machine-learning-based supervised strategies can employ traditional and deep learning classification methods.

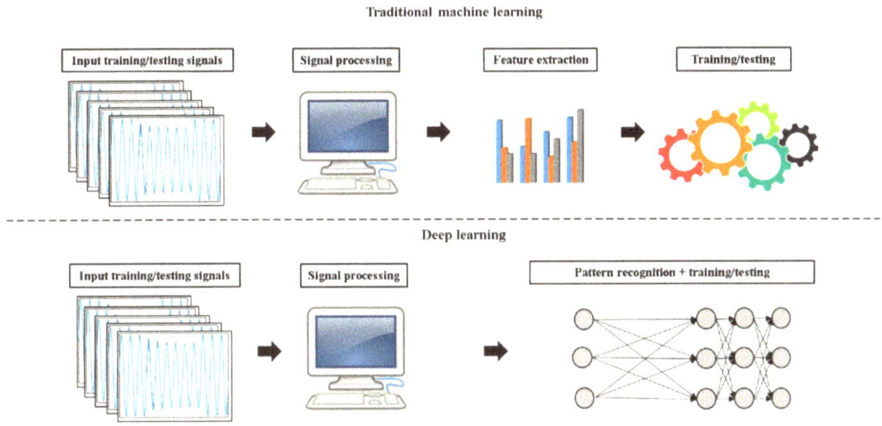

Figure 2.2 Methodology for traditional machine learning and deep learning methods in PQ disturbance classification.

2.1.2 Machine-learning-based strategies for the classification of PQ disturbances

Traditional supervised machine-learning-based (ML-based) classification strategies are commonly composed of three stages: signal processing, feature extraction, and classification [6], [7], [8]. During the signal processing stage, the raw signals are prepared by different techniques (that include, but are not limited to, signal decomposition, normalization, noise removal, etc.) for the extraction of the PQ disturbance features. Then, the feature extraction process gets relevant features that characterize the diverse PQ disturbances aiming to classify them. In this stage, a feature selection process can be added to get the most relevant features to improve the disturbance generalization in the training process. The final stage comprises firstly training the classification model to further test it with new disturbances.

The deep-learning-based methods automatically extract the disturbance features in a closed-loop process. This carries advantages related to the traditional machine learning methods because of its automatic pattern recognition, removing the need for a feature selection process and achieving higher accuracy [8]. With the complexity increasing of the monitored disturbances, a more complex deep neuronal network is also needed. The methodologies for the traditional machine learning methods and the deep learning methods are depicted in Fig. 2.2.

In Fig. 2.2, both methodologies share the stages of the input signal, signal processing, and the model training/testing need, although each methodology has its own pattern recognition and training process. Thus it can be said that the ML-based methods create classification models directly from the analysis of the input disturbance data, regardless of the methodology class.

Hence, the disturbance dataset used as input data is a key factor in the model training process for the accurate classification of the PQ disturbance. The training of the

Table 2.1 Pros and cons of the data types for the disturbance datasets.

Type of disturbance dataset	Advantages	Disadvantages
Synthetic	• Easy to develop • Low economic investment required • Access to a wide variety of disturbance classes • No class imbalance issue	• General disturbances generation • High dependency on disturbance parameters variation
Simulated	• Access to a wide variety of disturbance classes • No class imbalance issue • Faithful generation of disturbances dynamic • Nondependency to disturbances parameters variation	• Requires computational and modelling background to build benchmark system using professional software • High financial economic investment is required
Real-world	• Adapted to real DGSs • Nondependency to disturbances parameters variation • Real dynamic of the disturbances	• Limited disturbance classes • Class imbalance issue • Missing data issue • Expensive • Time consuming

classification model requires a large amount of quality data for the proper generalization of the disturbance classes aiming to get an outstanding classification performance.

2.2 Development of PQ disturbance datasets for training/assessing of the model

In PQ disturbance classification, the authors of published papers have trained and assessed the classification model based on three types of disturbance data, i.e., synthetic, simulated, and real-world datasets. Some advantages and disadvantages for the development of the different disturbance datasets are listed in Table 2.1.

On one hand, the development of real-world datasets from the DGSs requires distributed data acquisition schemes, with measurement devices, such as PMUs or μPMUs, at strategic sensing points to get a global PQ analysis. An alternative to the distributed data acquisition schemes for the development of real-world datasets is the generation of disturbance data through laboratory equipment [9]. However, the development of proper real-world disturbance datasets, represent a high economic investment and time-consuming activity [10]. Therefore there is a shortage of open real-world disturbance datasets in the literature, which makes difficult the direct comparison between classification methods [11]. In addition, the existing open datasets

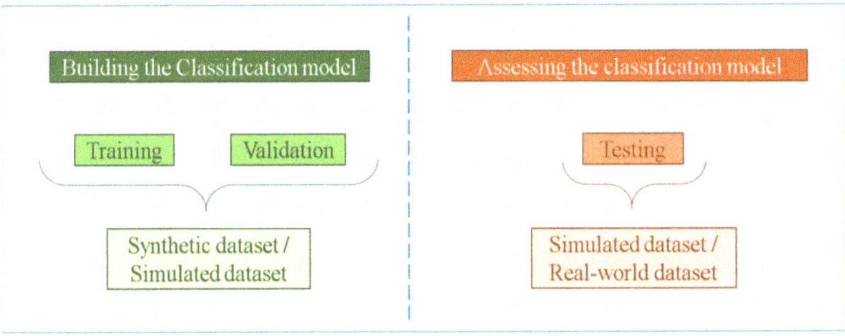

Figure 2.3 Commonly used dataset types in the classification model building (training/validation) and assessing (testing) processes.

have different limitations that hinder the training of the classification model, such as lack of disturbances classes or class imbalance issues.

To overcome the aforementioned issues, the synthetic and simulated data have been widely employed by the researchers for the training and assessing of the classification models [12]. The synthetic data is based on the computation of stated general numerical disturbance models, with parameters variations according to the current standards in PQ disturbance monitoring. For the dataset development of the simulated disturbances, the researchers employ professional software that contains advanced disturbance models, where the disturbance parameter variations are defined by the DGS elements.

Training and validation of the classification model before being tested are carried out in the literature by mainly using synthetic disturbance data, or simulated disturbance data. After the training, a real-world dataset is used to further assess the validated model, as depicted in the Fig. 2.3.

There is not an official benchmark for the development of different disturbance datasets. Therefore the proper comparison between classification methods is ambiguous, and the highlights of the proposed methods may be limited to specific situations.

For this reason, this chapter presents a practical guide for the development of PQ disturbance datasets based on the wide review of datasets commonly used in the classification of disturbances for the PQ monitoring of DGSs. The general procedure for the development of synthetic, simulated, and real-world datasets is presented below.

2.3 Generation and development of synthetic disturbance dataset

Synthetic disturbance generation is a dominant strategy for the development of disturbance datasets because of its key features in the classification model training. The principal highlights of the synthetic data generation are the controlled generation environment, the labeled data, large dataset generation, the varied data, and the balanced data control.

- ***Controlled generation environment.*** This feature refers to the full control in the generation process, allowing modification of the disturbance parameters to get variegated datasets.
- ***Labeled data.*** During the synthetic generation process, data can be labeled automatically avoiding the time-consuming data labeling hand process.
- ***Large dataset generation.*** With the controlled synthetic generation process of disturbances, the dataset can be as wide as required by the user.
- ***Varied data.*** The generation parameter control allows to get disturbances with a wide range of variations to build a varied dataset.
- ***Balanced data control.*** A main issue about disturbance dataset management is the unbalanced data that directly affects the classification process. Nevertheless, synthetic data generation avoids this issue due to the full disturbance generation control, giving place to a large and variated dataset.

The feasibility of the synthetic disturbance generation for PQ disturbance classification is related to the existence of well stated numerical models used in the literature. These numerical models allow to represent general behavior of PQ disturbances based on the variation of their generation parameters.

2.3.1 Numerical models for the generation of PQ disturbances

In the DGSs, there is plenty of single and combined disturbances. However, the disturbance generation starts with the analysis of the fundamental signal, corrupted by electrical variations, as follows:

$$y(t) = \sum_{j=1} c_j(t) + \eta, \tag{2.1}$$

where $y(t)$ is a multicomponent signal, mainly composed as the sum of the fundamental signal ($c_1(t) = Asin(\omega t)$, being $\omega = 2\pi f$), with different disturbances modeled as a mono-component signal ($c_j(t)$), and the stochastic noise denoted by η. In this sense, signal $y(t)$ can contain both, stationary and non-stationary components at different time-scale dynamics.

Thus, based on the multicomponent signal defined in (2.1), the representation of the PQ disturbance phenomena is carried out by the different disturbance numerical models for single and combined disturbances. The numerical models for the most common single and combined disturbances used in the literature for PQ monitoring are depicted in Table 2.2.

2.3.2 Typical parameters variation of disturbances according to PQ monitoring standards

Notwithstanding the numerical models described in the Table 2.2 allow to get the general behavior of the single and combined disturbances, they highly depended on the variation of their parameters. Hence, the proper parameter variation is crucial for the

Table 2.2 Numerical models for the most common single and combined disturbances [13], [14], [15].

Disturbance	**Numerical model**	**Parameters**
Undisturbed	$y(t)=A[sin(\omega t)]$	A (Amplitude) $\omega \rightarrow 2\pi f$
Sag	$y(t)=A[1-\alpha(U_1-U_2)]sin(\omega t)$	α (Sag amplitude) $U \rightarrow$ Ones vector
Swell	$y(t)=A[1+\alpha(U_1-U_2)]sin(\omega t)$	α (Swell amplitude)
Interruption	$y(t)=A[1-\alpha(U_1-U_2)]sin(\omega t)$	α (Interruption Amplitude)
Oscillatory transient	$y(t)=A\left[sin(\omega t)+\alpha e^{\frac{-(t-t_1)}{\tau}} sin(\omega_{ot}(t-t_1))(U_1-U_2)\right]$	α (o.t. amplitude) t_1 (Disturbance onset time)
impulse	$y(t)=A[sin(\omega t)+sign(sin(\omega t))[\alpha(U_1-U_2)]]$	α (Impulse amplitude)
Notch	$y(t) = A\,[sin(\omega t) - sign(sin(\omega t))$ $\left[\sum_{n=0}^{N} \alpha\{u(t-(t_1+0.02n)) - u(t-(t_2+0.02n))\}\right]]$	α (notch amplitude) t_1 (Initial time) t_2 (Ending time)
Harmonics	$y(t)=A\left[\sum_i \beta_i sin(i\omega t)\right]$	i (Harmonic) β_i (Amplitude)
Flicker	$y(t)=A\left[1+\alpha\left(sin(\omega_f t)\right)\right]sin(\omega t)$	α (Flicker Amplitude) ω_f (Flicker angular speed)
Sag + harmonics	$y(t)=A[1-\alpha(U_1-U_2)]\left[\sum_i \beta_i sin(i\omega t)\right]$	α (Sag amplitude) β_i (Harmonic amplitude)
Swell + harmonics	$y(t)=A[1+\alpha(U_1-U_2)]\left[\sum_i \beta_i sin(i\omega t)\right]$	α (Swell amplitude) β_i (Harmonic amplitude)
Interruption + harmonics	$y(t)=A[1-\alpha(U_1-U_2)]\left[\sum_i \beta_i sin(i\omega t)\right]$	α (Int. amplitude) β_i (Harmonic amplitude)
Flicker + harmonics	$y(t)=A\left[1+\alpha\left(sin(\omega_f t)\right)\right]\left[\sum_i \beta_i sin(i\omega t)\right]$	α (Flicker amplitude) β_i (Harmonic amplitude)
Flicker + sag	$y(t)=A\left[1+\alpha_f\left(sin(\omega_f t)\right)\right][1-\alpha_s(U_1-U_2)]sin(\omega t)$	α_f (Flicker amplitude) α_s (Sag amplitude)

continued on next page

Table 2.2 (*continued*)

Disturbance	Numerical model	Parameters
Flicker + swell	$y(t)=A\left[1+\alpha_f\left(sin(\omega_f t)\right)\right][1+\alpha_s(U_1-U_2)]sin(\omega t)$	α_f (Flicker amplitude) α_s (Sag amplitude)
Sag + oscillatory transient	$y(t)=A\left[1-\alpha_s(U_1-U_2)\right]\left[sin(\omega t)+\alpha_{ot}e^{\frac{-(t-t_1)}{\tau}}sin(\omega_{ot}(t-t_1))(U_1-U_2)\right]$	α_s (Sag amplitude) α_{ot} (Sag amplitude)
Swell + oscillatory transient	$y(t)=A\left[1+\alpha_s(U_1-U_2)\right]\left[sin(\omega t)+\alpha_{ot}e^{\frac{-(t-t_1)}{\tau}}sin(\omega_{ot}(t-t_1))(U_1-U_2)\right]$	α_s (Swell amplitude) α_{ot} (Sag amplitude)
Flicker + oscillatory transient	$y(t)=A\left[1+\alpha_f\left(sin(\omega_f t)\right)\right]\left[sin(\omega t)+\alpha_{ot}e^{\frac{-(t-t_1)}{\tau}}sin(\omega_{ot}(t-t_1))(U_1-U_2)\right]$	α_f (Flicker amplitude) α_{ot} (Sag amplitude)
Sag + harmonics + oscillatory transient	$y(t)=A\left[1-\alpha_s(U_1-U_2)\right]\left[\sum_i \beta_i sin(i\omega t)\right]\left[sin(\omega t)+\alpha_{ot}e^{\frac{-(t-t_1)}{\tau}}sin(\omega_{ot}(t-t_1))(U_1-U_2)\right]$	α_s (Sag amplitude) α_{ot} (Sag amplitude) β_i (Harmonic amplitude)
Swell + harmonics + oscillatory transient	$y(t)=A\left[1+\alpha_s(U_1-U_2)\right]\left[\sum_i \beta_i sin(i\omega t)\right]\left[sin(\omega t)+\alpha_{ot}e^{\frac{-(t-t_1)}{\tau}}sin(\omega_{ot}(t-t_1))(U_1-U_2)\right]$	α_s (Swell amplitude) α_{ot} (Sag amplitude) β_i (Harmonic amplitude)
Flicker + harmonics + oscillatory transient	$y(t)=A\left[1+\alpha_f\left(sin(\omega_f t)\right)\right]\left[\sum_i \beta_i sin(i\omega t)\right]\left[sin(\omega t)+\alpha_{ot}e^{\frac{-(t-t_1)}{\tau}}sin(\omega_{ot}(t-t_1))(U_1-U_2)\right]$	α_f (Flicker amplitude) α_{ot} (Sag amplitude) β_i (Harmonic amplitude)

realistic behavior of the generated disturbance. For this aim, the common range of the disturbance variations is defined according to the current PQ disturbance monitoring standards.

Standards provide a common point of reference in PQ monitoring. For this aim, international working groups provide guidelines with disturbance definitions, parameters, and variation limits for PQ monitoring. PQ indices (PQI) have been defined and presented by IEEE 1159 [16], IEC 61000-4-30 [17] and EN 50160 [18] standards. These standards characterized the different types of PQ event or disturbance. IEC 61000-4-30 standard is related to PQ testing and measurement methods, whereas EN 50160 categorized voltage characteristics in public distribution systems. On the other hand, IEEE standard (std.) 1159-2009 presents the recommended practice for PQ monitoring. This is the main standard for PQ parameter variation in PQ disturbance classification. The standard includes typical features of amplitudes, duration, and spectral content of the common PQ disturbances. Disturbance categories and their typical characteristics are defined in Table 2.2: Categories and typical characteristics of power system electromagnetic phenomena of the IEEE std. 1159-2009. Thus with the numerical models and parameters variation range defined, the synthetic PQ disturbances can be computed in software. The researchers usually generate the PQ disturbances in programming languages, such as Python or MATLAB®, computing the numerical models in Table 2.2, following parameter variations established in [16].

Despite the above methodology for the generation of synthetic disturbance datasets having been previously used in a wide number of research papers, this methodology has been barely formally studied. In this regard, [13] discusses the numerical models of different single and combined disturbances and introduces an integral mathematical model for the generation of PQ disturbances that is freely available in [19]. The mathematical model generates a controllable size dataset with 29 classes of PQ disturbances, synthetically generated through random variation of their building parameters, according to the PQ monitoring standard. An open-source dataset generator for power quality disturbances has recently been presented in [15]. The open source produces 16 types of standard disturbances widely used in the literature of deep learning classifiers. Besides, two main graphical user interfaces (GUIs) are included in the open-source. One GUI is a signal generator that synthesizes a disturbance signal based on the disturbance numerical model, whereas the other GUI creates a dataset with the 16 types of synthetic disturbances, variated according to the PQ monitoring standard. A practical example of the generation of PQ disturbances using an available mathematical model is presented in the Section 2.3.3.

2.3.3 Example of PQ disturbance generation using disturbances numerical models and Matlab programing

In the present example, the computation of the undisturbed fundamental voltage wave and different classes of PQ disturbances, including sag, oscillatory transient, and sag with oscillatory transient, among others, is carried out by using the MATLAB computing software and the mathematical model given in [19].

MATLAB is a numerical computing system that offers an integrated development environment (IDE), with its own programming language (M language). This language can be executed both in the interactive environment and through a script file (* .m files). This language allows vector and matrix operations, functions, lambda calculus, and object-oriented programming.

On the other hand, an algorithm presented in [13] allows to compute different numerical models for the generation of 29 classes of single mixed disturbances. The available algorithm includes a MATLAB function for automatic generation of synthetic PQ disturbances. Computation of synthetic PQ disturbances starts with the definition of global parameters according to [16], such as the number of signals, the sampling frequency, the system's frequency, the number of registered cycles, and the amplitude of the signal, as described in Box 2.1.

Next, provided function pqmodel is used with the defined parameters to generate a dataset containing 29 classes of disturbances.

Box 2.2 shows the function for disturbance generation, where output element "out" is a matrix of dimension (1000, 1600, 29), i.e., Number of samples per class, Number

Box 2.1 Global parameter set up.

```
ns = 1000;     % Number of signals
sf = 8000;     % Sampling frequency in Hz
f = 60;        % Simulation time in sec
n = 12;        % Number of registered cycles
A = 480;       % Amplitude of the signal
```

Box 2.2 Disturbance generator function.

```
[out] = pqmodel(ns, sf, f, n, A)
```

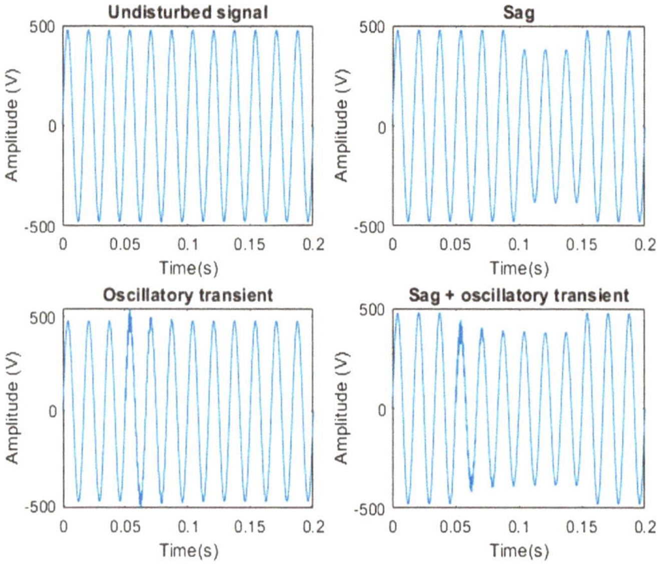

Figure 2.4 Disturbances generated using their numerical models and MATLAB programing.

of points per signal, Number of classes, for this example. Thus mathematical model in the provided function "*pqmodel*" automatically defines the disturbance parameters based on stochastic parameter variations, according to the recommended PQ monitoring practice [16].

Some signals, including the undisturbed wave and disturbed with sag, oscillatory transient, and sag with oscillatory transient, generated by the function "*pqmodel*," are shown in the Fig. 2.4. Unfortunately, even though the disturbances generated by numerical models present a behavior consistent to the real-world disturbance, the generic numerical model neglects the effect of some devices commonly installed in the real-world systems that might affect the disturbance's behavior, losing the full dynamic of the disturbance.

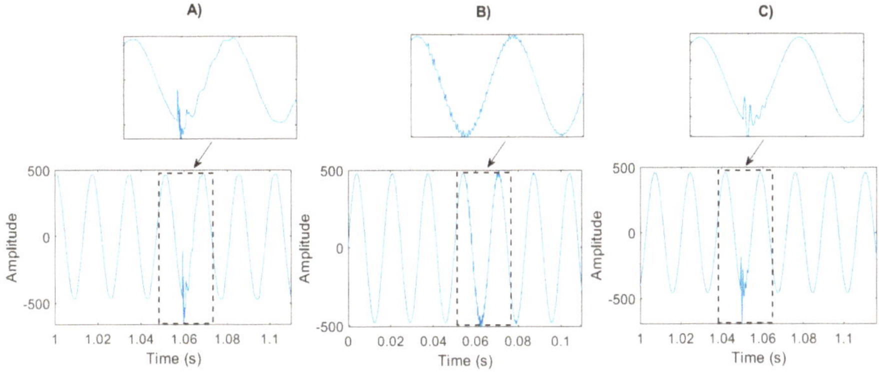

Figure 2.5 Wave behavior comparison for the oscillatory transient disturbance in A) the original dataset, B) the synthetic disturbance generation, C) obtained from the simulation software.

2.4 Development of PQ disturbance datasets with simulation software

Although the numerical models have been proved suitable and are highly used for the development of disturbance datasets in PQ monitoring, these are general models that critically depend on the selected parameters to generate the disturbances. Besides, on some occasions, it is necessary to generate disturbances with very particular variations, making necessary the use of more complex numerical models. A clear example where the use of the disturbance simulation approach is helpful, is when the testing system of interest can be built into the simulation software, allowing to get signals that better follow the dynamics of the original real-world disturbances obtained from the system, leading to a better training of the classification model.

2.4.1 Features of the simulation software

There are simulation softwares, such as PSCAD [20] or Simulink [21], that use detailed models for the realistic generation of disturbances through the connection of specifically modeled elements of DGSs. In addition, there are professional real-time (RT) simulators, such as OPAL-RT [22], Thyphoon-HIL, or RTDS [23], that interact with hardware of control and power. These RT simulators allow to analyze the real-time response of the DGSs during the presence of different PQ events.

Among the highlights for training disturbance classification models using disturbance simulation is the more realistic behavior in the simulated disturbances than in the synthetically generated disturbances. To exemplify the above statement, an oscillatory transient obtained from a collection of real-world measurements is considered. Then, the oscillatory transient disturbance simulation was carried out using a microgrid modeled in PSCAD (see Section 2.4.2).

Fig. 2.5 presents a comparison example that shows the difference in the wave behavior of an oscillatory transient disturbance of a real-world dataset (Fig. 2.5A), the synthetic disturbance generated with the disturbance numerical model (Fig. 2.5B), and the one obtained from the simulation software (Fig. 2.5C). In summary, Fig. 2.5 exemplifies that, although the synthetic oscillatory transient generated with its numerical model is suitable for the training of the classification model, it is evident that the transient obtained from the simulation, depicted in the Fig. 2.5C, presents a more realistic behavior visually comparable to the disturbance in the real-world dataset (Fig. 2.5A), which may enhance the model training to improve the classification performance for oscillatory transients.

2.4.2 Example of disturbances generation by simulation software for dataset development

In this section, PSCAD simulation software is used for the generation of different PQ disturbances. The test microgrid was proposed in [24] for the study of power management strategies in a DGS. Besides, testing microgrid has been employed in different power system studies on DGS, such as harmonic issues analysis in [25], islanding detection in [26], and PQ disturbance classification in [14], [9].

Setting up the initial simulation parameters and sample rate is a vital activity when generating disturbance with simulation software. Hence, for this example, 12 cycles of the event are sampled at 8 kHz as sampling frequency, according to [16], using 15 sensors located along the microgrid. This activity gives to the disturbance produced by the simulation a realistic behavior that depends on the event and its initial parameters. For the development of the simulated disturbance dataset, the considered events to produce the disturbance are diversified by the variation of the initial simulation parameters, as proposed below.

- **Momentary load increment**. It is considered load increments of 5%, 25%, and 50% in L1, L2, L3, L4, L5, and L6. The load increment starts at 0.02 s, 0.04 s, and 0.06 s. After the sixth, fourth, and second cycles, respectively, the load decreases by the same percentage as the increment.
- **Power bank switching**. This is initially is disconnected with a fixed power. The power bank is energized at 0.02 s, 0.04 s, and 0.06 s with a load increment of 5%, 25%, and 50% in L1, L2, L3, L4, L5, and L6.
- **Generator tripping.** It is considered a generator tripping in G1, G2, and G3 at 0.01 s, 0.02 s, 0.04 s, 0.06 s, during 0.12 s, 0.1 s, 0.06 s and 0.02 s, respectively.

The different signals that contain the PQ disturbances, produced by the considered events, are obtained from the voltage/current meters, at point of common coupling (PCC) and the rest of the buses connected to the PCC, through the computer interface to develop the dataset. A general view for the dataset development with simulated disturbances is shown in Fig. 2.6. At the end of the event simulation, disturbance signals obtained from simulation software are organized in time-depending disturbance arrays

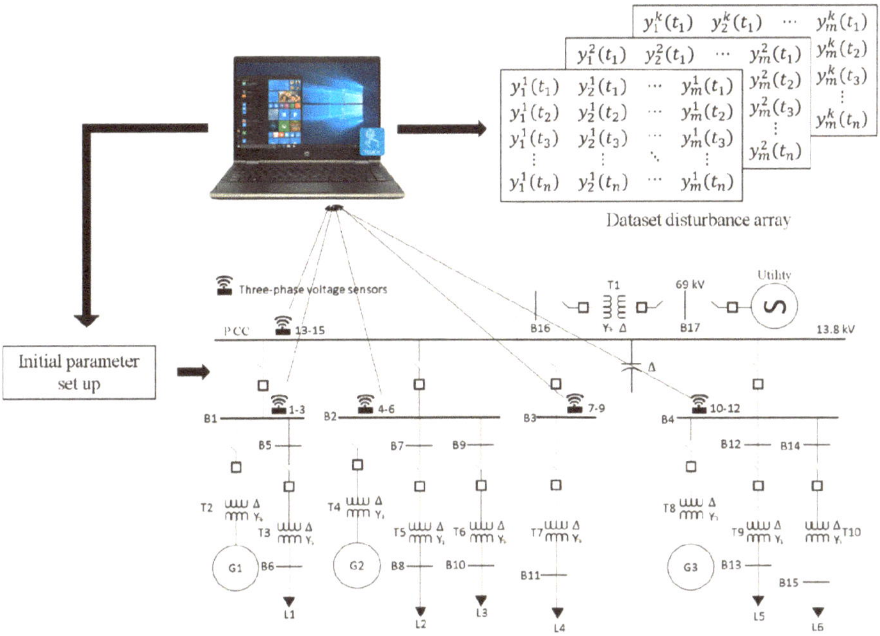

Figure 2.6 Disturbance dataset development using PSCAD software simulation.

as

$$X_k = \begin{bmatrix} y_1^k(t_1) & \cdots & y_m^k(t_1) \\ \vdots & \ddots & \vdots \\ y_1^k(t_n) & \cdots & y_m^k(t_n) \end{bmatrix}, \tag{2.2}$$

where the kth event simulation is stored in X_k containing m number of signals with n samples. Using varied considerations for the generation of the events allow to get a wide range of disturbances for the dataset. Nevertheless, the practical application of the classification models to the real-world systems usually demands more sophisticated data for its training to classify the disturbances with higher complex behavior.

2.5 Real-world datasets for PQ disturbance classification

Dissimilar to the idealized operation of the electronic devices considered in the analysis of synthetic and simulated disturbance, the operation of these devices is more complex in practice. Besides, external issues, such as human error, which is hardly possible to model lead to tangled disturbances. This highly complex behavior of the real-world disturbances has been well established in the literature for PQ monitoring [9], [27], [28], [29], [30]. Based on the above statement, despite the great advantages of the synthetic and simulated datasets for the training of classification models, it is not entirely clear if the validated accuracy achieved by the model training with these datasets is going to be sustained when classifying real-world disturbances. For this reason, it is crucial to include real-world disturbance data in the testing stage for the classification model assessing.

2.5.1 PQ monitoring recommendations for the development of real-world disturbance datasets

The development of real-world datasets requires distributed acquisition schemes and specialized measurement devices to be conducted. Besides, some considerations must be considered for the proper disturbance classification. In this regard, the IEEE std. 1159-2009 [16] not only works for defining the parameters of the disturbances, but also is a practical guide for the recommended practice of PQ monitoring, including the recommended monitoring location, the description of the monitoring instruments, and define some measurement thresholds, among other recommendations for PQ monitoring.

2.5.1.1 *Variables to be measured*

Voltage and current are the two main measurements for the analysis of the power quality phenomena. The selection of these measurements is because the most available instruments use analog-to-digital converters that easily sample and store the voltage and current waveforms. In addition, these electrical measurements allow us to compute and extract many desired features to detect and classify the disturbances, such as rms value, THD, or energy, to name a few features.

2.5.1.2 *Monitoring devices*

PQ analyzers are devises commonly employed for PQ monitoring. These analyzers can measure various electrical parameters, such as V, A, W, VA, VAr, power factor, phase, true rms, as well as fundamental harmonics, because of its incorporated algorithms. In addition, PQ analyzers are mobile, so they can be located at the point of monitoring interest, and can capture events with over 30 kHz of sampling frequency. However, PQ analyzers have high commercial cost and are not very flexible with respect to the computation algorithms they use. On the other hand, phasor measurements units (PMUs) and μPMUs have been recently widely used for the measurement of current and voltage due to their high precision synchronization time via GPS, which allows measurements comparison in a distributed data acquisition scheme using data concentrator. PMUs and μPMUs employ high sampling frequency around 30.7 kHz to capture points on the disturbance wave for further analysis through developed machine-learning-based algorithms.

2.5.1.3 Monitoring location

The location of the monitoring devices is crucial for the proper acquisition of the disturbance data. The initial location to install the monitoring devices depends on the aim of the survey. If the monitoring aim is to diagnose an equipment performance problem, then the monitor should be placed as close to the load as possible. This applies to performance problems with both sensitive electronic loads, such as computers and adjustable-speed drives (ASDs), and electrical distribution equipment, such as circuit breakers and capacitors.

2.5.1.4 Data acquisition

For the PQ disturbance monitoring, sophisticated phase-locked-loop sampling techniques are used to ensure that the first and subsequent samples of each cycle are in the same point-on-wave, even while monitoring highly distorted, reduced amplitude, and phase-shifted waveforms. The signals are sampled at 128, 256, or 512 samples/cycle. The sampling frequency attends to the Nyquist frequency for the respective disturbances to analyze. The values for voltage and current are obtained using fixed or sliding windows; 10 cycles is a common window size used in the literature, i.e., 200 or 166.6 ms for 50 Hz or 60 Hz, respectively.

2.5.2 Available open real-world datasets for PQ disturbance classification

A systematic analysis of available public datasets of PQ distortions until June 2020 was presented in [11]. In the study, two different datasets shared by IEEE were identified, one belonged to the IEEE 1159.2 working group [31], and the other to the IEEE 1159.3 working group [32]. Besides, employing popular scientific repositories of public datasets as IEEE data port, and Mendeley data, a dataset of sags entitled ``Real life power sags" [33], and other of transients ``Real life power quality transients" [34], were found and compared against the other real-world power distortions datasets.

From the comparative analysis of public datasets depicted in a comparative table in [11], it is easy to notice that there is a lack in public datasets due to the strong economic investment and time consumption that represent the development of real-world datasets. Besides, most of the datasets possess low number of disturbances with limited classes, and disturbance data is generally unbalanced. Hence, available public datasets are not entirely suitable for the training of the classification model, owing to training the model with an unbalanced dataset directly affects the training of the model, decreasing the PQ disturbance classification performance, representing real issue.

2.6 Effect of the class imbalance issue on machine learning models trained for PQ disturbance classification

Machine-learning-based (ML-based) methods have the ability to create classification models directly from the analysis of disturbance data. Therefore the training of the classification model requires a large amount of quality data for the proper generalization of the disturbance classes aiming to get an outstanding classification performance. Nevertheless, unbalanced datasets with class imbalance abound in practice for PQ disturbance due to anomalous disturbance.

A major feature of the unbalanced dataset is the existence of minority classes, giving rise to classes imbalance. During the training stage, the majority classes (the ones with most samples) as the minority classes (the ones with minimum samples) are given to the model aiming to learn the specific features of the disturbances for their classification. Nevertheless, after training, the model focuses on the classification of majority classes, putting aside the classification learning of minority classes. On one hand, the model has an outstanding performance classifying majority classes, but it has some troubles to classify minority classes.

Because of the unbalanced real-world dataset, in the general framework of PQ disturbance classification, researchers have generated synthetic disturbances for training the classification model. Still, the effect of the unbalanced dataset in the training of a model for disturbance classification is a crucial but limited-studied subject that needs to be attended. For this reason, an example that shows the influence of the unbalanced data-based training in the classification of PQ disturbance is presented below.

2.6.1 Example of PQ disturbances classification, influenced by unbalanced data-based training

In this example, three different ML-based supervised strategies are used to show the effect of unbalanced data on the classification model training for PQ disturbance classification. Study disturbance dataset has notable class imbalance for two classes, to clearly expose the effect of unbalanced data-based training in the classification performance.

2.6.1.1 Disturbance dataset description

The PQ disturbance dataset was obtained from the database given in [19]. Study dataset contains 10 different classes of single and mixed PQ disturbances (see the Table 2.3). For analysis purpose, the originally balanced dataset is advisedly unbalanced in the proportion showed in the Fig. 2.7.

Regarding the data pre-processing, histogram is a great descriptive statistics tool to detect class imbalance issue. In this sense, the disturbance class histogram in Fig. 2.7 shows an evident class imbalance in the study dataset. The unbalanced dataset contains 1000 examples per class, excluding the minority classes 7 and 8, which presents

Table 2.3 Disturbance classes in the dataset.

Class	Disturbance	Class	Disturbance
1	Pure sinusoidal	6	Harmonics
2	Sag	7	Sag with harmonics
3	Swell	8	Swell with harmonics
4	Interruption	9	Flicker
5	Oscillatory transient	10	Notch

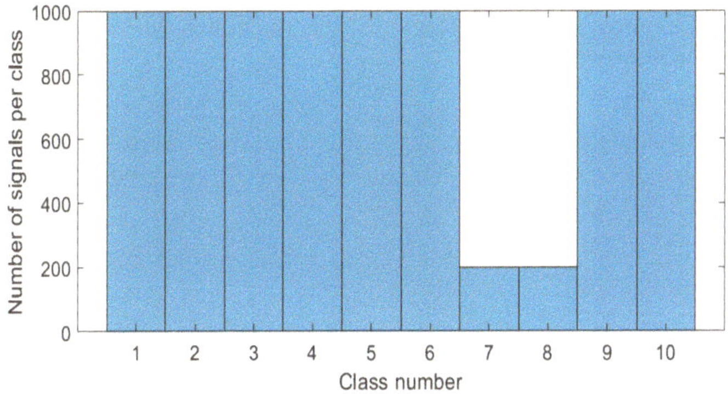

Figure 2.7 Disturbance class histogram showing the class imbalance in the dataset.

an 80% reduction in their number of examples. Supervised machine-learning based strategies stand out by the need for labeled data to train the classification model. In this regard, the clear unbalance in the class examples is expected to affect the training of the classification model, and brings classification problems to the trained model performance.

2.6.1.2 *Training of the classification models*

Supervised classification methods selected for disturbance classification based on the 10 disturbance classes of study dataset were SVM, KNN [35], and DCNN. SVM classifier employs a one vs. one structure [36]. On the other hand, 100 k-nearest neighbors were set for the classification by KNN. Moreover, the 1-D DCNN architecture is taken from [37], where it was used for the classification of different classes of single and mixed disturbances. The DCNN-based architecture has a sequential model with 6 1-D convolutional layers and 3 dense layers. First, the 2 convolutional layers with 32 filters are followed by a MaxPooling and dropout layer. Then, two more convolutional layers with 64 filters and a MaxPooling and dropout layers. Next, another 2 convolutional layers with 128 filters are followed by a GlobalMaxPooling and dropout layers, 2 dense layers with Relu activation function for 256 and 128 neurons, respectively; and a final dense layer with 10 neurons activated by Softmax function, corresponding to the number of classes with single and combined disturbances.

Traditional ML-based methods, such as SVM or KNN, carry out their training based on selected features extracted from the disturbance to generalize them. These features can be automatically selected from a batch of computed features using complex optimization techniques. However, optimization-based selection techniques are time-consuming, and preferred for highly multivariate data [14]. Hence, feature selection based on self-experience and prior knowledge is more commonly employed. On the contrary, the DCNN method has a deep learning-based (DL-based) architecture that performs the feature extraction for the training/classification in a closed-loop process, automatically defining features for the disturbance classification.

Computed features for the disturbance generalization on traditional ML-based methods are the following:

- Signal energy (E)
  Energy is a main feature of temporal series. It presents particular values for the PQ disturbances, allowing to separate them in similar behavior clusters.

$$E = \sum_t y^2 [t] . \tag{2.3}$$

  Signal energy is computed from the squared amplitude of the signal y at each t time.
- Root mean squared (*RMS*) voltage value
  The root mean square value is one of the most used indices for PQ classification.

$$RMS = \frac{1}{\sqrt{2}} \sqrt{\sum_t y^2 [t]} . \tag{2.4}$$

- Local maximum (l_{max})
  Local maximum of the signal defines its maximum amplitude.

$$l_{max} = \max (y [t]) \mid y [t-1] < y [t] > y [t+1] , \tag{2.5}$$

  where l_{max} founds the higher local maximum in the signal.
- Local minimum (l_{min})
  Local minimum of the signal defines its minimum amplitude.

$$l_{min} = \min (y [t]) \mid y [t-1] > y [t] < y [t+1] . \tag{2.6}$$

- Number local maximum (*Nlm*)
  Nlm obtains the number of local maximums in the signal.

$$Nlm = \sum y[t] \mid y [t-1] < y [t] > y [t+1] . \tag{2.7}$$

- Number of zero crossings (Zc)
  Number of zero crossings quantifies the number of zero crossings of the signal.

$$Zc = \sum |sgn (y \{t\}) - sgn (y \{t-1\})| \tag{2.8}$$

Thus traditional ML-based classification models are trained using a batch with the 6 computed features obtained from the signals in the study dataset, whereas the DCNN method uses the full original disturbance. DCNN model training is carried out using the sparse categorical cross entropy to calculate the losses during the training of the model. The loss function is defined as

$$H(y^k, \hat{y}^k) = -\sum_{m=1}^{M} y_m^k \log(\hat{y}_m^k) \tag{2.9}$$

where $m \in M$ is the total number of disturbance classes, and y^k and $\hat{y}^k$ are the actual and predicted labels for the $k-$esim disturbance in the dataset. Loss function on (2.9) is solved through backpropagation using the Adam optimizer.

2.6.1.3 Definition of the evaluation metrics

Although accuracy is the most used evaluation metric for assessing the performance of classification methods, when assessing classification performance of models trained with unbalanced-data it has limitations that prevent the deep and objective analysis of the results, covering the class imbalance issue. In consequence, to highlight the effect of unbalanced-data-based training on the disturbance classification performance, *F1-measure* is selected as evaluation metric. Computation of *F1-measure* is conducted by

$$F1 - measure = 2 \times \frac{1}{\frac{1}{Pre} + \frac{1}{Rec}}. \tag{2.10}$$

Pre and *Rec* correspond to the values of precision and recall, respectively. Precision and recall are two widely used evaluation metrics that quantify the classification errors based on false detections and nondetections. *Pre* and *Rec* are computed as

$$Rec = \frac{tp}{tp + fn} \times 100 \tag{2.11}$$

$$Pre = \frac{tp}{tp + fp} \times 100 \tag{2.12}$$

where tp are the true positives, fn are the false negatives, and fp are the false positives predictions. In essence, the combination of these evaluation metrics provide a clear vision of the general classification performance for the individual classes, and therefore would allow to identify the effect of the class imbalance in the classification process.

2.6.1.4 Classification performance assessment of trained models using the evaluation metrics

Typical performance assessment of classification methods is carried out using cross-validation. In this regard, k-fold cross-validation is a common tactic to estimate the skill of the ML-based trained model for classification of new data. Nevertheless, k-fold

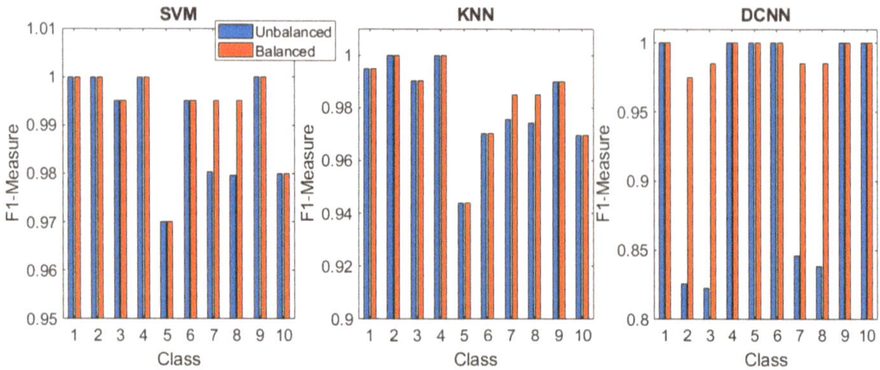

Figure 2.8 Results for the classification performance of the machine-learning-based methods.

separates k data examples per class designated for evaluation from the original training dataset, so the evaluation examples batch would preserve the unbalance in data, hiding the class imbalance issue. For this reason, the cross-validation in this study is carried out using a balanced validation batch, with 100 examples per class, intending to show the unbalanced-data-based training effect. Results of the trained methods performance for disturbance classification can be appreciated by way of Fig. 2.8.

The Fig. 2.8 shows the results for the classification performance of ML-based strategies trained with the unbalanced dataset described in the Section 2.6.1.1 and the originally balanced dataset, with the classes of Table 2.3. It is evident from Fig. 2.8 the superior performance of ML methods trained with balanced data, for the unbalanced classes 7 and 8. Besides, it is clear that class imbalance issue has major influence on DCNN, not only for the unbalanced classes, but also for balanced classes 2 and 3. This is due to deep learning-based methods requiring large amount of data to generalize the disturbance, and class imbalance issue reduces the data, impeding the DCNN from correct disturbance feature extraction. Therefore disturbance classes with similar characteristics as 2, 3, 7, and 8 are misclassified. On the other hand, results for traditional ML-based methods also show better classification performance for balanced training. However, since traditional ML methods depend on selected features, imbalance class issue only affects the classification of unbalanced classes.

2.6.2 Discussion about the influence of the unbalanced data-based training in machine learning methods for PQ disturbance classification

In the light of the obtained results, it is clear that unbalanced training affects the classification performance of ML-based methods. Nevertheless, class imbalance has different effect level on deep learning-based (DL-based) methods compared to traditional methods. In this regard, automatic feature extraction process of DL methods makes crucial the balanced training to avoid generalization errors, which produce misclassification on similar disturbance classes.

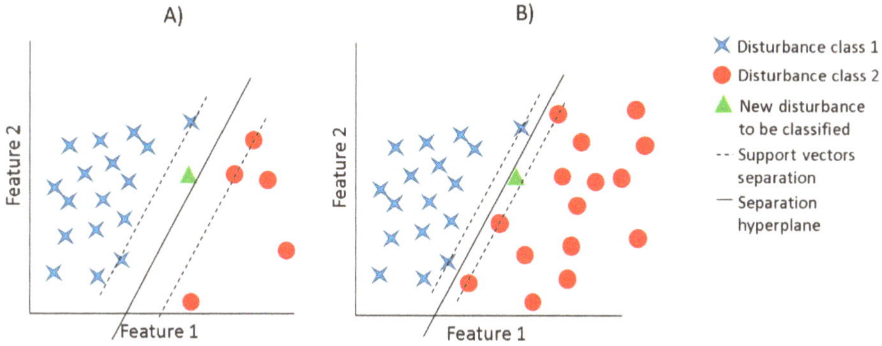

Figure 2.9 Example of separation hyperplane computation for SVM with A) unbalanced dataset, B) balanced dataset.

On the other hand, class imbalance has a significant effect in determining decision boundaries on feature space-based classification methods, such as SVM and KNN, producing classification errors on unbalanced classes. This statement is supported on the action principle of SVM, where class examples are used to compute the separation hyperplane among classes (see Fig. 2.9), also in KNN, where most proximate examples are used to determine the class of the new disturbance.

Fig. 2.9 shows a practical example of the class imbalance effect on the hyperplane computation for SVM. On the one hand, separation hyperplane computed with class imbalance in Fig. 2.9A indicates that new disturbance belongs to class 1. Nevertheless, training SVM with balanced class 2 in Fig. 2.9B, it can be appreciated that new disturbance presents similar features with examples on class 2, so is now classified as disturbance class 2, according to the computed separation hyperplane. Once the importance of balanced training for ML-based methods is highlighted, it is relevant to note that there are techniques to handle the class imbalance issue. These techniques can synthetically balance the classes through different methodologies. Among principal unbalanced data handling techniques are resampling and algorithm level methods.

Approaches, such as the resampling methods, manually balance the samples of the majority and minority classes by adding or removing samples. These approaches, modify the original distribution of the dataset causing information loss, or rise the computational burden with redundant information. In this regard, all misclassification errors are considered having equal cost in these approaches, which in practice is not suitable. On the contrary, algorithm-level approach is based on the modification of the classification algorithm without redistributing the original dataset. With the algorithm-level approaches, minority class samples are judged differently from the majority class samples with diverse strategies. For example, classification errors of minority class samples can acquire a higher weight during the loss calculation to force the classification algorithm to learn the distribution of the minority class samples and successfully classify them.

2.7 Chapter conclusions

A practical guide of the dataset development for the classification of PQ disturbances is presented in this chapter. The motivation of the chapter is the lack of an official benchmark for the generation of PQ disturbance dataset for the proper training and assessing of the classification models, especially for new researchers in the PQ monitoring area.

Three types of disturbance data generation are considered for the development of the PQ disturbance dataset, i.e., synthetic, simulated, and real-world disturbance. Evidently, real-world dataset would be the more suitable for the training of the model. However, class imbalance issue is common in public real-world disturbance dataset with low number of disturbance and limited classes. Since supervised machine-learning-based strategies stand out by the need for labeled data to train the classification model, unbalanced training data affect the training of the classification model, and bring classification problems to the trained model performance. Hence, evaluating the effect of class imbalance issue on ML-based classification methods was a key activity of the chapter. In this regard, results showed a clear influence of minority classes on the model training producing classification errors. Moreover, ML-based methods are influenced differently based on their action principle, given that DCNN is most affected by class imbalance issue.

Researchers empirical solution for class imbalance handling in PQ disturbance classification is the synthetic generation of disturbance for balanced training. Nevertheless, synthetic balancing of real-world dataset could be a better approach for the model training, letting better fit on real-world systems. Therefore using resampling methods or algorithm-level applications are promising solutions for the limited data and the class imbalance issue in real-world data.

Acknowledgments

The authors thank the project fordecyt-pronaces/1311344/2020 for the financial support to develop the present chapter.

References

[1] B. Hussain, S.M. Sharkh, S. Hussain, Impact studies of distributed generation on power quality and protection setup of an existing distribution network, in: SPEEDAM 2010, 2010, pp. 1243–1246.

[2] K. Suslov, N. Solonina, A. Smirnov, Distributed power quality monitoring, in: IEEE 16th International Conference on Harmonics and Quality of Power (ICHQP), 2014, pp. 517–520.

[3] A.F. Bastos, K.-W. Lao, G. Todeschini, S. Santoso, Accurate identification of point-on-wave inception and recovery instants of voltage sags and swells, IEEE Transactions on Power Delivery 34 (2019) 1–9.

[4] Y. Wang, L.-F. Deng, M.H.J. Bollen, X.-Y. Xiao, Calculation of the point-on-wave for voltage dips in three-phase systems, IEEE Transactions on Power Delivery 35 (2020) 1–11.
[5] G. Wu, Q. Zhong, Q. He, Z. Xu, Evaluation method and probabilistic index of voltage sag severity considering point-on-wave, Journal of Modern Power Systems and Clean Energy 9 (2021) 633–638.
[6] G.S. Chawda, A.G. Shaik, M. Shaik, S. Padmanaban, J.B. Holm-Nielsen, O.P. Mahela, P. Kaliannan, Comprehensive review on detection and classification of power quality disturbances in utility grid with renewable energy penetration, IEEE Access 8 (2020) 146807–146830.
[7] S. Khokhara, A. Asuhaimi, A. Safawi, M. Pesara, A comprehensive overview on signal processing and artificial intelligence techniques applications in classification of power quality disturbances, Renewable and Sustainable Energy Reviews 51 (1) (2015) 1650–1663.
[8] K. Poras, M.T. Madan, A critical and comprehensive review on power quality disturbance detection and classification, Sustainable Computing: Informatics and Systems 28 (2020) 1–11.
[9] O. Cortes-Robles, E. Barocio, A. Obushevs, P. Korba, F.R. Segundo Sevilla, Fast-training feedforward neural network for multi-scale powerquality monitoring in power systems with distributed generationsources, Measurement 170 (2021) 1–12.
[10] Y. Yuan, Y. Guo, K. Dehghanpour, Z. Wang y, Y. Wang, Learning-based real-time event identification using rich real PMU data, IEEE Transactions on Power Systems 36 (2021) 1–12.
[11] S. Dominguez-Gimeno, R. Igual, C. Medrano, Analysis of public datasets of power quality distortions, in: 18th International Conference on Renewable Energies and Power Quality (ICREPQ'20), vol. 18, 2020, pp. 321–326.
[12] R.K. Beniwal, M.K. Saini, A. Nayyar, B. Qureshi, A. Aggarwal, A critical analysis of methodologies for detection and classification of power quality events in smart grid, IEEE Access 9 (2021) 83507–83534.
[13] R. Igual, C. Medrano, F. Arcega, G. Mantescu, Integral mathematical model of power quality disturbances, in: 18th International Conference on Harmonics and Quality of Power (ICHQP), 2018, pp. 1–6.
[14] O. Cortes-Robles, E. Barocio, J. Segundo, D. Guillen, J.C. Olivares-Galvan, A qualitative-quantitative hybrid approach for power quality disturbance monitoring on microgrid systems, Measurement 154 (2020) 1–14.
[15] R. Machlev, A. Chachkes, J. Belikov, Y. Beck, Y. Levron, Open source dataset generator for power quality disturbances with deep-learning reference classifiers, Electric Power Systems Research 195 (2021) 1–7.
[16] IEEE Recommended Practice for Monitoring Electric Power Quality, Revision of IEEE Std 1159-1995, 2009, pp. 1–94.
[17] I.E.C.S.I 61000-4-30, Testing and Measurement Techniques-Power Quality Measurement Methods, 2003.
[18] E.S.E. 50160, Voltage Characteristics of Electricity Supplied by Public Distribution System, 2010.
[19] R. Igual, C. Medrano, F.J. Arcega, G. Mantescu, Mathematical model of power quality disturbances, Mendeley Data (2017).
[20] PSCAD, pscad.com, [Online]. Available https://www.pscad.com/, (Accessed 2021).
[21] MathWorks, mathworks.com, [Online]. Available https://la.mathworks.com/products/simulink.html, (Accessed 2021).
[22] OPAL-RT, opal-rt.com, [Online]. Available https://www.opal-rt.com/, (Accessed 2021).

[23] RTDS, rtds.com, [Online]. Available https://www.rtds.com/, (Accessed 2021).
[24] F. Katirarei, M.R. Iravani, Power management strategies for a microgrid with multiple distributed generations units, IEEE Transactions on Power Systems 21 (4) (2006) 1821–1831.
[25] G. Agundis, J. Segundo, R. Peña, N. Visairo, C. Núñez, F. Guerrero, M. Savagh, Harmonic issues assessment on PWM VSC-based controlled microgrids using Newton methods, IEEE Transactions on Smart Grid 9 (2) (2018) 1002–1011.
[26] F. Katiraei, M.R. Iravani, P.W. Lehn, Micro-grid autonomous operation during and subsequent to islanding process, IEEE Transactions on Power Delivery 20 (1) (2005) 248–257.
[27] O.F. Eikeland, I.S. Holmstrand, S. Bakkejord, M. Chiesa, F.M. Bianchi, Detecting and interpreting faults in vulnerable power grids with machine learning, IEEE Access 9 (2021) 150686–150699.
[28] M. Pavlovski, M. Alqudah, T. Dokic, A.A. Hai, M. Kezunovic, Z. Obradovic, Hierarchical convolutional neural networks for event classification on PMU measurements, IEEE Transactions on Instrumentation and Measurement 70 (2021) 1–13.
[29] M.R. Alam, F. Bai, R. Yan, T.K. Saha, Classification and visualization of power quality disturbance-events using space vector ellipse in complex plane, IEEE Transactions on Power Delivery 36 (3) (2021) 1380–1389.
[30] Z. Li, H. Liu, J. Zhao, T. Bi, Q. Yang, A power system disturbance classification method robust to PMU data quality issues, IEEE Transactions on Industrial Informatics 18 (1) (2022) 130–142.
[31] IEEE 1159.2 Working Group: Test waveforms, IEEE Power and Energy.
[32] IEEE 1159.3 PQDIF Task force, IEEE Power and Energy Society.
[33] Real-Life Power Quality Sags, IEEE Dataport (2017).
[34] Real-Life Power Quality Transients, IEEE Dataport (2017).
[35] T. Cover, P. Hart, Nearest neighbor pattern classification, IEEE Transactions on Information Theory 13 (1) (1967) 21–27.
[36] S. Knerr, L. Personnaz y, G. Dreyfus, Single-layer learning revisited: a stepwise procedure for building and training a neural network, Neurocomputing (1990).
[37] S. Wang, H. Chen, A novel deep learning method for the classification of power quality disturbances using deep convolutional neural networks, Applied Energy 235 (2019) 1126–1140.

Advances in compression algorithms for PMU and Smart Meter data based on tensor decomposition

3

Betsy Sandoval Guzmán[a], *Emilio Barocio Espejo*[a], *Petr Korba*[b], *and Felix Rafael Segundo Sevilla*[b]
[a]University of Guadalajara, Guadalajara, Mexico, [b]Zurich University of Applied Sciences, Winterthur, Switzerland

3.1 Introduction

Clean energy that does not burn limited fossil fuels and its use does not generate excessive carbon emissions, requires not only modern technologies to exploit, convert, transfer, and store renewable energy as electricity, but also an electric grid that ensures, through smart technologies, that the grid is responsive and robust [5]. Since one of the pillars for the smart grid transformation is the information exchanged, the monitoring system, constituted by smart meters (SMs) and synchronized phasor measurement units (PMUs), becomes an essential part of the transformation to the smart grids, leading to the use of clean energies. On the one hand, the SMs capture the electricity consumption across the power grid, improving operational efficiency [6]. On the other hand, the PMU data allow monitoring the network conditions [7]. Resulting in an increasing global trend to implement more PMUs and SMs worldwide, the following lists the numbers of those devices installed in different parts of the world:

- *USA and Mexico. SMs in the USA*: 86'844,157 [8]. *SMs in Mexico:* By the end of 2016, 772 thousand devices were installed [9]. *PMUs:* The number of synchrophasors installed in the North-east of the USA is approximately 400 in the Midwest, 150 in the South, 120 in Texas, 500 in WECC, and at least 300 PMUs in Mexico [10].
- *Europe. SMs:* As of 2018, 33.83% of all electricity metering points were equipped with a SM, i.e., 99 million [11]. *PMUs:* About 300 were installed in the Continental European Electrical Power System and about 100 PMUs in Scandinavia and the Baltic countries [1].
- *China. SMs:* By the end of 2015, China had installed 340 million *SMs* [2]. *PMUs:* as the mid of 2017, approximately 4100 PMUs had been installed [3].

It must be emphasized that both SM and PMU data are high-dimensional time series. The SM measurements report rates of 3 s–1 hr [12] for a total number of 28,800 to 24 samples per day, generating about 23.7 TB of information monthly per million

Monitoring and Control of Electrical Power Systems Using Machine Learning Techniques
https://doi.org/10.1016/B978-0-32-399904-5.00009-0

of SMs [13]. While the PMUs are sensing at a sampling rate of 35–120 Hz [14] collecting from 2,592,000 to 10,368,000 measurements per day from one single stream, 100 PMUs with 20 measurements at 30 Hz can produce 50 GB of data per data, and higher sampling rates (60–120 Hz) increase the volume of data to 100–200 GB [4]. Consequently, despite the enormous benefits of monitoring through SMs and PMUs, one of the main challenges is the required storage space since measurements must be stored for long periods, as pointed in 2020 North American Synchro Phasor initiative (NASPI) [15], and the multiple research works addressing the big data problems caused by PMU and SM data [6], [16–18].

This storage overhead challenge opens the opportunity to explore data compression techniques for SM and PMU data. Inspired by the lack of techniques that achieve a high compression ratio (CR) and a low reconstruction error for both SM and PMU measurements, this chapter has compiled the advances in data compression techniques based on tensor compositions. Given that the tensor decompositions are dimensional reduction techniques, these approaches allow achieving high compression ration and low reconstruction error. Tensors and their decompositions are a well-documented and established area that has been successfully applied to different disciplines, such as Neuroscience [19], Psychometric [20], Chemometrics [21], and recently in Electrical Power System applications [22], [23]. Tensor decomposition for data compression was first introduced in image processing [24], [25]; however those works do not address how to work with time series; moreover, data compression in the Electrical Power Systems area has not been fully exploited. Therefore this chapter represents the first overview of techniques for data compression using dimension reduction based on tensor decompositions, archiving a high compression ratio and low reconstruction error.

The key contributions of this chapter are as follows:

- *Innovative conceptualization*: The problem of compression is reformulated using three-dimensional arrays, instead of classical matrices in two dimensions. The new conceptualization is not trivial and allows the user to have more flexibility when handling vast volumes of data.
- *High compression ratios* (*CRs*): It is demonstrated how the memory size of the files containing the measurements can be significatively compressed. This measure is quantified using the metric compression ratio (CR).

The chapter is organized as follows: Section 3.2 discusses the state of the art of compression algorithms for both SM and PMU data. The traditional data compression based on two-dimensional reduction is explained in Section 3.3. Section 3.4 introduces the mathematical background of data compression through three-dimensional reduction using tensor decompositions and the advantages and limitations of each presented decompositions. Section 3.5 provides the main results for both SM and PMU data. Finally, conclusions are discussed in Section 3.6.

3.2 State of the art

In general terms, data-compression methods are categorized in lossless compression and lossy compression methods [26]. In lossless compression methods, the reconstructed data are identical to the original data; while lossy compression methods is more popular to preserve the relevant information by allowing a small recosntruction error, thus achieving a higher compression ratio and extracting useful information during the process. It is important to mention that both approaches exploit the correlation and redundancy in the data. On the one hand, the lossy compression techniques use the correlation to create statistic models that measure the frequency of a specific symbol treating the input data as generic files without differentiating among text, images, or time series. On the other hand, the lossy techniques exploit the spatial and temporal correlation present in the data and create a lower-dimensional representation.

The compression performance in lossy and lossless techniques is evaluated through the compression ratio (CR) index, which measures the ratio between the size of the original data S_o and the size of the compressed data S as follows:

$$\mathrm{CR} = S_o / S, \tag{3.1}$$

where the higher the value of CR, the smaller the magnitude of the compressed file is.

To measure the reconstruction error in lossy techniques, there is not a standard metric; however, normalized mean absolute error (MAE) and normalized root mean squared error (RMSE) have been selected in this chapter, calculated as

$$\mathrm{MAE} = \frac{\frac{1}{N}\sum_{i=1}^{N} \left|\hat{x}_i - x_i\right|}{\frac{1}{N}\sum_{i=1}^{N} x_i} \tag{3.2}$$

$$\mathrm{RMSE} = \frac{\sqrt{\frac{1}{N}\sum_{i=1}^{N} \left|\hat{x}_i - x_i\right|^2}}{\frac{1}{N}\sum_{i=1}^{N} x_i} \tag{3.3}$$

where x_i is the original value, $\hat{x}_i$ is the estimated value, and N represents the total number of samples in the time series.

The following subsections show the main results reported in the literature for both compression classifications applied to SM and PMU data.

3.2.1 Compression data techniques for SM data

Three of the most popular lossless methods to compress SM data are A-XDR [25], DEGA [10], and LZMH [26]. The comparison among those three algorithms was stated in [10], where the CRs achieved were ~1.8, ~4.7, and 4.0, respectively. However, in the same work, the authors conclude that fine granularity (high time sampling) is required to achieve a higher CR, for instance, intervals of up to 3 s–30 s. However, since most domestic electricity consumption standards require 15-minute intervals, lossy approaches represent a better option for achieving higher CR in the SM data compression.

Table 3.1 Brief summary of compression techniques for SM data.

Technique	Type of compression	Reconstruction error	CR	Reference
A-XDR	Lossless	–	~1.8	[12]
DEGA	Lossless	–	~4.7	[29]
LZMH	Lossless	–	4	[30]
SVD	Lossy	RMSE 0.70	6	[26], [27]
FLDC	Lossy	MPPE 5.57	55.71	[28]

Among the most important lossy compression techniques for SM data are those based on singular value decomposition (SVD), in [26] and [27] a CR of 6 and RMSE of 0.70 was achieved. Recently in [28], a feature-based load data compression method (FLDC) was proposed, archiving a CR of 55.71 and a reconstruction error MPPE equal to 5.57. However, although the high CR reported especially for the lossy techniques, their reconstruction error is high. Consequently, the development of compression algorithms for SM datasets that achieve high CR and low compression error is still an open challenge. Table 3.1 summarizes the lossless and lossy compression techniques for SM data.

3.2.2 Compression data techniques for PMU data

This state-of-the-art review for PMU data compression techniques focuses only on voltage measurements. This is because most of the reported techniques in the literature achieve the lowest CR for voltage measurements. Therefore the compression of voltage measurements is one of the biggest challenges in PMU data compression.

In [31], five lossless methods for PMU data compression were compared, namely gzip, bzip2, 7-zip, szip, and szip-i, where szip-i achieved the largest CR (2.77) for the measurements of two different PMUs. A variation of the delta-encoding approach referred to as slack referenced encoding (SRE) in combination with three standard compression tools Deflate, BZip2 and LZMA was proposed in [4]; the value of the CR achieved was 2.45 for the measurements from 1 PMU. The Erwan–Ranganathan compression version (ER1c) algorithm in combination with UHA was investigated in [32], where a high CR value of 100.89 was achieved for the measurements of 100 PMUs.

Under the principle that the measurements from different PMUs are correlated and redundant [33], the data can be stored in low-dimensional representations. In other words, lossy compression algorithms can be used to achieve high CR value and low reconstruction error. Then, most of the lossy techniques are based on dimensional reduction approaches. In [33], the authors exploit the temporal and spatial correlation using principal components analysis (PCA), discrete wavelet cosine transform (DWCT), and discrete wavelet transform (DWT); datasets of 4 PMUs from a local area are compressed, achieving a CR of 8.93. In [34] is shown an application based on singular value decomposition (SVD) with a CR of 92.3. The authors do not address the number of PMUs, but the data come from 50 substations to assume at least the same number of PMUs. An improved SVD-based approach was recently developed in [16],

Table 3.2 Brief summary of compression techniques for PMU data.

Technique	Type of compression	No. PMU	Reconstruction error	CR	Reference
sz-i	Lossless	2	–	2.77	[31]
SRE	Lossless	1	–	2.45	[4]
ER1c+ zipx	Lossless	100	–	100.89	[32]
PCA+DWCT + DWT	Lossy	4	MADE 0.456×10^{-3}	8.93	[33]
SVD	Lossy	50 (substations)	MPE: 4.15/MAE:0.036	92.3	[34]
Improved SVD	Lossy	13	MAE 0.003	8.94	[18]
MSPCA	Lossy	193	NMSE 5.35×10^{-9}	18.223	[35]

where CR of 8.94 for the voltage of 13 PMU was achieved. Multiscale principal component analysis (MSPCA) was proposed in [35] to compress the measurements of 193 PMUs, obtaining a CR of 18.22. Table 3.2 shows a brief summary of the mentioned lossless and lossy compression techniques for PMU voltage data.

Despite the high compression ratios obtained with the previous lossless and lossy approaches for PMU data, it must be highlighted that to obtain high CR, both approaches (lossless and lossy) employ the principle introduced in [33], which is to benefit from the properties such as correlation and redundancy inherent on the data from different PMUs. This property was demonstrated in [36], meaning that the more correlation and redundancy on the measured data, the higher the compression ratio achieved; therefore the number of bits required to store the data grows sublinearly with the density of the PMUs in the power grid. It is worth noticing that up to now, methods that achieve large CRs depend on these on the above statements, and they fail when these particular conditions are not met. This can be easily visualized in Table 3.2, where the CR increases as the number of PMUs also increases, regardless of the techniques used. Note that in [35] the results seem to contradict the previous statement; however, the low reconstruction error should be noted, which suggests that the authors preferred to obtain a smaller reconstruction error, rather than a high CR. As a result, the development of compression algorithms for PMU datasets without apparent correlation and redundancy is still an open challenge, e.g., historical data of a single PMU in power systems.

3.3 Two-dimensional reduction for data compression based on singular value decomposition (SVD)

The power system is interconnected in its different voltage levels through the transmission and distribution lines. This induces a spatial and temporal correlation in the measurements collected by different SMs and PMUs close to each other. Then, under this principle, most of the lossy compression algorithms for SM and PMU data are based on dimension reduction using singular value decomposition (SVD) or one of its variants. This section shows how to arrange and compress the data into two-dimensional (2D) arrays based on the SVD. Furthermore, this 2D approach represents a good framework for understanding the tensor-based approach and its advantages.

Given a time series $\boldsymbol{x}_i$ $\{\boldsymbol{x}_i = [x_{i1}, \ldots, x_{im}, \ldots, x_{iM}]\} \in \mathbb{R}^{1 \times M}$ with M time samples, I different time series can be arranged into a matrix as follows:

$$\mathbf{X} = \begin{bmatrix} x_{11} & \cdots & x_{1m} & \cdots & x_{1M} \\ \vdots & \ddots & \vdots & \ddots & \vdots \\ x_{i1} & \cdots & x_{im} & \cdots & x_{iM} \\ \vdots & \ddots & \vdots & \ddots & \vdots \\ x_{I1} & \cdots & x_{Im} & \cdots & x_{IM} \end{bmatrix} \in \mathbb{R}^{I \times M};$$

$$\text{with } i = 1, 2, \ldots, I; \; m = 1, 2, \ldots, M, \tag{3.4}$$

where the *i-th* row holds the *i-th* time series $\boldsymbol{x}_i$ (dashed line), and the columns represent the I measurements at the *m-th* time instance (dotted line). Then, the size in bytes of the matrix $\mathbf{X} \in \mathbb{R}^{I \times M}$ can be calculated as

$$S = 8\,(I \times M)\,, \tag{3.5}$$

where the number 8 represents the storage size in bytes necessary to represent a double-precision value [37], and the product $(I \times M)$ calculates the number of elements of the matrix $\mathbf{X} \in \mathbb{R}^{I \times M}$. Then, to compress the data, SVD can be applied, where the $\mathbf{X}$ matrix with the time series is factorized as the product of three matrices, which have been truncated in their first σ singular values:

$$\hat{\mathbf{X}} = \mathbf{U}^{\sigma}\, \boldsymbol{\Sigma}^{\sigma}\, \mathbf{V}^{\sigma T}, \tag{3.6}$$

where the matrix $\hat{\mathbf{X}}$ represents the approximation of the original matrix $\mathbf{X}$; σ represents the number of selected singular values, according to the variance explained calculated as $\upsilon_{SVD} = \frac{\sigma_r^2}{\sum_{r=1}^{R} \sigma_r^2} * 100$ [22]; the matrices $\mathbf{U}^{\sigma}$ and $\mathbf{V}^{\sigma}$ hold the first σ left and right singular vectors, respectively; and the matrix $\boldsymbol{\Sigma}^{\sigma}$ keeps the first σ singular values. After that, the data of the matrix $\mathbf{X}$ can be described by the matrices $\boldsymbol{\Sigma}^{\sigma}$, $\mathbf{U}^{\sigma}$, and $\mathbf{V}^{\sigma}$, which represents the compressed version of the original data; and the size of this compressed version S_{SVD} is given by the following expression:

$$S_{SVD} = 8\,(U_s + \Sigma_s + V_s)\,; \quad \text{with } U_s = I \times \sigma;\, \Sigma_s = \sigma \times \sigma;\, V_s = M \times \sigma, \tag{3.7}$$

where the number 8 represent the storage size in bytes necessary for defining a double-precision value [37], U_s is the stored space required for the matrix $\mathbf{U}^{\sigma}$, and Σ_s and V_s are the stored spaces for the matrices $\mathbf{\Sigma}^{\sigma}$ and $\mathbf{V}^{\sigma}$, respectively. From Eq. (3.7), it must be noted that in this traditional 2D approach, most of the final size of the compressed data will be weighted by 1) the number of rows (if $I > J$) or by 2) the number of columns (if $J > I$), since $\sigma \ll J, I$.

3.4 Tensor decomposition for data compression through three-dimensional reduction

In this section the compression through three-dimensional reduction using tensor decomposition is stated. To do so, the tensor background is presented first, as well two tensor decomposition approaches are explained, parallel factor model (PARAFAC) and Tucker. Finally, a numerical example is given for every decomposition to understand the compression process.

3.4.1 Tensor background

A tensor is a multidimensional array, and its order is the number of dimensions, also known as modes [38]. In this work, a three-dimensional (3D) tensor is used to arrange SM and PMU data. Therefore following the standardized notation for tensor decomposition introduced in [39], scalars and vectors are designated by lowercase italics x and lowercase italics in bold $\boldsymbol{x}$, matrices by uppercase letters in bold $\mathbf{X}$, and a 3D tensor is represented by bold underlined letters $\underline{\mathbf{X}}$. The $i-th$ input of vector $\boldsymbol{x} \in \mathbb{R}^{I}$ is denoted as x_i, the element i, j from the matrix $\mathbf{X} \in \mathbb{R}^{I \times J}$ is represented as x_{ij}, whereas the element i, j, k from the tensor $\underline{\mathbf{X}} \in \mathbb{R}^{I \times J \times K}$ is denoted as x_{ijk}. In the following sections the subarrays of a 3D-tensor called slices, and the unfolding process are used; both terminologies are explained below.

3.4.1.1 Slices of a 3D tensor

The slices are subarrays of a tensor and are computed fixing one index and shifting the other two indexes (see Fig. 3.1). Based on this, three different slices can be computed for a 3D tensor: horizontal (Fig. 3.1.a), fixing only the index "i" $\left(\underline{\mathbf{X}}(i,:,:)\right)$; lateral (Fig. 3.1.b), where the index "j" remain constant $\left(\underline{\mathbf{X}}(:,j,:)\right)$; and frontal (Fig. 3.1.c), where only the index "k" is fixed $\left(\underline{\mathbf{X}}(:,:,k)\right)$.

3.4.1.2 Unfolding process for a 3D tensor

The unfolding process reorganizes the elements of a tensor into a matrix, having three different cases for a 3D tensor: unfolding mode-1 $\underline{\mathbf{X}}_{(1)}$ (Fig. 3.2.a), unfolding mode-2 $\underline{\mathbf{X}}_{(2)}$ (Fig. 3.2.b), and unfolding mode-3 $\underline{\mathbf{X}}_{(3)}$ (Fig. 3.2.c).

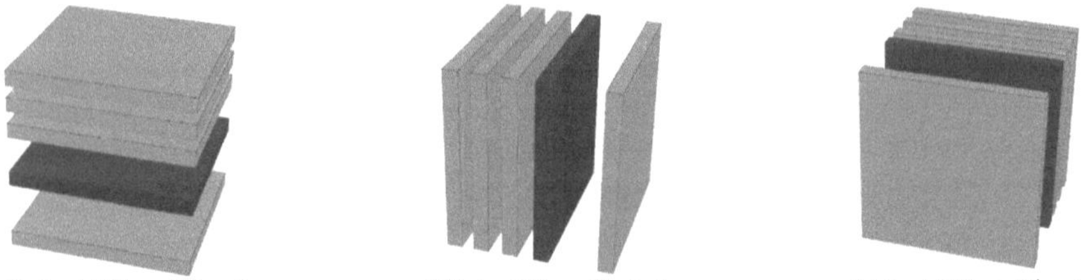

(a) Horizontal Slices: $\underline{\mathbf{X}}(i,:,:)$ (b) Lateral Slices: $\underline{\mathbf{X}}(:,j,:)$ (c) Frontal Slices: $\underline{\mathbf{X}}(:,:,k)$

Figure 3.1 The three different types of slices for a 3D tensor.

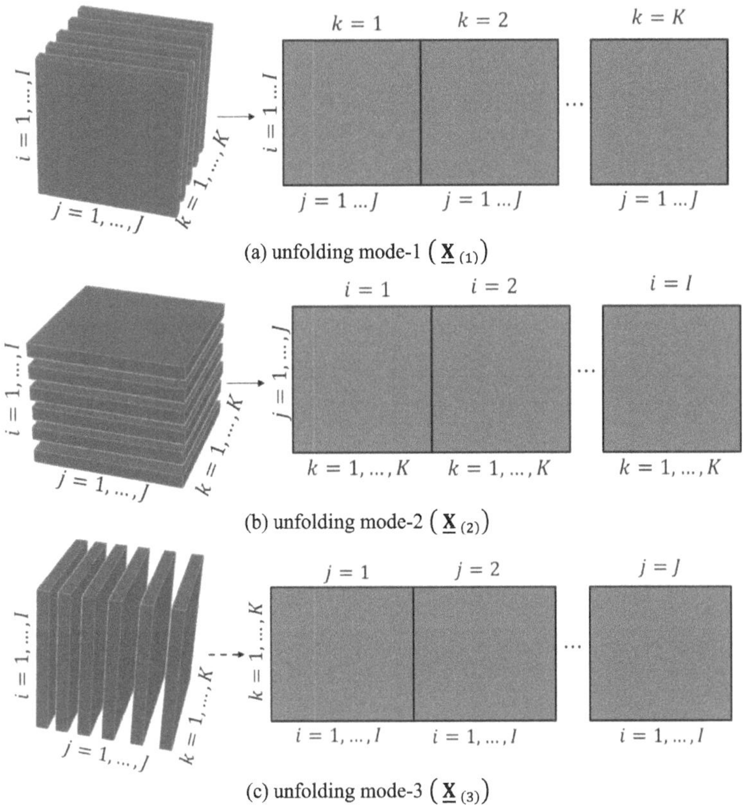

Figure 3.2 The three cases of the unfolding process for a 3D tensor.

3.4.2 Tensor representation

In this section, the way of arranging the data into a 3D tensor is explained. Let us assume a time series $\boldsymbol{x}_i$ $\{\boldsymbol{x}_i = [x_{i1}, \ldots, x_{iJ}, \ldots, x_{iJ}]\}$ that has been measured for K different conditions; then the information for I different time series can be organized into a three-dimensional array $\underline{\mathbf{X}} \in \mathbb{R}^{I \times J \times K}$ (3D-tensor) as follows:

$$\underline{\mathbf{X}}(:,:,1) = \begin{bmatrix} x_{111} & \cdots & x_{1j1} & \cdots & x_{1J1} \\ \vdots & \ddots & \vdots & \ddots & \vdots \\ x_{i11} & \cdots & x_{ij1} & \cdots & x_{iJ1} \\ \vdots & \ddots & \vdots & \ddots & \vdots \\ x_{I11} & \cdots & x_{Ij1} & \cdots & x_{IJ1} \end{bmatrix}, \cdots,$$

$$\underline{\mathbf{X}}(:,:,K) = \begin{bmatrix} x_{11K} & \cdots & x_{1jK} & \cdots & x_{1JK} \\ \vdots & \ddots & \vdots & \ddots & \vdots \\ x_{i1K} & \cdots & x_{ijK} & \cdots & x_{iJK} \\ \vdots & \ddots & \vdots & \ddots & \vdots \\ x_{I1K} & \cdots & x_{IjK} & \cdots & x_{IJK} \end{bmatrix},$$

$$\text{with } \underline{\mathbf{X}}(:,:,k) \in \mathbb{R}^{I \times J}; i = 1, 2, \ldots, I; j = 1, 2, \ldots, J; k = 1, 2, \ldots, K \quad (3.8)$$

where every condition has been allocated in one frontal slice, i.e., the *k-th* condition is allocated into the *k-th* frontal slice $\underline{\mathbf{X}}(::, k)$ (see Fig. 3.3). Note that a condition corresponds to a third variable that has been considered, for instance, the geographic position where the measurements were taken. Then, the size of the data contained in the tensor can be calculated with the following expression:

$$S = 8\,(I \times J \times K)\,, \quad (3.9)$$

where the number 8 represents the storage size in bytes for a double-precision value, I is the number of time series, and J and K the number of time samples and the number of conditions, respectively.

3.4.3 PARAFAC tensor decomposition for data compression

The parallel factor model (PARAFAC) tensor decomposition was first introduced by Hitchcock in 1927 [40]. Given a 3D tensor $\underline{\mathbf{X}} \in \mathbb{R}^{I \times J \times K}$, PARAFAC factorizes it into R triads of vectors (see Fig. 3.3) defined as

$$min \left\| \underline{\mathbf{X}} - \hat{\underline{\mathbf{X}}} \right\|_F^2 \text{ with } \hat{\underline{\mathbf{X}}} = \sum_{r=1}^{R} \boldsymbol{a}_r \circ \boldsymbol{b}_r \circ \boldsymbol{c}_r, \quad (3.10)$$

where $\|\cdot\|_F$ denotes the Frobenius norm, $\hat{\underline{\mathbf{X}}}$ is the calculated approximation (which minimizes the cost function in terms of least square), R is called the rank of the decomposition, "$\circ$" represents the outer vector product, and the vectors $\boldsymbol{a}_r$, $\boldsymbol{b}_r$, and $\boldsymbol{c}_r$ belong to the load matrices $\mathbf{A} \in \mathbb{R}^{I \times R}$, $\mathbf{B} \in \mathbb{R}^{J \times R}$, and $\mathbf{C} \in \mathbb{R}^{K \times R}$, respectively, as shown in Eq. (3.11).

$$\begin{aligned} \mathbf{A} &= \left[\boldsymbol{a}_1, \ldots, \boldsymbol{a}_p, \ldots, \boldsymbol{a}_R\right] \in \mathbb{R}^{I \times R}, \boldsymbol{a}_p = \left[a_{1r}, \ldots, a_{ir}, \ldots a_{Ir}\right]^T \in \mathbb{R}^I \\ \mathbf{B} &= \left[\boldsymbol{b}_1, \ldots, \boldsymbol{b}_q, \ldots, \boldsymbol{b}_R\right] \in \mathbb{R}^{J \times R}, \boldsymbol{b}_q = \left[b_{1r}, \ldots, b_{jr}, \ldots b_{Jr}\right]^T \in \mathbb{R}^J \end{aligned} \quad (3.11)$$

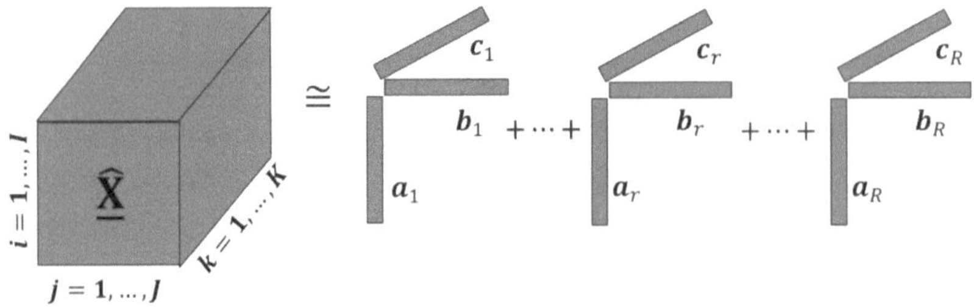

Figure 3.3 PARAFAC decomposition for a 3D tensor.

$$\mathbf{C} = [\boldsymbol{c}_1, \ldots, \boldsymbol{c}_r, \ldots, \boldsymbol{c}_R] \in \mathbb{R}^{K \times R}, \boldsymbol{c}_r = [c_{1r}, \ldots, c_{kr}, \ldots c_{Kr}]^T \in \mathbb{R}^K$$

To carry out the solution of Eq. (3.10), the alternating least squares (ALS) algorithm is used [41], which iteratively estimates the load matrices. After the decomposition, the original data are now in a low-dimensional representation since $R \ll I, J, K$, i.e., in a compressed version. The size of compressed data using PARAFAC expressed in bytes using double precision is given by the following expression:

$$S_P = 8\,(A_P + B_P + C_P)\,; \text{ with } A_P = I \times R;\; B_P = J \times R;\; C_P = K \times R, \quad (3.12)$$

where the number 8 represents the storage size in bytes necessary for representing a double-precision value, A_P is the stored space required to store the matrix **A**, and B_P and C_P represent the stored space required to store the matrix **B** and **C**, respectively. Then, to retrieve the data, the element x_{ijk} of the tensor $\hat{\underline{\mathbf{X}}}$ can be reconstructed using the information contained in the matrices **A**, **B** , and **C** as follows:

$$\hat{x}_{ijk} = \sum_{r=1}^{R} a_{ir} b_{jr} c_{kr} \; for \; i = 1, 2, \ldots, I, j = 1, 2, \ldots, J, k = 1, 2, \ldots, K, \quad (3.13)$$

where the element $\hat{x}_{ijk}$ represent the approximation of x_{ijk}.

3.4.3.1 *Analysis of the PARAFAC tensor decomposition and the "principle of parallel proportional profiles"*

To gain more in-depth analysis of PARAFAC tensor decomposition, the principle of parallel proportional profiles [20] is first explained. This principle states that the same set of vectors $(\boldsymbol{a}_r, \boldsymbol{b}_r, \boldsymbol{c}_r)$ that describes a data set can describe other data sets. For instance, let us call one data set to one frontal slice of the tensor from Eq. (3.8), which corresponds to the *k-th* condition ($\underline{\mathbf{X}}\,(:,:,k)$). Then, after applying PARAFAC, the reconstruction of this slice is given by the following expression:

$$\hat{\underline{\mathbf{X}}}\,(:,:,k) = \mathbf{A}\; diag\,(\boldsymbol{c}_k)\, \mathbf{B}^T, \quad (3.14)$$

where $diag\,(\cdot)$ is an algebraic operation that transforms a vector into a diagonal matrix, and the row vector $\boldsymbol{c}_k$ $\{\boldsymbol{c}_r = [c_{k1}, \ldots, c_{kr}, \ldots c_{kR}]^T \in \mathbb{R}^R\}$ obtained from **C** contains the c_{kr} values for the $k - th$ slices of the tensor. From Eq. (3.14), it can be noted that the K total data sets (frontal slices) of the tensor can be described with the same vectors $\boldsymbol{a_r}$ and $\boldsymbol{b_r}$ from the matrices $\mathbf{A}$ and $\mathbf{B}$ only weighted by the vectors $\boldsymbol{c_k}$ in different proportions. Therefore every data set from Eq. (3.8) will be reconstructed using the matrices $\mathbf{A}$ and $\mathbf{B}$ and weighted by its respective row of the matrices $\mathbf{C}$. Note that the principle states the properties of PARAFAC decomposition, establishing that the data sets (slices) must have correlation across the modes; therefore it is not possible to fit data in which the correlation between slices changes over time; otherwise it might lead to degenerated solutions. It must be noted that since the PARAFAC model is symmetric in its three modes, the principle applies for all the three modes.

After explaining the relationship between PARAFAC and the principle of parallel proportional profiles, the following points can be concluded:

(i) To use the PARAFAC model to factorize a tensor, the power system data stored in the tensor must follow the principle of parallel proportional profiles, otherwise the PARAFAC decomposition would lead to a degenerated solution.

(ii) This means that the data must be "homogeneous", i.e., there must be a strong temporal and spatial correlation among the data of different data sets (different slices).

3.4.3.2 An illustrative example of PARAFAC decomposition for data compression

To gain insight on the PARAFAC decomposition used as a compression technique, an illustrative example is presented in this subsection.

Let us assume a 3D tensor $\underline{\mathbf{X}} \in \mathbb{R}^{8\times4\times2}$ composed as follows:

$$\underline{\mathbf{X}}(:,:,1)\begin{bmatrix} 0.4885 & 0.3922 & 0.2981 & 0.4742 \\ 0.5272 & 0.4266 & 0.3365 & 0.5228 \\ 0.4430 & 0.3568 & 0.2755 & 0.4339 \\ 0.0348 & 0.1125 & 0.3852 & 0.3074 \\ 0.0040 & 0.0803 & 0.3340 & 0.2531 \\ -0.0192 & 0.0762 & 0.3824 & 0.2776 \\ 0.4931 & 0.3961 & 0.3021 & 0.4795 \\ 0.0419 & 0.1228 & 0.4093 & 0.3291 \end{bmatrix}$$

$$\underline{\mathbf{X}}(:,:,2)\begin{bmatrix} 0.7002 & 0.5634 & 0.4332 & 0.6841 \\ 0.7751 & 0.6108 & 0.4806 & 0.7478 \\ 0.6348 & 0.5119 & 0.3974 & 0.6237 \\ 0.0380 & 0.1077 & 0.3552 & 0.2864 \\ -0.0051 & 0.0661 & 0.2990 & 0.2221 \\ -0.0405 & 0.0509 & 0.3341 & 0.2304 \\ 0.7067 & 0.5689 & 0.4383 & 0.6912 \\ 0.0474 & 0.1194 & 0.3791 & 0.3092 \end{bmatrix},$$

$$\text{with } \underline{\mathbf{X}}(:,:,k) \in \mathbb{R}^{8\times4} \tag{3.15}$$

where the size of the data is equal to 512 bytes calculated through Eq. (3.5) as $8(8 \times 4 \times 2) = 512$ bytes.

Then, selecting a rank with a value equal to $R = 2$, the tensor is factorized into two triads of vectors $([\boldsymbol{a}_1, \boldsymbol{b}_1, \boldsymbol{c}_1], [\boldsymbol{a}_2, \boldsymbol{b}_2, \boldsymbol{c}_2])$ as follows:

$$\mathbf{A} = \begin{vmatrix} \boldsymbol{a}_1 & \boldsymbol{a}_2 \\ 1.4879 & -0.0207 \\ 1.6028 & 0.0046 \\ 1.3483 & -0.0093 \\ 0.0365 & 0.6642 \\ -0.0512 & 0.6064 \\ -0.1339 & 0.7218 \\ 1.5016 & -0.0188 \\ 0.0541 & 0.7003 \end{vmatrix} \in \mathbb{R}^{8\times 2}, \quad \mathbf{B} = \begin{vmatrix} \boldsymbol{b}_1 & \boldsymbol{b}_2 \\ 0.5744 & 0.0464 \\ 0.4642 & 0.2094 \\ 0.3638 & 0.7691 \\ 0.5675 & 0.6019 \end{vmatrix} \in \mathbb{R}^{4\times 2},$$

$$\mathbf{C} = \begin{vmatrix} \boldsymbol{c}_1 & \boldsymbol{c}_2 \\ 0.5724 & 0.7389 \\ 0.8199 & 0.6737 \end{vmatrix} \in \mathbb{R}^{2\times 2}.$$

Now, the original data is represented by the three matrices, **A**, **B**, and **C**; and the size of this compressed version can be calculated using Eq. (3.12) as $S_P = 8((8 \times 2) + (4 \times 2) + (2 \times 2)) = 224$ bytes.

To evaluate the compression performance, firstly, the compression ratio is calculated using Eq. (3.1), archiving a value of 2.28 (512/224). Then, to calculate the reconstruction error, all the tensor elements are reconstructed using Eq. (3.13). For illustrative purposes, the process of the reconstruction of the elements x_{111} and x_{121} from the slice $\underline{\mathbf{X}}(:,:,1)$, and the elements x_{112} and x_{122} from the slice $\underline{\mathbf{X}}(:,:,2)$ is shown below:

$\hat{x}_{111} = a_{11} \cdot b_{11} \cdot c_{11} + a_{12} \cdot b_{12} \cdot c_{11}$
$\hat{x}_{121} = a_{11} \cdot b_{12} \cdot c_{11} + a_{12} \cdot b_{22} \cdot c_{12}$
$\hat{x}_{111} = 1.4879 \cdot 0.5744 \cdot 0.5724 + -0.0207 \cdot 0.0464 \cdot 0.7389$
$\hat{x}_{121} = 1.4879 \cdot 0.4642 \cdot 0.5724 + -0.0207 \cdot 0.2094 \cdot 0.7389$
$\hat{x}_{121} = 0.3921$
$\hat{x}_{111} = 0.4884$
$\hat{x}_{122} = a_{11} \cdot b_{12} \cdot c_{21} + a_{12} \cdot b_{22} \cdot c_{22}$
$\hat{x}_{112} = a_{11} \cdot b_{11} \cdot c_{21} + a_{12} \cdot b_{12} \cdot c_{22}$
$\hat{x}_{112} = 1.4879 \cdot 0.5744 \cdot 0.8199 + -0.0207 \cdot 0.0464 \cdot 0.6737$
$\hat{x}_{122} = 1.4879 \cdot 0.4642 \cdot 0.8199 + -0.0207 \cdot 0.2094 \cdot 0.6737$
$\hat{x}_{112} = 0.7001$
$\hat{x}_{122} = 0.5634$

The results for this compression example are summarized in Table 3.3.

3.4.4 Tucker tensor decomposition for data compression

The Tucker model decomposition was proposed in 1963 by Tucker [42]. Given a 3D tensor, the Tucker decomposition factorized it into a core tensor multiplied by a matrix

Table 3.3 Compression results for the illustrative example using PARAFAC.

$\mathbf{A} \in \mathbb{R}^{8\times 2}$	$\mathbf{B} \in \mathbb{R}^{4\times 2}$	$\mathbf{C} \in \mathbb{R}^{2\times 2}$	**Total size**	**CR**	**Reconstruction error** $\left(\left\|\underline{\mathbf{X}}-\hat{\underline{\mathbf{X}}}\right\|_F^2\right)$
128	64	32	224	2.2857	4.069×10^{-15}

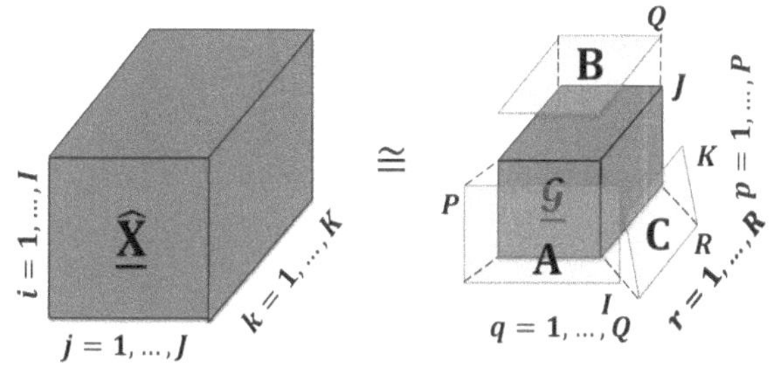

Figure 3.4 Tucker decomposition for a 3D tensor.

along each of its modes (see Fig. 3.4), as described Eq. (3.16).

$$\min_{\hat{\underline{\mathbf{X}}}} \left\|\underline{\mathbf{X}}-\hat{\underline{\mathbf{X}}}\right\| \text{ with } \hat{\underline{\mathbf{X}}} = \underline{\mathcal{G}} \times_1 \mathbf{A} \times_2 \mathbf{B} \times_3 \mathbf{C}$$
$$\text{subject to } \underline{\mathcal{G}} \in \mathbb{R}^{P\times Q\times R}, \mathbf{A} \in \mathbb{R}^{I\times P}, \mathbf{B} \in \mathbb{R}^{J\times Q}, \mathbf{C} \in \mathbb{R}^{K\times R}, \tag{3.16}$$

where "$\times_n$" denotes the *n-mode* product of a tensor with a matrix; $\underline{\mathcal{G}} \in \mathbb{R}^{P\times Q\times R}$ is the core tensor and shows the level of interaction between the different components; $\mathbf{A} \in \mathbb{R}^{I\times R}$, $\mathbf{B} \in \mathbb{R}^{J\times Q}$, and $\mathbf{C} \in \mathbb{R}^{K\times R}$ are non-singular matrices and are called the load matrices, which are composed of the vector $\boldsymbol{a}_r$, $\boldsymbol{b}_r$, and $\boldsymbol{c}_r$, as depicted in Eq. (3.17).

$$\begin{aligned} \mathbf{A} &= \left[\boldsymbol{a}_1, \ldots, \boldsymbol{a}_p, \ldots, \boldsymbol{a}_R\right] \in \mathbb{R}^{I\times P}, \boldsymbol{a}_p = [a_1 \ldots a_I]^T \in \mathbb{R}^P \\ \mathbf{B} &= \left[\boldsymbol{b}_1, \ldots, \boldsymbol{b}_q, \ldots, \boldsymbol{b}_R\right] \in \mathbb{R}^{J\times Q}, \boldsymbol{b}_q = [b_1 \ldots b_J]^T \in \mathbb{R}^Q, \\ \mathbf{C} &= [\boldsymbol{c}_1, \ldots, \boldsymbol{c}_r, \ldots, \boldsymbol{c}_R] \in \mathbb{R}^{K\times R}, \boldsymbol{c}_r = [c_1 \ldots c_K]^T \in \mathbb{R}^R \end{aligned} \tag{3.17}$$

where P, Q, and R are the number of columns in the factor matrices components, also called ranks, and represent the number of components of **A**, **B**, and **C**, respectively. Then, if $R, Q, R \ll I, J, K$, the Tucker decomposition is called the truncated core version.

After the decomposition, the original data are now in a compressed version, and the size S_{Tuk} expressed in bytes using double precision is given by the following expression:

$$S_{Tuk} = 8\left(\underline{\mathcal{G}}_s + A_s + B_s + C_s\right); \qquad \text{with } A_P = I \times R; B_P = J \times R; C_P = K \times R, \tag{3.18}$$

where represents $\underline{\mathcal{G}}_s$ is the stored spaces required to store the core $\underline{\mathcal{G}}$; and A_s, B_s, and C_s are the stored spaces required to store the matrices **A**, **B**, and **C**, respectively.

After the compression, the tensor can be reconstructed using the element wise Tucker form to retrieve the compressed information, as shown Eq. (3.19).

$$\hat{x}_{ijk} = \sum_{p=1}^{P}\sum_{q=1}^{Q}\sum_{r=1}^{R} g_{pqr}a_{ip}b_{jq}c_{kr}$$
$$for\ i = 1, 2, \ldots, I,\ j = 1, 2, \ldots, J,\ k = 1, 2, \ldots, K, \tag{3.19}$$

where the element $\hat{x}_{ijk}$ represent the approximation of x_{ijk}.

3.4.4.1 Analysis of the Tucker tensor decomposition

In principle, the Tucker mode can decompose any data set into a tensor core and the load matrices. Mainly because Tucker allows combining elements of different ranks to reconstruct a tensor element. That can be noted from Eq. (3.19), where to compute the element $\hat{x}_{ijk}$ of the tensor, the P elements of the $i-th$ row from the matrix $\mathbf{A} \in \mathbb{R}^{I\times P}$ are combined with the Q elements of the $j-th$ row from the matrix $\mathbf{B} \in \mathbb{R}^{J\times Q}$, and with the R elements of the $k-th$ row from the matrix $\mathbf{C} \in \mathbb{R}^{K\times R}$, as well. Then, every element is reconstructed by a total of $P \times Q \times R$ factors, which also corresponds to the total number of elements in the core tensor $\underline{\mathcal{G}} \in \mathbb{R}^{P\times Q\times R}$. After the analysis of the Tucker tensor decomposition, the following points can be concluded:

(i) In principle, Tucker decomposition can decompose any tensor, which means any dataset can be factorized using Tucker.

(ii) In the case of electrical power systems, this means that the data can be "heterogeneous," i.e., there can be variations between different datasets. In other words, data may be weakly correlated or not correlated at all.

3.4.5 An illustrative example of Tucker decomposition for data compression

To gain insight into the Tucker decomposition for data compression, in this subsection an illustrative example is given. To do so, the test tensor presented in Eq. (3.15) is decomposed through Eq. (3.16). The computed core tensor and load matrices are shown below:

$$\mathbf{A} = \begin{array}{cc} \boldsymbol{a}_1 & \boldsymbol{a}_2 \end{array} \left|\begin{array}{cc} -0.4697 & -0.1693 \\ -0.5128 & -0.1635 \\ -0.4280 & -0.1467 \\ -0.1793 & 0.4592 \\ -0.1366 & 0.4280 \\ -0.1394 & 0.5170 \\ -0.4745 & -0.1693 \\ -0.1940 & 0.4825 \end{array}\right|, \quad \mathbf{B} = \begin{array}{cc} \boldsymbol{b}_1 & \boldsymbol{b}_2 \end{array} \left|\begin{array}{cc} 0.5198 & 0.6226 \\ 0.4436 & 0.2463 \\ 0.4304 & -0.7107 \\ 0.5896 & -0.2154 \end{array}\right|,$$

$$\mathbf{C} = \begin{vmatrix} 0.6013 & -0.7989 \\ 0.7989 & 0.6013 \end{vmatrix} \quad (\text{columns } c_1, c_2)$$

$$\underline{\mathcal{G}}(:,:,1) = \begin{vmatrix} -3.1187 & -0.0102 \\ 0.0030 & -0.8437 \end{vmatrix}, \qquad \underline{\mathcal{G}}(:,:,2) = \begin{vmatrix} -0.0375 & -0.0553 \\ -0.2270 & 0.1383 \end{vmatrix}$$

$$\quad g_{11} \quad g_{21} \qquad\qquad g_{12} \quad g_{22}$$

Now, the data is represented by the three matrices, **B**, **C**, and the tensor core $\underline{\mathcal{G}}$, and the size using Eq. (3.17) is $S_P = 8\left((2 \times 2 \times 2) + (8 \times 2) + (4 \times 2) + (2 \times 2)\right) =$ 288bytes. To calculate the reconstruction error, firstly, all the elements of the tensor are reconstructed using Eq. (3.19). For illustrative purposes, the process of the reconstruction of the elements x_{111} and x_{121} from the slice $\underline{\mathbf{X}}(:,:,1)$, and the elements x_{112} and x_{122} from the slice $\underline{\mathbf{X}}(:,:,2)$ are shown below:

$$\begin{aligned}\hat{x}_{111} = {} & g_{111} \cdot a_{11} \cdot b_{11} \cdot c_{11} + g_{211} \cdot a_{12} \cdot b_{11} \cdot c_{11} + g_{121} \cdot a_{11} \cdot b_{12} \cdot c_{11} \\ & + g_{221} \cdot a_{12} \cdot b_{12} \cdot c_{11} + g_{112} \cdot a_{11} \cdot b_{11} \cdot c_{12} + g_{212} \cdot a_{12} \cdot b_{11} \cdot c_{12} \\ & + g_{122} \cdot a_{11} \cdot b_{12} \cdot c_{12} + g_{222} \cdot a_{12} \cdot b_{12} \cdot c_{12}\end{aligned}$$

$$\begin{aligned}\hat{x}_{111} = {} & -3.1187 \cdot -.4697 \cdot 0.5198 \cdot 0.6013 + 0.0030 \cdot -0.1693 \cdot 0.5198 \cdot 0.6013 \\ & + -0.0102 \cdot -0.4697 \cdot 0.6226 \cdot 0.6013 + -0.8437 \cdot -0.1693 \\ & \cdot 0.6226 \cdot 0.6013 + -0.0375 \cdot -.4697 \cdot 0.5198 \cdot -0.7389 + -0.2270 \\ & \cdot -0.1693 \cdot 0.5198 \cdot -0.7389 + -0.0553 \cdot -0.4697 \cdot 0.6226 \cdot -0.7389 \\ & + 0.1383 \cdot -0.1693 \cdot 0.6226 \cdot -0.7389\end{aligned}$$

$$\hat{x}_{111} = 0.4885$$

$$\begin{aligned}\hat{x}_{121} = {} & g_{111} \cdot a_{11} \cdot b_{21} \cdot c_{11} + g_{211} \cdot a_{12} \cdot b_{21} \cdot c_{11} + g_{121} \cdot a_{11} \cdot b_{22} \cdot c_{11} \\ & + g_{221} \cdot a_{12} \cdot b_{22} \cdot c_{11} + g_{112} \cdot a_{11} \cdot b_{21} \cdot c_{12} + g_{212} \cdot a_{12} \cdot b_{21} \cdot c_{12} \\ & + g_{122} \cdot a_{11} \cdot b_{22} \cdot c_{12} + g_{222} \cdot a_{12} \cdot b_{22} \cdot c_{12}\end{aligned}$$

$$\hat{x}_{121} = 0.3922$$

$$\begin{aligned}\hat{x}_{112} = {} & g_{111} \cdot a_{11} \cdot b_{11} \cdot c_{21} + g_{211} \cdot a_{12} \cdot b_{11} \cdot c_{21} + g_{121} \cdot a_{11} \cdot b_{12} \cdot c_{21} \\ & + g_{221} \cdot a_{12} \cdot b_{12} \cdot c_{21} + g_{112} \cdot a_{11} \cdot b_{11} \cdot c_{22} + g_{212} \cdot a_{12} \cdot b_{11} \cdot c_{22} \\ & + g_{122} \cdot a_{11} \cdot b_{12} \cdot c_{22} + g_{222} \cdot a_{12} \cdot b_{12} \cdot c_{22}\end{aligned}$$

$$\hat{x}_{112} = 0.7002$$

$$\begin{aligned}\hat{x}_{122} = {} & g_{111} \cdot a_{11} \cdot b_{21} \cdot c_{21} + g_{211} \cdot a_{12} \cdot b_{21} \cdot c_{21} + g_{121} \cdot a_{11} \cdot b_{22} \cdot c_{21} \\ & + g_{221} \cdot a_{12} \cdot b_{22} \cdot c_{21} + g_{112} \cdot a_{11} \cdot b_{21} \cdot c_{22} + g_{212} \cdot a_{12} \cdot b_{21} \cdot c_{22} \\ & + g_{122} \cdot a_{11} \cdot b_{22} \cdot c_{22} + g_{222} \cdot a_{12} \cdot b_{22} \cdot c_{22}\end{aligned}$$

Table 3.4 Compression Results for the illustrative example using Tucker.

$\mathcal{G} \in \mathbb{R}^{2\times2\times2}$	$A \in \mathbb{R}^{8\times2}$	$B \in \mathbb{R}^{4\times2}$	$C \in \mathbb{R}^{2\times2}$	**Total size**	**CR**	**Reconstruction error** $\left(\left\|\underline{X}-\hat{\underline{X}}\right\|_F^2\right)$
64	128	64	32	288	1.7777	2.1371e-15

$\hat{x}_{122} = 0.534$

The compression performance is evaluated using the compression ratio index and the reconstruction error $\left(\left\|\underline{X}-\hat{\underline{X}}\right\|_F^2\right)$; Table 3.4 summarizes the results.

From Tables 3.3 and 3.4 it can be observed that the PARAFAC decomposition achieved lower CR than the Tucker decomposition (2.28>1.77); however, Tucker achieved lower reconstruction error ($2.1371\times10^{-15} < 4.069\times10^{-15}$).

3.5 Applications using power system data

This section presents the effectiveness of the proposed approach using the two presented tensor decompositions on power system data. First, electrical homogeneous electrical load profiles of smart meters from the ERCOT database in the USA are used as data input to show the performance of PARAFAC. Then, heterogeneous three-phase voltage measurements of PMUs from *ENTSO-E* in Europe are compressed using the Tucker decomposition.

3.5.1 *Data compression for homogeneous smart meter data using PARAFAC decomposition*

The theoretical background of PARARAF decomposition for homogeneous data compression has been presented. In this section, the results of applying the proposed methodology to SM data are presented. To do so, the repository open-source data provided on the website of the Electrical Reliability Council of Texas (ERCOT) is used as input [43]. The historical data is collected from eight different geographic zones displayed in Fig. 3.5.a: 1) Coast, 2) East, 3) Far West, 4) North Central, 5) North, 6) South Central, 7) South, and 8) West. In this chapter, the summers (92 days) from 2000 to 2006 of the business low load profile type (BLLP type) are selected to be compressed, and the load profiles are displayed in Fig. 3.5.b. Note that those load profiles were selected since they belong to the same type of profile and the geographical regions are close to each other; therefore they represent a good example of multivariable homogeneous SM data. The load profiles were collected using 15 minutes sampling time to complete 24 hrs. The information is originally stored in a 2D array $\mathbf{X} \in \mathbb{R}^{(5152)\times96}$ in the form of comma-separated values (CVS) file, where the subscript "5152" rep-

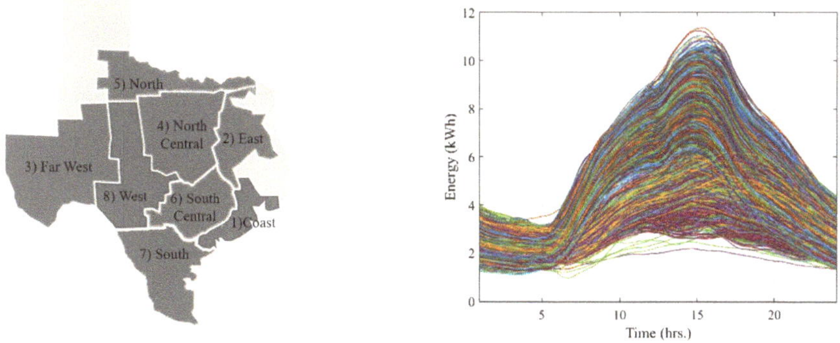

(a) Geographical zones of ERCOT

(b) Load profiles of the summers from 2000 to 2006 of the Business Low Load Profile (BLLP) type)

Figure 3.5 ERCOT Smart Meter data.

resents the total number of load profiles for the summer, considering the seven years and the eight geographic areas (92×7×8) (see Fig. 3.6.a).

3.5.1.1 SM data compression based on three-dimensional reduction using PARAFAC decomposition

To apply PRAFAC tensor decomposition, the data must be arranged into a tensor. However, note that since the data are multivariable (days × time samples × geographical areas), they can be organized naturally into a 3D tensor of dimensions 644 × 96 × 8, as shown in Fig. 3.6.b, where "644" indicates the number of load profiles for the seven years (92 × 7), "96" represents the number of samples per day (24 × 15), and "8" is the number of geographical areas in the ERCOT system. The total size of the data is 565.248 kB, which was calculated using Eq. (3.5). Note that this size is the same for the tensor arrangement and the matrix arrangement.

To compress the tensor constructed with the information, the PARAFAC decomposition of $\underline{\mathbf{X}} \in \mathbb{R}^{644\times96\times8}$ is calculated with Eq. (3.10), using $R = 3$ selected using CORCONDIA algorithm [44]. After the decomposition of $\underline{\mathbf{X}}$ has been computed, the matrices $\mathbf{A} \in \mathbb{R}^{644\times2}$, $\mathbf{B} \in \mathbb{R}^{96\times2}$, and $\mathbf{C} \in \mathbb{R}^{8\times2}$ hold the compresses version of the data.

Table 3.5 summarizes the results of the compression obtained using PARAFAC decomposition, where the final size is 17.953 kB (Eq. (3.12)), and the compressed file is about 31 (CR=31.484) times smaller than the original file.

To complete the results of Table 3.5, Fig. 3.7.a displays the original load profiles of three different days that were randomly chosen, namely 1st, 20th, and 50th day of the summer of the year 2000 (solid line in black) over their reconstruction (dotted line in red). It can be observed that both traces in Fig. 3.7.a follow the same pattern. Moreover, Fig. 3.7.b displays the box-whisker plot of the metrics MAE and RMSE for all the load profiles (5152), with a mean of 0.045 and 0.038, respectively. Thus demonstrating that in addition to a high CR, PARAFAC also achieves a low reconstruction error.

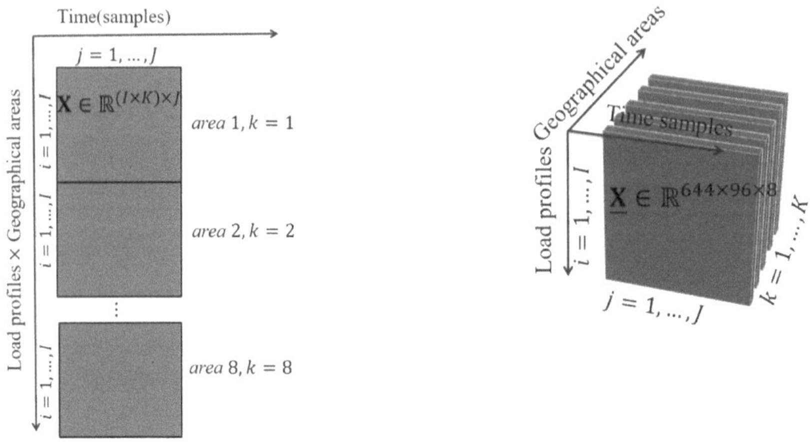

(a) 2D array $\mathbf{X} \in \mathbb{R}^{5152\times96}$ b) 3D array $\underline{\mathbf{X}} \in \mathbb{R}^{644\times96\times8}$

Figure 3.6 Different ways to store SM data.

Table 3.5 Compression Results for SM data using Parafac.

Method	Size of matrices			Total kB	CR
PARAFAC	$\mathbf{A} \in \mathbb{R}^{644\times3}$	$\mathbf{B} \in \mathbb{R}^{96\times3}$	$\mathbf{C} \in \mathbb{R}^{8\times3}$	17.953	31.484
$\underline{\mathbf{X}} \in \mathbb{R}^{644\times96\times8}$	10.304 kB	1.536 kB	0.128 kB		

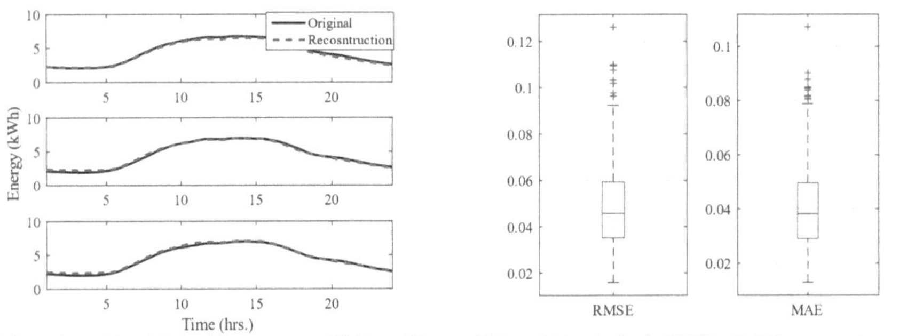

(a)Comparison of the original and the reconstructed SM data of the days 1st, 20th, 50th of the summer of 2000.

(b) Box-whisker plot for the RMSE and MAE reconstruction errors for SM data using PARAFAC.

Figure 3.7 Result of the reconstruction after the data compression based on Tensor decomposition using PARAFAC for the smart meter measurements.

3.5.1.2 SM data compression based on two-dimensional reduction using SVD decomposition

To compare the results, the data in its original 2D representation is compressed using SVD, since this approach has been widely used for SM data compression. Then, the matrix $\mathbf{X} \in \mathbb{R}^{5152\times96}$, which represents the original data arrangement, is factorized using SVD using Eq. (3.6), where the number of singular values needed to represent the 90% of the variance explained was taken from Fig. 3.8, thus $\sigma = 2$. It must be

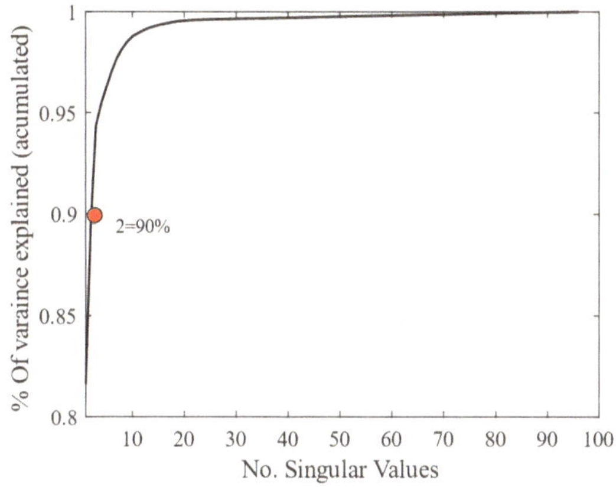

Figure 3.8 Variance explained (%) associated to the matrix $\mathbf{X} \in \mathbb{R}^{5152\times 96}$.

Table 3.6 Compression Results for SM data Using SVD.

Method	Size of matrices			Total kB	CR
SVD	$\mathbf{U} \in \mathbb{R}^{5152\times 2}$	$\mathbf{\Sigma} \in \mathbb{R}^{2\times 2}$	$\mathbf{V} \in \mathbb{R}^{96\times 2}$	84.000	6.7291
$\mathbf{X} \in \mathbb{R}^{5152\times 96}$	82.432 kB	0.032 kB	1.536 kB		

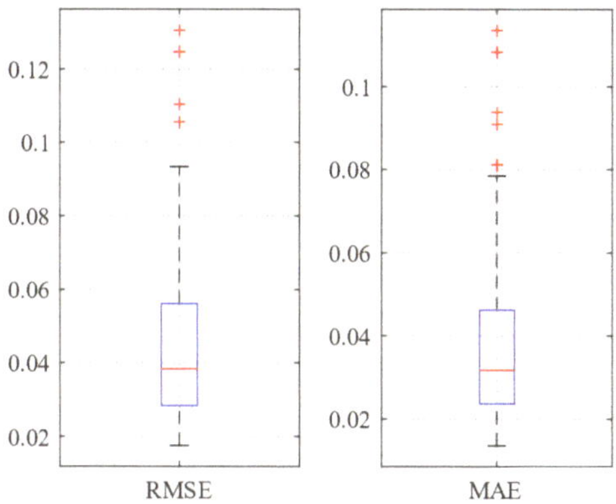

Figure 3.9 Box-whisker plot for the RMSE and MAE reconstruction errors for SM data using SVD.

highlighted that this value was selected because it represents a goof trade-off between the error reconstruction and the CR.

The results after the compression using SVD are shown in Table 3.6, and Fig. 3.10. The results show that the total size of the file is 84 kB, representing only 6.7 smaller than the original size; whereas the means of the reconstruction errors are similar to those obtained with PARAFAC, 0.039 and 0.35, respectively.

From Table 3.5 it can be noted that the CR obtained through the proposed approach using PARAFAC is considerably larger than the obtained with SVD (31.48>6.72), whereas Fig. 3.9 shows that the mean of reconstruction errors are similar for both approaches. (See Figs. 3.7.b and 3.9.)

3.5.2 Data compression for heterogeneous PMU data using Tucker decomposition

The theoretical background of Tucker decomposition for heterogeneous data compression has been presented. This section presents the results of applying the proposed methodology to PMU data. To do so, the voltage measurements from the European Network of Transmission System Operators (ENTSO) collected by the power system grid lab of Zurich University of Applied Science (ZHAW) in Switzerland are used as input. All phasor measurements (provided by ABB's RES 670 and National Instruments PMUs) were collected to a sampling rate of 25 Hz. Each measurement point consists of three measurement channels generating three-phase voltage phasors. The PMU voltage data is originally stored in an Excel file in comma-separated values (CSV) format, i.e., into a 2D array $\mathbf{X} \in \mathbb{R}^{30\times 2160000}$, as shown in Fig. 3.10.a, where the subscript "30" is the total number of voltage measurements corresponding to the three-phase system for the ten days (10×3), and "2160000" is the number of samples per day. The size of the generated data is equal to 518.400 MB.

3.5.3 PMU data compression based on three-dimensional reduction using Tucker decomposition

To understand the advantage of storing the data into a tensor to compress, it must be highlighted that if the data is compressed using this 2D representation the final weight will be weighted mostly by the number of time samples ($M = 2160000$) since $M << I$ (see Eq. (3.7)). Therefore, to reduce the final size of the original file, the total M samples per day can be represented as the sum of K shorter time series J. Then M can be replaced with the product of $J \times K$, with J representing the number of time samples, and K indicating the number of time windows. Now, the matrix $\mathbf{X} \in \mathbb{R}^{I\times M}$ is renamed $\mathbf{X} \in \mathbb{R}^{I\times(J\times K)}$, and these data can now be stored in the tensor $\underline{\mathbf{X}} \in \mathbb{R}^{I\times J\times K}$, where every window k was collocated in one frontal slice. Note that the process of rearranging the data into a tensor not only reduces substantially the storage space required, but also capitalizes the correlation between the slices of minutes, i.e., K windows; thus achieving a high CR for a single PMU, even when there is apparently no correlation in the data.

For this work, the number of time samples per window was set at $J = 1500$, note that this value is the number samples per minute (25×60), and it was selected since a minute time window represents a good trade-off between the CR and the reconstruction error. Then the dimension of the tensor are $30 \times 1500 \times 1400$, where the superscript "30" indicates the total number of voltage measurements corresponding to the three-phase system for the ten days, "1500" is the number of time samples in one minute ($30 \times 60 \times 60$), and "1400" is the total number of minutes in one day. The tensor constructed tensor is displayed in Fig. 3.10.b.

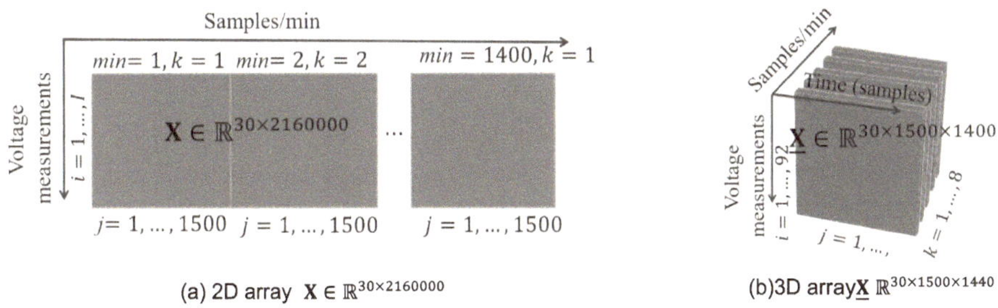

Figure 3.10 Different ways to store PMU data.

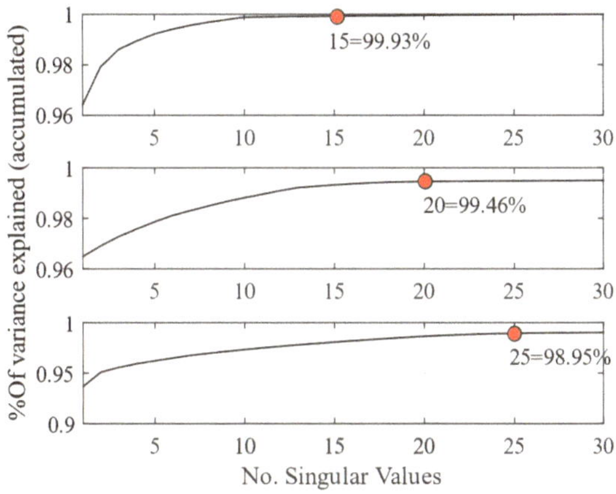

Figure 3.11 Percentage of variance explained associated at unfolding mode-1 $\underline{\mathbf{X}}_{(1)} \in \mathbb{R}^{30\times(1440\times1500)}$, unfolding mode-2 $\underline{\mathbf{X}}_{(2)} \in \mathbb{R}^{1400\times(1500\times30)}$, and unfolding mode-3$\underline{\mathbf{X}}_{(3)} \in \mathbb{R}^{1500\times(30\times1400)}$.

Then, to apply Tucker decomposition to the created tensor, the three input values of the ranks P, Q, and R, are calculated using the singular values of the unfolding mode-1, mode-2, and mode-3, respectively [45]. Fig. 3.11 shows their values and their percentage of variance explained, where the selected values are $P = 15$, $Q = 20$, and $R = 25$.

After the decomposition of $\underline{\mathbf{X}} \in \mathbb{R}^{30\times1440\times1500}$, the data are now in their compressed version represented by the matrices $\mathbf{A} \in \mathbb{R}^{30\times15}$, $\mathbf{B} \in \mathbb{R}^{1440\times20}$, $\mathbf{C} \in \mathbb{R}^{1500\times25}$, and the core tensor $\underline{\mathcal{G}} \in \mathbb{R}^{15\times20\times25}$; where the final achieved size was 591.6 kB, representing a reduction size of 876 times smaller than the original file (CR of 876.267) (see Table 3.7).

Fig. 3.12a depicts the comparison of the original and reconstructed PMU voltage measurements for the phase A (V_a) voltage from the January 1st, 5th, 10th, 2019, for the first 5 hours of each day. To complete the results, Fig. 3.12b shows the reconstruction error in the box-whicker for all the voltage time series, with a mean of 4.9×10^{-4} and 3.35×10^{-4} for MAE and RMSE, respectively. From Fig. 3.12.a, it can be ob-

Table 3.7 Compression Results for PMU data Using Tucker.

Method	Size of matrices				Total kB	CR
Tucker	$\mathcal{G} \in \mathbb{R}^{15\times20\times25}$	$\mathbf{A} \in \mathbb{R}^{15\times30}$	$\mathbf{B} \in \mathbb{R}^{20\times1500}$	$\mathbf{C} \in \mathbb{R}^{25\times1440}$	591.600	876.267
$\underline{\mathbf{X}} \in \mathbb{R}^{30\times1500\times1440}$	60 kB	3.6 kB	240 kB	288 kB		

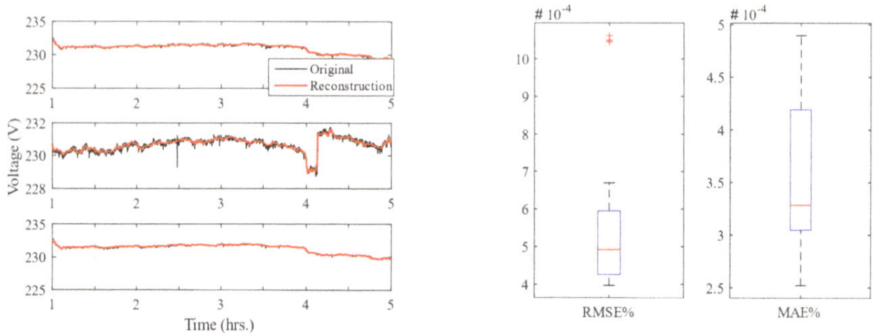

Comparison of the original and reconstructed PMU data of the five first hours of January 1st, 5th, 10th 2019 using Tucker

(a) Box-whisker plot for the RMSE and MAE reconstruction errors for PMU data using Tucker

Figure 3.12 Evaluation of the compress data using Tucker decomposition.

Table 3.8 Compression Results for PMU using SVD.

Method	Size of matrices			Total kB	CR
SVD	$\mathbf{U}\in\mathbb{R}^{30\times15}$	$\mathbf{\Sigma}\in\mathbb{R}^{15\times15}$	$\mathbf{V}\in\mathbb{R}^{2160000\times15}$	259,205.4	1.999
$\mathbf{X}\in\mathbb{R}^{30\times2160000}$	3.6 kB	1.800 kB	259200 kB		

served that the original signal (solid black line) and the reconstructed (dotted red line) follow the same pattern.

3.5.3.1 PMU data compression based on three-dimensional reduction using SVD decomposition

To have a broader overview of the effectiveness of the proposed compression using Tucker, the results are compared again against SVD. Note that to compute SVD the number of singular values can be taken from the unfolding mode-1 in Fig. 3.11 since $\mathbf{X} \in \mathbb{R}^{I\times(J\times K)} = \underline{\mathbf{X}}_{(1)} \in \mathbb{R}^{I\times(J\times K)}$, thus $\sigma = 20$. The results after compressing the data using SVD are summarized in Table 3.8 and Fig. 3.13.

From Table 3.8 and Table 3.7, it can be observed that Tucker achieves a significantly larger CR (876.267) in comparison with SVD (1.999). From Fig. 3.12.b and Fig. 3.13.b, it can be noted that the reconstruction error of Tucker is slightly greater than the obtained with SVD. However, it must be highlighted that this is because Tucker works as a filter, reducing the noise in the voltage measurement (see Fig. 3.12.a and Fig. 3.13.a). Therefore it is evident that the comparison between the original measurements and the filtered reconstruction will impact the error calculation. Note that the approach based on Tucker archived high CR and low reconstruction error for our

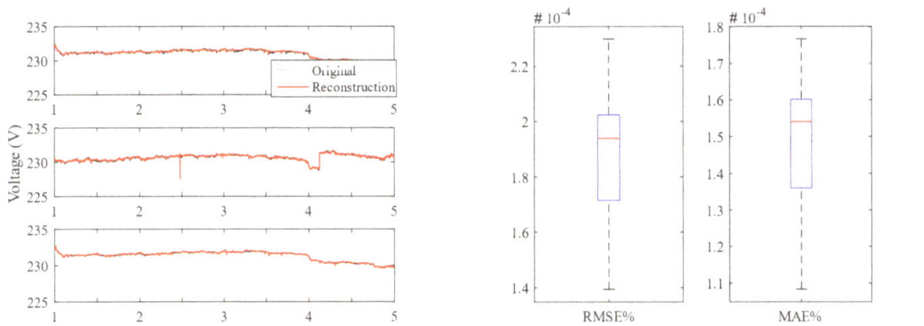

(a)Comparison of the original and reconstructed PMU data of the five first hours of January 1st ,5th, 10th 2019 using SVD

(b) Box-whisker plot for the RMSE and MAE reconstruction errors for PMU data using SVD

Figure 3.13 Results of the compression PMU voltage using SVD decomposition.

test data coming from a single PMU; therefore the proposed approach is a good compression tool for uncorrelated data.

3.6 Conclusions

In this chapter, advances in compressing SM and PMU data based on tensor decomposition were presented. First, the mathematical framework of the two main tensor decompositions used for compression was shown. Then, illustrative examples of both decompositions were presented to illustrate the compression process, understand the properties of each decomposition, and the type of input data that can be compressed.

Two case studies were presented to test the theoretical guarantee of the proposed tensor approaches. First, homogeneous electric load profiles from SMs of Electric Reliability Council of Texas (ERCOT) were compressed using PARAFAC tensor decomposition, showing the way to build a tensor directly from multivariable data. Then, the compression of heterogeneous voltage measurements from the European Network of Transmission System Operators (ENTSO-E) was presented using the Tucker tensor decomposition. In this second application, the tensor was built from data that was originally arranged into a 2D matrix, thus demonstrating that the data may not be multidimensional to be compressed with the tensor decomposition approaches.

To underline the ability of the proposed methodology to achieve higher compression ratios, the results were compared against a traditional low-dimensional approach, such as SVD.

The results confirm the advantage of the proposed tensor dimensional reduction approach over the 2D dimensional approach, achieving a higher CR and a low error reconstruction. Moreover, the main contribution of the methodology presented is the formulation of arranging the time series into a tensor representation, which is a higher-order matrix, thus achieving higher compression ratios and, at the same time, preserving the main features of the original data.

Appendix 3.A

The illustrative examples for the PARAFAC and Tucker decompositions in this chapter were computed using the N-way toolbox for Matlab® [46]. The codes used to run the routine are shown below.

```
%Routine for the illustrative examples (PARAFAC and Tucker)
%This routine computes the PARAFAC toy example chapter
%To run the routine, the N-WAY tool box is necessary, (see ref.[42])
% ------INPUT--------
% X      3D tensor
% Rp      No. of rank for PARAFAC tensor decomposition
% Pt      No. of rank for Tucker decomposition for matrix A
% Qt      No. of rank for Tucker decomposition for matrix B
% Rt      No. of rank for Tucker decomposition for matrix C
%---------OUTPUT-----
%Ap,Bp,Cp   Load matrices PARAFAC
%At,Bt,Cp   Load matrices Tucker
%ep        Reconstruction error PARAFAC
%et        Reconstruction error Tucker
%Xp        Reconstructed Tensor PARAFAC
%Xt        Reconstructed Tensor Tucker

X(:,:,1)=[0.4885 0.3922 0.2981 0.4742; 0.5272 0.4266 0.3365 0.5228
          0.4430 0.3568 0.2755 0.4339; 0.0348 0.1125 0.3852 0.3074
          0.0040 0.0803 0.3340 0.2531;-0.0192 0.0762 0.3824 0.2776
          0.4931 0.3961 0.3021 0.4795; 0.0419 0.1228 0.4093 0.3291];

X(:,:,2)=[0.7002 0.5634 0.4332 0.6841; 0.7751 0.6108 0.4806 0.7478
          0.6348 0.5119 0.3974 0.6237; 0.0380 0.1077 0.3552 0.2864
         -0.0051 0.0661 0.2990 0.2221;-0.0405 0.0509 0.3341 0.2304
          0.7067 0.5689 0.4383 0.6912; 0.0474 0.1194 0.3791 0.3092];

Rp=2; Pt=2; Qt=2; Rt=2;

[Factors_p] = parafac(X,Rp);  Ap=Factors_p{1,1}; Bp=Factors_p{1,2};
Cp=Factors_p{1,3};
[Factors_t,G]=tucker(X,[Pt Qt Rt]); At=Factors_t{1,1};
Bt=Factors_t{1,2}; Ct=Factors_t{1,3};
%TENSOR RECOSNTRUCTION
%PARAFAC
    for i=1:length(Ap(:,1))
       for j=1:length(Bp(:,1))
           for k=1:length(Cp(:,1))
              for r=1:Rp;
```

```
                xp(r)=Ap(i,r)*Bp(j,r)*Cp(k,r) ;
            end
            Xp(i,j,k)=sum(xp);
         end; end; end
 %Tucker
       for i=1:length(At(:,1))
          for j=1:length(Bt(:,1))
              for k=1:length(Ct(:,1))
                   for p=1:Pt
                      for q=Qt
                          for r=Rt
                   xt(p,q,r)=G(p,q,r)*At(i,p)*Bt(j,q)*Ct(k,r);
                          end
                      end
                   end
               Xt(i,j,k)=sum(sum(sum(xt)));
       end; end; end

%Reconstruction Error
ep=sum(sum(sum(X-Xp)));  %PARAFAC
et=sum(sum(sum(X-Xt)));  %Tucker

%SIZE OF THE TENSOR (S)
I=length(X(:,1,1)); J=length(X(1,:,1)); K=length(X(1,1,:)); S=8*(I*J*K);
%RESUL
%PARAFAC
Ap=I*Rp; Bp=J*Rp; Cp=K*Rp; Sp=8*(Ap+Bp+Cp);CRp=S/Sp;
%Tucker
Gt=Pt*Qt*Rt; At=I*Pt; Bt=J*Qt; Ct=K*Rt; St=8*(Gt+At+Bt+Ct);CRt=S/St;
```

References

[1] European Commission DG Energy, European Smart Metering Benchmark, Belgium, 2019.

[2] A. Shivakumar, et al., Smart energy solutions in the EU: state of play and measuring progress, Energy Strateg. Rev. 20 (January 2018) 133–149.

[3] Sarma Nuthalapati, Power System Grid Operation Using Synchrophasor Technology, Springer International Publishing, Vancouver, WA, USA, 2019.

[4] R. Klump, P. Agarwal, J.E. Tate, H. Khurana, Lossless compression of synchronized phasor measurements, in: IEEE PES Gen. Meet. PES 2010, 2010, pp. 1–7.

[5] B. Pilkington, The role of semiconductors in clean energy, [Online]. Available https://www.azocleantech.com/article.aspx?ArticleID=965.

[6] Y. Wang, Q. Chen, T. Hong, C. Kang, Review of smart meter data analytics: applications, methodologies, and challenges, IEEE Trans. Smart Grid 10 (3) (2019) 3125–3148.

[7] M.U. Usman, M.O. Faruque, Applications of synchrophasor technologies in power systems, J. Mod. Power Syst. Clean Energy 7 (2) (2019) 211–226.
[8] A.U.E. Information, How many smart meters are installed in the United States, and who has them?, [Online]. Available https://www.eia.gov/tools/faqs/faq.php?id=108&t=3. (Accessed 1 April 2020).
[9] L.L. Ireri, M. Arturo, R. Enrique, C.G. Marx, G.J. Antonio, Smart Grids en México: Situación actual, retos y propuesta de implementación Smart Grid in Mexico: Current situation, challenges and implementation, Ingeniería, investigación y tecnología 20 (2) (2019).
[10] A.G. Phadke, T. Bi, Phasor measurement units, WAMS, and their applications in protection and control of power systems, J. Mod. Power Syst. Clean Energy 6 (4) (2018) 619–629.
[11] Tractebel, European smart metering benchmark – Benchmarking Smart Metering Deployment in the EU-28, June 2019, pp. 1–128.
[12] A. Unterweger, D. Engel, M. Ringwelski, The effect of the data granularity on load data compression, in: S. Gottwalt, L. König, S. Hartmut (Eds.), Energy Informatics, Springer, Cham, 2015, pp. 69–80.
[13] N. Yu, S. Shah, R. Johnson, R. Sherick, M. Hong, K. Loparo, Big data analytics in power distribution systems, in: 2015 IEEE Power Energy Soc. Innov. Smart Grid Technol. Conf. ISGT 2015, 2015, pp. 1–5.
[14] NASPI, Time Synchronization in the Electric Power System, March 2017, p. 59.
[15] N. American, S. Initiative, T. Report, NASPI 2020 Survey of Industry Best Practices for Archiving Synchronized Measurements Acknowledgments, 2020.
[16] B.P. Bhattarai, et al., Big data analytics in smart grids: state-of-the-art, challenges, opportunities, and future directions, IET Smart Grid 2 (2) (2019) 141–154.
[17] M. Ghorbanian, S.H. Dolatabadi, P. Siano, Big data issues in smart grids: a survey, IEEE Syst. J. 13 (4) (2019) 4158–4168.
[18] J. Zhao, Y. Ye, H. Yu, P. Li, J. Wu, C. Wang, Improved SVD-based data compression method for synchronous phasor measurement in distribution networks, Int. J. Electr. Power Energy Syst. 129 (February 2021) 106877.
[19] F. Cong, Q.H. Lin, L.D. Kuang, X.F. Gong, P. Astikainen, T. Ristaniemi, Tensor decomposition of EEG signals: a brief review, J. Neurosci. Methods 248 (2015) 59–69.
[20] R.B. Cattell, Parallel proportional profiles and other principles for determining the choice of factors by rotation, Psychometrika 9 (4) (1944) 267–283.
[21] C.J. Appellof, E.R. Davidson, Strategies for analyzing data from video fluorometric monitoring of liquid chromatographic effluents, Anal. Chem. 53 (13) (1981) 2053–2056.
[22] B. Sandoval, B. Emilio, P. Korba, F.R. Segundo Sevilla, Three-way unsupervised data mining for power system applications based on tensor decomposition, Electr. Power Syst. Res. 187 (March 2020) 106431.
[23] B. Sandoval, P. Korba, F.R.S. Sevilla, E.B. Espejo, A tensor decomposition approach for contingency screening and coherency identification in power systems, in: 2021 IEEE Madrid PowerTech, PowerTech 2021 – Conf. Proc., 2021.
[24] M.A.O. Vasilescu, D. Terzopoulos, Multilinear analysis of image ensembles: tensorfaces, Lect. Notes Comput. Sci. (including Subser. Lect. Notes Artif. Intell. Lect. Notes Bioinformatics) 2350 (2002) 447–460.
[25] A. Karami, M. Yazdi, G. Mercier, Hyperspectral image compression based on Tucker decomposition and wavelet transform, in: Work. Hyperspectral Image Signal Process. Evol. Remote Sens., 2011, pp. 1–4.

[26] Y. Wang, Q. Chen, C. Kang, Q. Xia, M. Luo, Sparse and redundant representation-based smart meter data compression and pattern extraction, IEEE Trans. Power Syst. 32 (3) (2017) 2142–2151.
[27] Y. Wang, et al., Residential smart meter data compression and pattern extraction via non-negative K-SVD, in: IEEE Power Energy Soc. Gen. Meet. 2016-Novem no. 51325702, 2016, pp. 1–5.
[28] Y. Wang, Q. Chen, C. Kang, Smart meter data analytics: Electricity consumer behavior modeling, aggregation, and forecasting, 2020.
[29] A. Unterweger, D. Engel, Resumable load data compression in smart grids, IEEE Trans. Smart Grid 6 (2) (2015) 919–929.
[30] M. Ringwelski, C. Renner, A. Reinhardt, A. Weigel, V. Turau, The Hitchhiker's guide to choosing the compression algorithm for your smart meter data, in: 2012 IEEE Int. Energy Conf. Exhib. ENERGYCON 2012, 2012, pp. 935–940.
[31] P. Top, J. Breneman, Compressing phasor measurement data, in: 45th North Am. Power Symp. NAPS 2013, 2013, pp. 1–4.
[32] E. Olivo, M. Campion, P. Ranganathan, Data compression for next generation phasor data concentrators (PDCs) in a smart grid, J. Inf. Secur. 07 (05) (2016) 291–296.
[33] P.H. Gadde, M. Biswal, S. Brahma, H. Cao, Efficient compression of PMU data in WAMS, IEEE Trans. Smart Grid 7 (5) (2016) 2406–2413.
[34] J.C.S. De Souza, T.M. Lessa Assis, B.C. Pal, Data compression in smart distribution systems via singular value decomposition, IEEE Trans. Smart Grid 8 (1) (2017) 275–284.
[35] G. Lee, D.I. Kim, S.H. Kim, Y.J. Shin, Multiscale PMU data compression via density-based WAMS clustering analysis, Energies 12 (4) (2019).
[36] S. Kirti, Z. Wang, A. Scaglione, R. Thomas, On the communication architecture for wide-area real-time monitoring in power networks, in: Proc. Annu. Hawaii Int. Conf. Syst. Sci., 2007.
[37] D. Xue, Linear Algebra and Matrix Computations with MATLAB, Illustrate, De Gruyter STEM, Shenyang, China, 2020.
[38] T.G. Kolda, B.W. Bader, Tensor decompositions and applications, SIAM Rev. 51 (3) (2009) 455–500.
[39] H.A.L. Kiers, Towards a standardized notation and terminology in multiway analysis, J. Chemom. 14 (3) (2000) 105–122.
[40] F.L. Hitchcock, The expression of a tensor or a polyadic as a sum of products, J. Math. Phys. 6 (1–4) (1927) 164–189.
[41] T.G. Kolda, B.W. Bader, Tensor decompositions and applications, SIAM Rev. 51 (3) (2009) 455–500.
[42] L. Tucker, Some mathematical notes on three-mode factor analysis, Psychometrika 31 (3) (1996).
[43] Electrical Reliability Council of Texas, ERCOT Repository of Load Profile Data, [Online]. Available https://www.ercot.com/mktinfo/loadprofile.
[44] R. Bro, H.A.L. Kiers, A new efficient method for determining the number of components in PARAFAC models, J. Chemom. 17 (5) (2003) 274–286.
[45] B. Sandoval, E. Barocio Espejo, P. Korba, F.R.S. Sevilla, Data Compression for Advance Sensing and Communication Technology in Smart Grids based on Tensor Decomposition, Sustain. Energy, Grids Networks.
[46] C.A. Andersson, R. Bro, The N-way Toolbox for MATLAB, Chemom. Intell. Lab. Syst. 52 (1) (2000) 1–4.

Machine learning and digital twins: monitoring and control for dynamic security in power systems

4

Christoph Brosinsky[a], *Mert Karaçelebi*[b], *and Jochen L. Cremer*[b]
[a]Technische Universität Ilmenau, Ilmenau, Germany, [b]Delft University of Technology, Delft, Netherlands

Indexes

i	System state
j	Control set-point

Symbols

$\hat{y}_i$	Prediction of time discrete sample i
u_j	Values of control set-point j
$\overline{x}$	Constrained differential state
x_i	Values describing the system state i
N	Horizon length (number of samples)
p	Model parameter set
y, y^m	Output signal obtained by measurements

Acronyms

AC	Alternating current
AI	Artificial Intelligence
ANN	Artificial neural network
AVR	Automatic voltage regulator
DAE	Differential algebraic equations
DSA	Dynamic security assessment
DSO	Distribution system operator
DT	Digital twin
EMS/DMS	Energy management system / Distribution management system
FPR	False positive rate
hf	high fidelity
HVAC	High voltage alternating current
ICT	Information and Communications Technology
KF	Kalman filter
lf	low fidelity
MHE	Moving horizon estimation
ML	Machine learning
MLP	Multi-layer perceptron
PE	Parameter estimation
ROC	Receiver operator characteristics
SSA	Static security assessment
TPR	True positive rate
TSO	Transmission system operator

Monitoring and Control of Electrical Power Systems Using Machine Learning Techniques
https://doi.org/10.1016/B978-0-32-399904-5.00010-7

VSC-HVDC Voltage source converter high voltage direct current transmission system
WAMS Wide area monitoring system

4.1 Introduction

This chapter connects techniques from machine learning (ML) and digital twins (DTs) to provide novel insights into monitoring and control of (dynamic) security for electrical power systems. DTs are validated and verified high-fidelity (hf) models of high simulation accuracy. They can be applied to simulate the supervised process (e.g., power system operation) and provide synthetic data whenever measurement data is scarce. However, such hf simulation models are not always appropriate for real-time applications in monitoring and control as they correspond to high computational effort. The computational effort, i.e., the time required to perform simulations can be significantly reduced by applying surrogate models, which may be mandatory for some applications executed in real-time. There, ML can create an application-specific, low-fidelity (lf) approximation of the hf digital twin. These trained lf models are promising in real-time applications, where the time to react is scarce and lf information is sufficient. This chapter aims at providing a conceptual overview of combining the advantages of hf and lf models.

The combined framework of hf DT and lf ML is illustrated in Fig. 4.1. The two models, hf DT and lf ML, collect data from the power system (measurements and topology) and from the system operator. The hf model processes and generates the required data for the lf model, where the lf model feeds predictions and relevant study

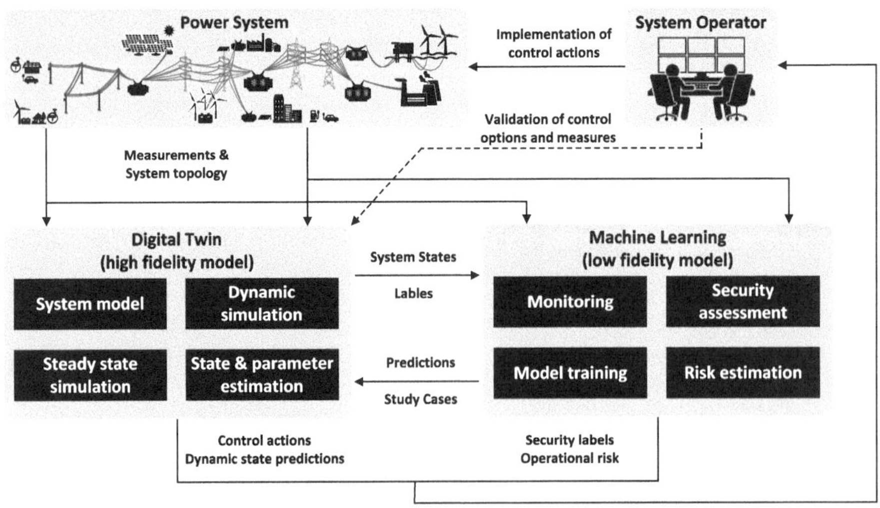

Figure 4.1 Digital twin and machine learning framework.

cases for the hf DT. Combining the two can provide helpful information for system operators to enhance operational security.

The chapter is split into two parts. The first part introduces the individual concepts from ML and DT, and the second part focuses on combining the concepts to predict dynamic security and analyze operational risks. First, the concept of hf DTs for online power system studies and their corresponding model parameter tuning from measurements is introduced. Subsequently, ML approximated lf models and their corresponding training frameworks are introduced. The first part concludes with a concept to apply data from hf models for the training of lf models.

The second part of this chapter introduces concepts for combining hf and lf models for purpose-driven real-time power system dynamic security assessment (DSA) applications. Concepts of conventional DSA are reviewed, and real-time DSA with ML methods are discussed. Finally, both hf and lf models are applied for probabilistic risk analysis in a didactic example.

4.2 Machine learning and digital twins for power system analysis

New techniques applying ML or DTs have individual strengths offering opportunities and limitations for power system analysis. Therefore the two techniques are introduced, and their advantages and disadvantages are discussed.

4.2.1 Machine learning for power system analysis

With the wide availability of measurement data and other information in electric power grids, learning models from this data with ML is the logical continuation to classical rule-based knowledge inference in power system management. The promising applications of ML techniques for power system analysis have been investigated in the last decades [1–4]. These are embedded in so-called decision support, assistant or expert systems, which comprise automatic fault diagnosis, isolation, and evaluation [5], alarm prioritization, fault switching schedules, safety checking, routine switching schedules, automatic switching, and network optimization [6,7].

Recent developments in computing hardware enable the application of these concepts combined with novel ML techniques [8]. These provide the foundation for power system engineers to augment their experience and heuristic knowledge with ML predictions [3]. However, the major limitation of ML is its generalizability, which leads to low accuracies and may make ML unsuitable. Generalization means that a trained ML model may be suitable for a specific trained data, but it may be challenging to apply it beyond that specific data. Thus in case the topology of the observed system changes or the operating conditions are different, the ML model must be retrained [9]. In addition, sometimes ambiguity exists about the ML results and their accuracy. Hence, the trust, accuracy, robustness, and reliability of ML models are of particularly high importance when developing ML-based task applications; otherwise, operators will circumvent the tools and consequently make conservative decisions to avoid security limit violations [10]. Additionally, the interpretability of some ML and the explanation of the predicted results may be difficult, e.g., with deep neural networks [11].

4.2.2 Digital twins for power system analysis

"Digital Twins (DT) are software-based abstractions of complex physical systems or objects which are connected via a communication link to the real object through a continuous data flow from the real world" [12]. Although other approaches are applicable to create a DT [13], the ideal DT model should be of analytic and deterministic character, i.e., physics-based (first principles), and sufficiently accurate while capable in real-time [14]. Thus a major challenge arises to determine the required detail-level. Whereas a simplified model does not unveil the DT value, a highly accurate approach may cause unbearable complexity and exhaustive computation times [12].

The term DT epitomizes the general trend of converging information and operational technology [15], and has been recently identified as a key concept by several industries [16]. It evolved from the broad concept of cyber-physical systems (CPS), i.e., physical systems or processes, which integrate computational applications and information communications technology (ICT). Fig. 4.2 (adapted from [17]) illustrates the difference to the predecessors of DTs. A digital model does not exchange data or information automatically. This exchange is usually done manually. A *digital shadow* is created by telemetry interrogation, but does not automatically report back to the physical world. The counterpart of a *digital shadow* is a *digital generator* [18], which can be applied to design systems (e.g., offline planning studies or compliance tests of control schemes prior to commissioning), or to train ML tools. The digital twin communicates bidirectionally and automatically reports to the physical world (closed loop). Thus the main characteristic that distinguishes a digital model and a digital twin is exchanging data automatically in a bidirectional way with the physical object.

Depending on the purpose of the application, a DT can consist of many executable models, especially when it comes to a high level of complexity, i.e., considering different timeframes or nonlinearities at runtime. These models can provide insights into the system and adapt to the system by causal (first principle) connections. Considering power system models, a high accuracy depends on a hf model. Furthermore, a high granularity in model components allows flexibility to model different power systems, topologies, operating conditions, and time-domains. From the user perspective, a high system model accuracy raises the confidence in the decisions made upon the DT model or the subsequently applied decision support functions [10]. Taking power system models as an example, increasing the level of detail in terms of considered time frames, such as models, would comprise a steady state representation, a phasor or average value model representation for electromechanical interactions, and a model including all necessary components to reflect the electromagnetic timeframes up to the microsecond scale. A suitable metamodel or interface is required to describe these interrelated dependencies to avoid redundancy of the model causalities and model parameters. This seamless integration requires a modular design approach, and suitable a framework for consistent and flexible modeling.

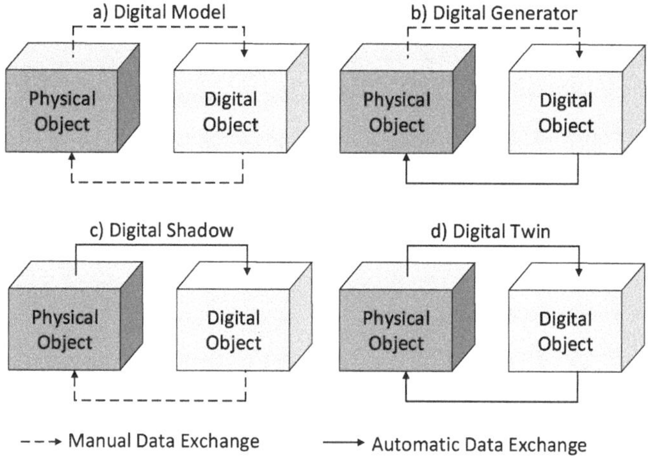

Figure 4.2 Difference between digital model (a), digital generator (b), digital shadow (c), and digital twin (d), adapted from [17].

Limiting factors to the DT approach include high demands in the computation hardware, i.e., the high computation effort respecting the real-time requirements in operational environments (e.g., online DSA for large systems). In particular, the execution of multiple simulations to find an optimal operating state requires long computing times. Another limiting factor is the expertise required from dedicated power system engineers and ICT specialists to create a real-time capable, high-fidelity DT. Tackling these challenges is promising, as DTs in operational systems enable novel automation routines, and automated system control functions up to autonomous systems [19].

4.2.3 Application of digital twins for online power system studies

A DT can estimate a state value, which is not directly observable by sensors, and report this to the real world. This fundamental property of a DT appears to be a similarity to the concept of an observer, which is often applied in control engineering (see Fig. 4.3). According to the control theory, observers apply a mathematical model of the original system and a function of the measurement noise (i.e., a correction term) to estimate the system state from measurements [20]. As an ideal, DT represents the case of a high-fidelity model without uncertainties, in practical application the estimated state is expected to converge to the true state of the observed entity. These properties of the DT concept are illustrated in Fig. 4.3. Depending on the applied DT model, the input signal u_j, the measured output y_j, and the corresponding model response $\hat{y}_j$ is generic and may comprise a single or an array of values. When the model parameter set p is unknown, a suitable model parameter identification method is required to ob-

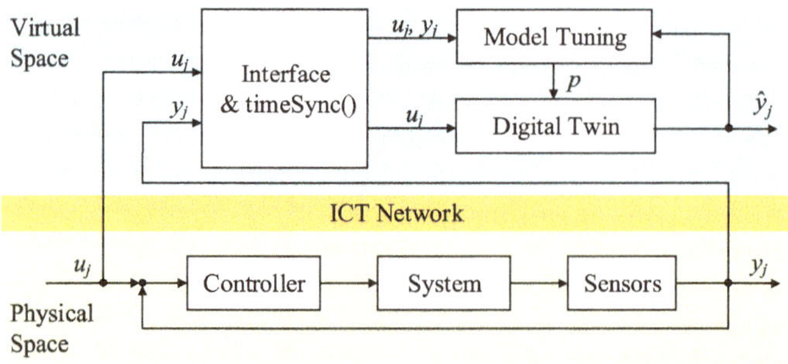

Figure 4.3 Flowchart illustration of the creation or training of a generic digital twin.

tain a valid model response (see subsection on parameter estimation for continuous digital twin model tuning).

Apparent fields of application for power system DTs include:

- Continuous system analysis and anomaly detection (e.g., malfunctions or cyber-attacks)
- Enhanced predictability of future operating conditions, allowing coordinated system operation among distribution system operators (DSOs) and transmission system operators (TSOs)
- Increased power system observability
- Consideration of system dynamics in network operation to prove control actions
- Reduction of unplanned outages by continuous equipment monitoring
- Provision of a realistic operator training simulator
- Implementation of a higher degree automation schemes and assistant systems for power system operation

As illustrated in Fig. 4.4, the study model is derived from a steady state real-time snapshot by conventional state estimation. Thus today's power system control applications do not provide a validated time-domain model. Furthermore, the modeling results derived by screening security-related scenarios, and contingencies are not yet stored in conventional operational environments (e.g., EMS/DMS long term archives). An orderly archiving can help to apply ML methodologies to evaluate critical system states faster and to analyze subordinate mechanisms better. Consequently, DTs provide the basis for advanced operator assistance systems and decision support functions for online system operation.

Fig. 4.4 illustrates the state of the art of modern EMS/DMS environments. The real-time view on the left side of Fig. 4.4 comprises the SCADA, and sometimes WAMS data, applied to derive a steady-state estimation of the power system state. A snapshot from the actual state estimation result is applied by subsequent analytic functions (e.g., contingency analysis, security assessment, or redispatch). These functionalities depend on the scenario-based models illustrated on the right side of Fig. 4.4. Today, the model result and the observed dynamic system response are not considered for model validation. The proposed DT concept, as illustrated in Fig. 4.3, comprises a validated model to close this applicational gap.

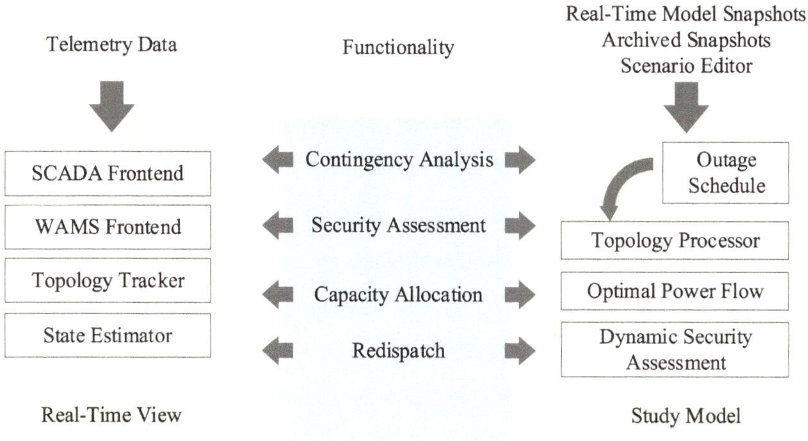

Figure 4.4 Differences between real-time model and study model.

4.2.4 Combining machine learning and digital twins for power system analysis

When combining novel ML and DT concepts, their strength can be fully unfolded for system operation. A DT is a detailed replication (twin) of the real, physical system and can simulate the system in real-time with a high level of accuracy. The strength of DT is to simulate the system accurately. Taking the system state x_i (obtained from sensor measurements) and control set-points u_j as an input, future system states x_{i+1}, x_{i+2}, ..., x_{i+n} are predictable. A limiting factor is the computational time, which depends on the amount of data, the system size, and the dimensionality of investigated study cases.

In contrast, the strength of ML is a quick prediction at the cost of a lower accuracy score (e.g., resulting from training biases). ML methods train a model from data without prior knowledge of possible model parametrizations. Therefore one concept to combine the two individual strengths by using the DT to generate training data (when computational time is abundant), and then use the data to train the ML model. The advantage is that a DT can generate data, which is rarely observable, and which can only be acquired with great effort, or is unobtainable in the real world. Thus combining the strengths of ML and the DT concept can support real-time power system analysis, whereas ML predictions are fast, but of lower accuracy, DTs can correct the ML prediction.

Furthermore, ML predictions can be proven or validated by the DT model considering only a few in-depth studies and suggested control-set points. One possible application combining ML and DT is a power system perturbation, e.g., caused by an equipment fault. Then, the ML can immediately predict the power system's dynamical response. As illustrated in Fig. 4.5, ML can suggest ad hoc control responses after the occurrence of the fault at time t_1. Subsequently, the proposed primary control actions can be taken considering the time delay between sending the control signal and the pri-

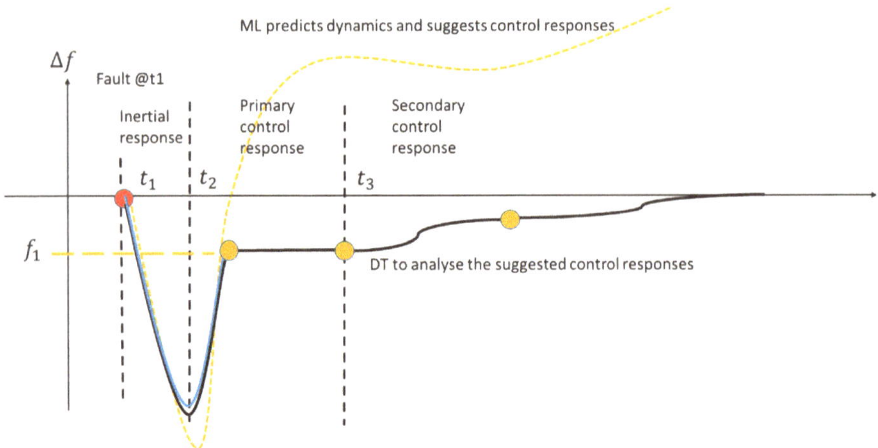

Figure 4.5 Combined benefits from ML and DT for accurate and fast control responses to faults. ML predicts immediately at fault t_1 the possible dynamical trajectory and suggests a control action at t_2. Between time t_1 and t_2, the DT simulates the control action in detail and confirms the suitability of the action (modified from [21]). *Black line* (DT) is available at t_2, *yellow line* (ML) is available at t_1. *Blue line* is the actual system response.

mary control actions at time t_2. In the meantime, the DT can start to execute a detailed dynamic simulation with higher accuracy. When the hf DT simulation is available, it can replace the lf ML prediction to improve the system stability and take secondary control actions. The ML ability to provide a rough prediction of the initial dynamic trajectory very fast, in combination with the advantage of DT to simulate the dynamic trajectory accurately to prove the suggested control action in a hf model, raises the confidence in the control action and maximizes the stabilizing effect of the response to the fault.

The capability of the power system to anticipate, absorb, adapt, or recover from faults (and other events), comprises the intra-day operation (handling of disruptions in real-time operation) and the operational planning (optimized topology and structural reinforcements). Combining the benefits at their full potential, DT and ML support the development of novel control schemes. Applying control schemes that are both corrective and preventive enables operational planning closer to the power system security boundary, i.e., allowing a higher asset utilization. The control schemes derived from the combination of ML and DT have the potential to be fast and sophisticatedly robust at the same time. They may even provide decision support for cascading failures by successive and systematic assessments. As shown in Fig. 4.5, following a disturbance at t_1, the ML predicts the system dynamics (yellow dashed line) and the following control action (red points), while in the meantime the DT computes the dynamic trajectory (black line), to correct the control action according to the observed system state (blue line) towards a full system recovery in Fig. 4.6.

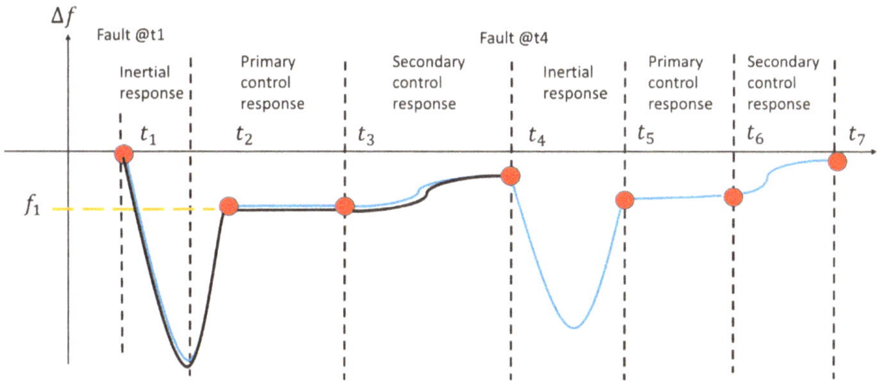

Figure 4.6 The system response with the implemented control actions. Computing control actions in nearly real-time has the potential to even consider cascading failures in real-time (modified from [21]).

4.2.4.1 Parameter estimation for continuous digital twin model tuning

As an accurate and validated high-fidelity dynamic power system model is required, model validation and parameter estimation methods are necessary. Parametric time-domain models contain internal parameters, which describe the physical behavior of the modeled object or control system. These parameters are often unknown and need to be estimated. This is typically done by fitting the mathematical models to measured time series data. Some common techniques for parameter estimation from online measurements are the Kalman filter (KF) and its derivatives (e.g., extended KF, unscented KF, or cubature KF) [22].

Another suitable method is the moving horizon estimation (MHE) [23]. The MHE approach can handle constraints, which is essential to raise the accuracy and validity of the estimated states and parameters of nonlinear system models. The general concept of the moving horizon estimation (MHE) methodology is illustrated in Fig. 4.7. Instead of applying the full information available as input, it only takes a sample of N values. Thus the MHE approach combines the advantages of full information estimation with a tractable online computation [23]. The MHE's computational power is as demanding as the KF and its derivatives. However, in comparison, it can deal with highly nonlinear estimation problems in the presence of state constraints $\overline{x}$. The general form to describe an MHE is given in [23], whereas methods for implementation are discussed in [24].

The fundamental concept of the MHE is to not fit the parameters to the complete data available at time k, i.e., $0, 1, \ldots, k$, but rather to only the previous n time steps, i.e., $k - N; k - N + 1, \ldots, k$. This concept is illustrated in Fig. 4.7. As the time frame moves along the captured time series, the objective function as given in [23], is minimized to estimate the parameters that fit the model response to the captured values. Starting from an initial parameter set, the model parameters are varied in such a way that the objective function reaches a minimum. At each time step k, an optimization problem is solved to minimize the objective function. It penalizes the deviation of the

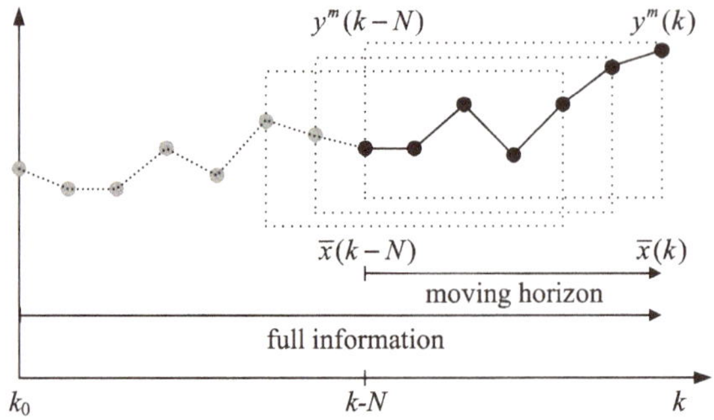

Figure 4.7 Parameter estimation with MHE scheme (adapted from [23]).

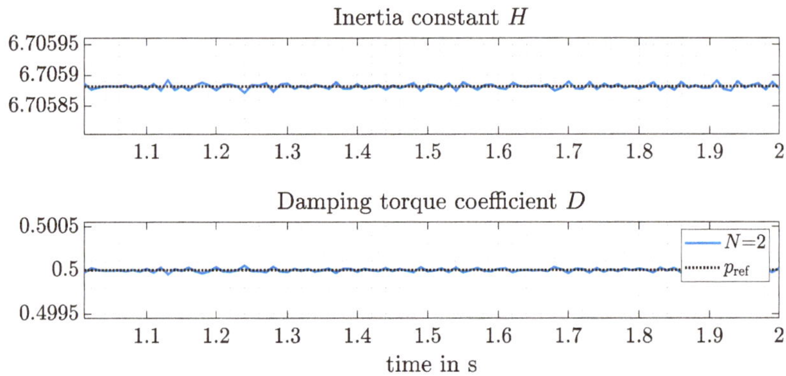

Figure 4.8 MHE example with horizon length $N = 2$.

computed output y from the measured output y^{m}, and the deviations from the system dynamics. The state and parameter guess is based on the previous time step (as illustrated in Fig. 4.7). An example illustrating a MHE parameter estimation result with $N = 2$ for the synchronous machine inertia constant H, and the damping coefficient D is illustrated in Fig. 4.8.

A simplified graphical representation of an analytical time-domain model of a sixth-order synchronous machine model, based on the equivalent circuit, is given in Fig. 4.9. The model comprises a dynamic state vector x, including a generic representation of its controllers (see also Fig. 4.9), and state names can be found in Table 4.1. In the study model applied for the didactic example (see section 4.5), the excitation system IEEE AC4A and a simple speed governor IEEE HyGOV are applied. These internal states, which are hard to obtain by sensors and are not observable from a power system control room, add valuable information for the ML training procedure, as this information can be related to system stability.

4.2.4.2 Training the ML model from data

The supervised ML technique can derive purpose-driven surrogate models from data. These ML models are applicable for classification and regression, and take the feature

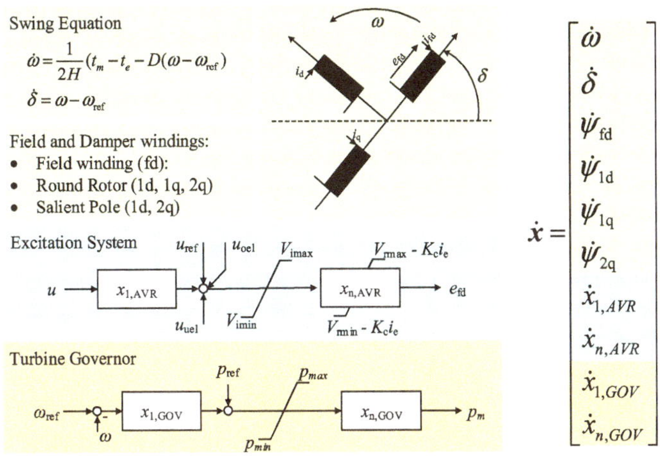

Figure 4.9 Simplified analytical model and the dynamic state vector of a synchronous machine model including governing and excitation system.

Table 4.1 Dynamic system states corresponding to Fig. 4.9.

Symbol	System state
δ	Rotor angle (rad)
ω	Angular frequency (rad s^{-1})
t_e	Electrical torque (p.u.)
t_m	Mechanical torque (p.u.)
ψ_{fd}	Excitation flux (p.u.)
ψ_{1d}	Flux in 1d-axis damper winding (p.u.)
ψ_{1q}	Flux in 1q-axis damper winding (p.u.)
ψ_{2q}	Flux in 2q-axis damper winding (p.u.)
u	Control set point
x_{AVR}	Automatic voltage regulator state
x_{GOV}	Governor state

vector x_i as an input to apply the trained function

$$\hat{y}_i = \hat{f}(x_i)$$

for predicting the output data $\hat{y}_i$. While for the purpose of classification, the function $\hat{f}$ predicts discrete values, the values for regression are continuous $\in R$.

To train the function $\hat{f}$, a data set of Ω value pairs (x, y) is required. One sample corresponds to the pair $(x_i, \ y_i)$. As shown in Fig. 4.10, the data should be preprocessed before training the model. This includes the imputation of missing data, and normalization i.e., the data within each feature's vector is scaled, so each feature is between [0, 1], as the features may have different dimensions. This can be done by

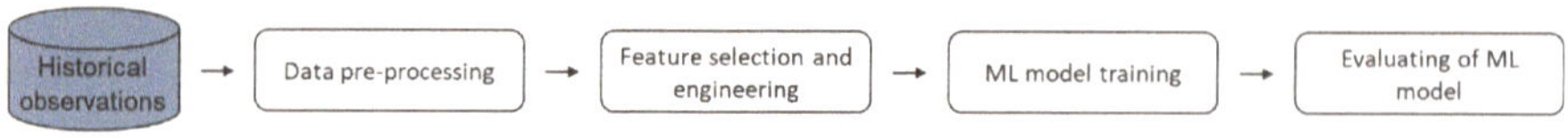

Figure 4.10 Framework to train a ML model.

applying

$$x_{ij}^{scaled} = \frac{x_{ij} - \min(x_j)}{\max(x_j) - \min(x_j)},$$

where $\min(x_j) = \min(x_{0j}, x_{1j}, \ldots, x_{\Omega j})$ and $\max(x_j)$ denote the smallest and largest value among all data withing Ω, considering each respective feature j. The following expression assumes a normalized x and drops the superscript "*scaled*." Subsequently, the typical objective of the training procedure is to minimize the training loss, i.e., the squared difference between predicted and actual labels

$$loss = \frac{1}{\Omega}\sum_i \left(y_i - \hat{f}(x_i)\right)^2.$$

The function $\hat{f}$ is a suitable candidate to represent the data when this loss is low. However, in order to infer the suitability of this function, it is important to consider that the function may overfit or underfit the data and that the initial candidate parameterizations for $\hat{f}$ must be selected.

For instance, the engineer may decide on an initial way to parameterize the function $\hat{f}$, and the decision could be to select a polynomial function,

$$\hat{f} = \theta_0 + \theta_1 x^1 + \theta_2 x^2 + \theta_3 x^3 + \cdots + \theta_n x^n,$$

which has n terms with n parameters to fit. In case, that n is suboptimally selected, the function tends to under- and over fitting. An exhaustive overview on selecting appropriate loss functions, including the parameterization of function $\hat{f}$, and the consideration of feature selection, is given in [25]. Different algorithms are applicable to minimize the loss function . For instance, an iterative algorithm is the gradient descent. Assuming $n = 1$, the function $\hat{f} = \theta_0 + \theta_1 x^1$ has only two parameters. Then the loss function becomes

$$loss = \frac{1}{\Omega}\sum_i \left(y_i - (\theta_0 + \theta_1 x_i^1)\right)^2,$$

and the partial derivatives with respect to θ_1 and θ_0 are

$$D_{\theta_1} = \frac{-2}{\Omega}\sum_i x_i^1 \left(y_i - \left(\theta_0 + \theta_1 x_i^1\right)\right),$$

$$D_{\theta_0} = \frac{-2}{\Omega}\sum_i y_i - \left(\theta_0 + \theta_1 x_i^1\right).$$

The gradient descent algorithm updates the parameters accordingly to these gradients:

$$\theta_k = \theta_k - LD_{\theta_k} \quad \forall k = \{0, 1\}.$$

These updated parameters result in an updated candidate function $\hat{f}$, and the loss of this candidate function can be computed. This procedure to update the parameters is repeated iteratively, until the termination criterion is met. An effective design of ML training and the performance of the trained ML model involves engineering and selecting the most valuable features and the objective to evaluate and train an effective model. This feature engineering transforms the data $x^{engineered} = g(x)$ into a more meaningful representation. For instance, the data may comprise the current x_{ic} and the voltage x_{iv} and the objective is to predict overheating of a device. So an effective technical feature here can be the power, since it correlates strongly with overheating, hence the engineered feature may be $x^{engineered} = x_{ic}\, x_{iv}$. Feature selection can also enhance the performances and can be based on statistical methods. Many approaches exist to select features, and their approach can be either to forward or backward select features. In backward selection, the ML model is trained with all features, and then the least-relevant feature is removed to retrain the ML model. This iteration repeats until the ML model does not improve the selected performance metrics. The forward selection starts with only randomly selected features and adds one feature in each iteration until the performance metrics for the ML model do not improve anymore.

The approximation capability of ML is determined by the structure of function $\hat{f}$. To illustrate this by an example, a model based on an artificial neural network (ANN) is described. The ANN constructs the function $\hat{f}$ out of many chained functions inspired by neurons in the human brain. This results in a multi-layer perceptron (MLP), i.e., a feed-forward type network with a unidirectional information flow. MLP architectures start with features x_i at the input layer, and end at the output layer which contains the model prediction $\hat{y}_i$. An example MLP is illustrated in Fig. 4.11. There, neurons are located in the hidden layers between input and output layers. Each neuron contains a weight vector w and a corresponding scalar bias b. The neurons act like a linear map $g(X) = W^T X + B$. However, linear functions can only construct a linear relationship between features, and the activation functions generate the nonlinear interactions. The activation functions σ are applied to the resulting output of each neuron. Popular activation functions are sigmoid, rectified linear unit (ReLU) or hyperbolic tangent.

Assume a neural network $\hat{f}(x)$ approximates an unknown function $f(x) = y$ applying 6 neurons. These are distributed equally to 2 hidden layers with an activation function σ, where $x, y \in R$. The prediction $\hat{y}$ then results in

$$\hat{f}(x) = \hat{y} = w_y(\sigma\left(W_2^T \sigma\left(W_1^T x + b_1\right) + b_2\right)) + b_y.$$

These neuron weights and biases are optimized during the training. The most basic parameter update algorithm is the gradient descent algorithm applied to the predefined loss function. Many variants and improvements (e.g., Momentum, RMSProp, Adam, etc.) to gradient descent were proposed, such as stochastic gradient descent [26].

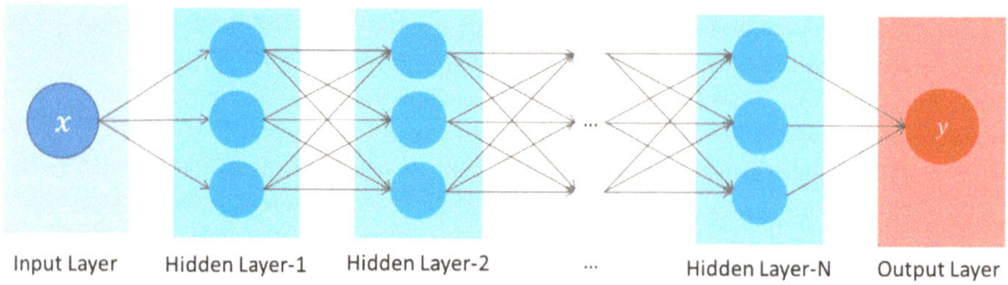

Figure 4.11 Example architecture of a multi-layer perceptron.

There are many other metrics to evaluate the training of the ML model beyond the mean squared difference. For instance, when the ML task is binary classification ($y_i = \{0, 1\}$) and not regression, the error may be suitable, which refers to the ratio of inaccurately predicted values

$$error = \frac{1}{\Omega} \sum_i | y_i - \hat{y}_i | .$$

On the other hand, the error might not be suitable for classification problems with an unbalanced dataset. For example, historical data regarding a power system security assessment problem usually contains more secure cases over insecure cases. The precision metric calculates the ratio of correctly predicted actual values ($\hat{y}_i = y_i = 1$) over all true predictions ($\hat{y}_i = 1$). Similarly, the recall metric or the true positive rate (TPR) calculates the ratio of correctly predicted true values over all true values $y_i = 1$. The metric F_1 calculates the weighted average of precision and recall.

$$F_1 = 2 \times \frac{Precision \times Recall}{Precision + Recall}.$$

The computational complexity of the ML training depends on the selected model complexity and size of the training data. The overall training time is in the order of the computational time of hf DT simulations, but this process can be done offline, without time limitations. However, the prediction speed is very fast and almost constant, as the trained ANN model predicts the output with simple algebraic operations.

Besides the model and metrics described in this section, many more ML models can be applied in combination with DTs.

4.2.4.3 Generating ML training data by digital twin models

The training of ML models is typically performed offline when a DT of the power system examines possible operating scenarios an related data, which then can be used to train the ML model. The relevant data typically involves an operating condition as input feature, and a possible output label, which the ML model later in during aims to predict, where the output labels could be control actions, a fault analysis, or a classification of an expected risk-level. The data archive can contain an initial database of the input and output pairs of historical observations. However, a DT can generate data

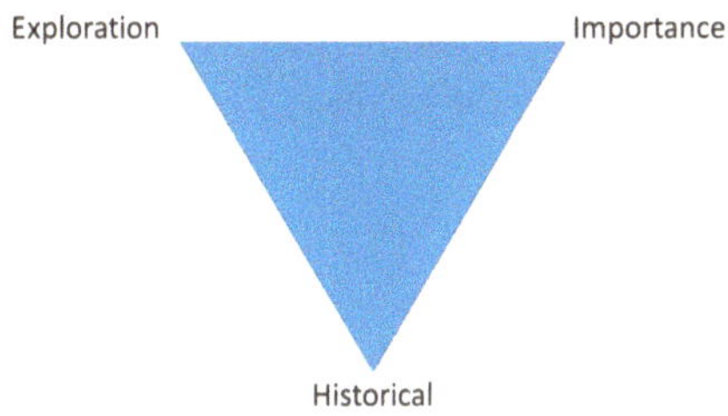

Figure 4.12 The three dimensions of an ideal ML model training database.

beyond these observed and recorded historical system conditions. Thus, it can enrich the database to enhance generalization capability by extrapolating important data samples and examine specific scenarios of interest. A suitable approach to sample the data from a DT balances (i) the data importance with (ii) exploration needs, and (iii) with data similar to historical observations. Therefore the ideal training database combines all three dimensions as shown in Fig. 4.12.

The idea behind the generation of data similar to historical observations is that these consider operating conditions most likely to occur. To generate such similar data, typically, probability distributions are fitted to the historical data, then Monte Carlo sampling techniques are applied to sample data that follow this fitted probability distribution [27]. Suitable probability distributions need to capture the complex dependency for many variables, which can be nonlinear. Therefore, copula models can be applied to model such dependencies [28]. For example the canonical vines (C-vines) copula models, are determined by the following probability density function

$$f(x_1, \ldots, x_m) = \left(\prod_{i=1}^{m} f_i(x_i)\right) \times c_{1\ldots m}\left(F_1(x_1, \ldots, F_m(x_m)\right),$$

$c_{1\ldots d} : [0,1]^m \rightarrow \mathbb{R}$ is an m-dimensional copula, and $U = \{U_1, U_2, \ldots, U_m\} = \{F_1(X_1), F_2(X_2), \ldots, F_m(X_m)\}$ are uniform marginal distributions. Then, the product of its marginal distribution and the multivariate copula density functions builds the joint density function.

The Monte Carlo sampling method can be applied to sample up the historical data using C-vine copulas. Generated conditions define the input space (x_i) for the DT simulation. Subsequently, the DT emulates the system behavior against the predefined disturbance scenarios with new operating conditions to obtain the desired labels (y_i) for ML training.

Obtaining an information-rich knowledge base is essential for ML applications. Therefore, additional samples can be obtained from certain samples with higher information value to improve the prediction performance. For instance, samples close to the decision boundary in a classification problem provide more information, and thus can improve the prediction performance. In power systems security assessment, the decision boundary corresponds to the system security boundary, which is illustrated in Fig. 4.13, with respect to two generators' power output. However, often the majority of cases in Fig. 4.13(a) [29] share the same class with their direct neighbors so their contributing information to training is low. However, Fig. 4.13(b) shows also cases located near the security boundary, which are more relevant, and the generation of these new samples from the interested region improves the prediction performance of ML models.

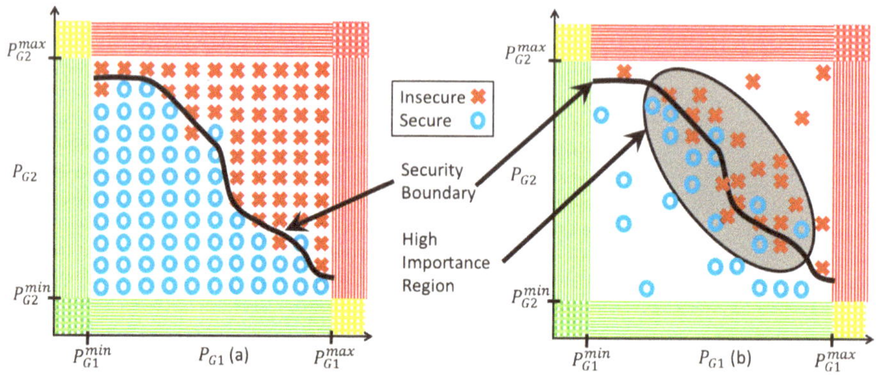

Figure 4.13 Uniform random sampling (a) and importance sampling in (b), details in [29], [30].

One method to obtain influential samples from the original dataset is importance sampling, as shown in Fig. 4.14. The sample importance is measured by the trained classifiers to filter out the points with low information value. The DT conducts dynamic simulations for the reduced dataset, hence the overall simulation dimensionality, i.e., the computational time is reduced without compromising the ML performance. Following this procedure, simulated samples are added to the training database, aiming for a higher prediction or classification accuracy.

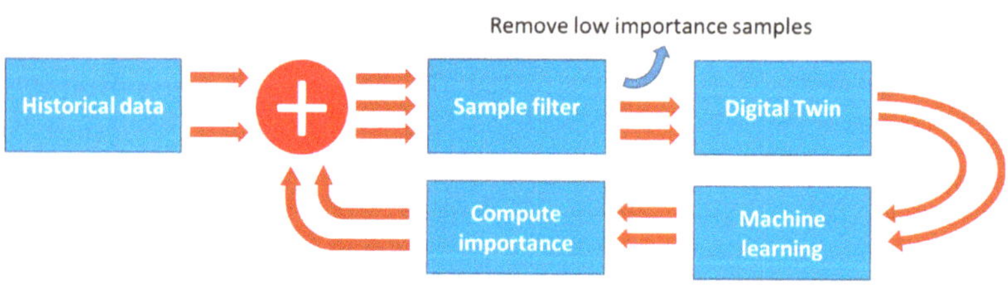

Figure 4.14 Example of an importance sampler for DSA applications.

This importance of the samples can be computed using ensemble learning methods, such as Adaboost, extreme trees, or random forests.

The ensemble models construct multiple learners $k \in \Omega^E$. The prediction score $s(x_i)$ is computed from the hypotheses of the individual learners $h_k(x_i) \in \{0, 1\}$ for sample x_i as

$$s(x_i) = \frac{1}{|\Omega^E|} \sum_{k \in \Omega^E} h_k(x_i).$$

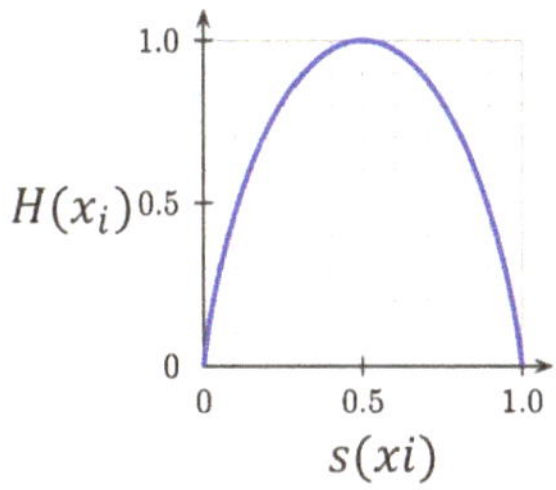

Figure 4.15 Example plot of the entropy vs. score.

The sample importance $I(x_i)$ is calculated as the measurement of disagreement, expressed as the entropy, as illustrated in Fig. 4.15, by the following formula:

$$I(x_i) = -\frac{1}{\log_2 2}\left[s(x_i)\log_2 s(x_i) + (1 - s(x_i))\log_2(1 - s(x_i))\right].$$

4.2.4.4 *Summary of advantages and disadvantages considering the interaction of digital twins and ML techniques*

Table 4.2 summarizes the advantages and disadvantages of traditional modeling and seeking solutions with assistance by ML methods. When it is possible to formulate the relationships between the physical variables using mathematical equations, building a coherent hf DT simulation model of reality has many advantages. On the other hand, a well-trained neural network can also find correlations that are difficult to investigate by conventional modeling techniques. Thus to combine both techniques is promising for decision support systems, where both accuracy and prediction speed are required.

Table 4.2 Comparison of model-based and data driven methodologies.

	Digital Twin (hf model)	**Machine Learning (lf model)**
Advantages	A simulation model is explainable, extendable, and reproducible by experts Verified models are applicable outside explicitly tested areas High trust in results	Detailed knowledge of the modeled system or process is not required ML has the potential to recognize pattern or correlations in data that are not perceptible by a human Fast execution times
Disadvantage	The creation of validated models is time-consuming and requires highly skilled domain experts System states that do not correspond to the normal operating condition are hardly reproducible with reasonable effort High execution time	To obtain high quality training data in the required quantity of data is often difficult The model can lead to inaccurate decisions The solution approach of a neural networks is not traceable by humans and the implementation of solutions require human supervision Trust in results not always given

4.3 Purpose driven surrogate models

Following the previous section, where the functionality and possible interactions between hf DT and lf ML models have been described, this section gives an example for real-time power system dynamic security applications. Therefore, the previously discussed concepts are utilized to obtain a purpose-driven surrogate model for DSA applications.

4.3.1 *Dynamic security assessment with digital twins*

DSA studies of electric power systems are mainly based on time-domain simulations. These mathematical models often contain stiff or moderately stiff nonlinear differential algebraic equation (DAE) systems of high order. The solution of these systems of equations involves a relatively high computational effort. Additionally, the applicable numerical methods can only approximate the system dynamics. Consequently, a validation procedure involving real process data is required. One major obstacle for meaningful DSA implementations in practice is to gather enough data for such validation, especially for security-related situations (e.g., critical contingencies, dynamic security boundary violation, severe disturbances). Thus, two main obstacles for online DSA can be identified:

- Obstacle 1: DAE solver-based solutions for screening of many scenarios in large power system models must be real-time capable to support online operation.
- Obstacle 2: Validation data sets for security-related conditions are lacking (e.g., suitable disturbance records).

4.3.1.1 Security assessment

Following the definition of security given in [31,32], security assessment tools evaluate the ability of the power system to survive perturbations and to work reliable under adverse conditions.

To analyze power system security, either static or dynamic security assessment tools are applied [37]. Static security assessment (SSA) tools evaluate post-disturbance conditions, i.e., violation equipment ratings or voltage constraints based on power flow calculations. Thus, the dynamic transition from pre- to post-contingency is not considered. Dynamic security assessment (DSA) methodologies also evaluate the phase of dynamic transition. When the system returns to a secure steady state after the disturbance, the system is considered as secure. As the conventional concept of SSA, cannot sufficiently guarantee the system integrity in all situations [38], it is mandatory for TSOs to assess dynamic stability to identify stability limits [39]. The criteria applied by DSA systems to evaluate contingencies and their impact on system security include rotor angle stability, voltage stability, and frequency stability; as well as the damping of oscillatory modes within the interconnected power system [33,34]. They must comply with boundary definitions from grid codes or other specified thresholds. Some indices considered in DSA applications for transmission systems are given in [35,36].

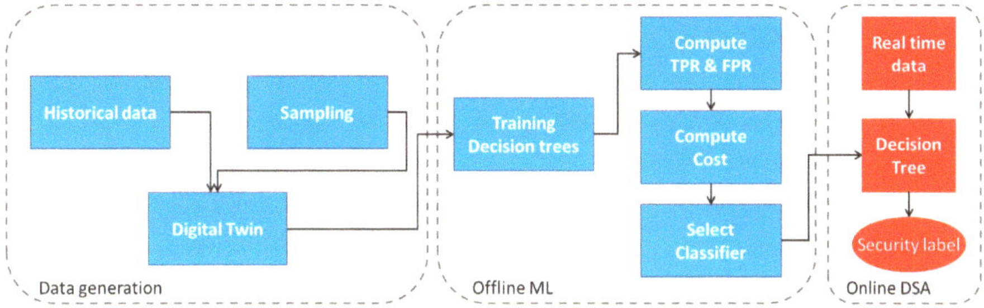

Figure 4.16 Approach of using ML for real-time DSA.

We furthermore distinguish between offline and online DSA. Though offline DSA is often part of power system planning studies, these offline studies inherit several uncertainties, as these do not consider all possible operational conditions [40]. These uncertainties can be eliminated by implementing online DSA. By including real-time process information, i.e., the actual point of operation and previously unexpected operational conditions can be considered. As online DSA can support decisions during real-time operation, the ENTSO-E recommends a continuous validation and fine tuning of system models (e.g., by WAMS) to avoid invalid results and, in the worst case, wrong decisions [38]. As the system modeling requirements vary for DSA studies, these need to be balanced in terms of model detail and required computation time [38]. This is one of the main arguments in favor of using a digital twin, as DT models are suitable for both online and offline DSA studies.

4.3.2 Real-time DSA with ML

Real-time dynamic security assessment (DSA) for system operation becomes important in the future as power system operation is done closer to operating limits. However, online DSA has its limits for real-time operation. One idea for an approach to real-time DSA is to apply a trained ML model to predict the outcome of the simulation.

Offline training requires an extensive dataset, which is generated by DT, as described in the section entitled "Generating ML training data by digital twin models." There, the operating condition x_i is defined by active and reactive power at the generator node , and voltage magnitude and phase angles given by PMUs. Subsequently, the security of the operating condition is predicted with the trained ML model for when the contingency c occurs. As shown in Fig. 4.16, the training of the ML model includes various $x \times c$ combinations within the function $f(x)$ to predict security label $\hat{y}$.

The decision tree shown in Fig. 4.16 is a common, nonparametric ML model. The model name "decision tree" comes from the upside tree structure that has a single root node, many terminals (leaf), and internal nodes. Starting from the root node, the training data is split into 2 subsets based on the decision rule that generates the maximum homogeneity between subsets. A decision rule is a simple comparison of selected fea-

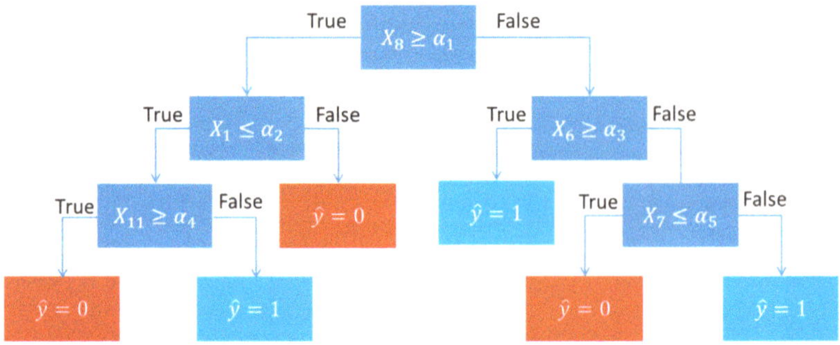

Figure 4.17 Decision tree structure for binary classification.

ture X_i to the optimized threshold value α_i. The homogeneity is measured by the impurity functions as an value of entropy or the Gini impurity. The Gini impurity $H(\theta_k)$ is formulated from the data proportion (p_{km}) of class$m$ and θ_k is the subset at node k as

$$p_{km} = \frac{1}{N_k} \sum_{y \in \theta_k} 1(y = m)$$

$$\text{Gini}: H(\theta_k) = \sum_m p_{km}(1 - p_{km}).$$

The decision tree represents an approximation of a high-dimensional piecewise constant function. Therefore the decision tree can define nonlinear interactions between features. Besides the predicted class, the prediction score $s(x) \in [0, 1]$ can be retrieved from the decision tree, which is equal to the ratio of the same class training samples over all samples in the prediction leaf. The binary classification tree model predicts the class (0–1) by comparing the output score $s(x_i)$ with the decision threshold Z. For the security assessment problem, the operation condition x_i is classified as secure $(\hat{f}(x_i) = \hat{y}_i = 1)$ when the corresponding prediction score $s(x)$ outperforms Z, i.e., $s(x) \geq Z$. The default threshold value $Z = 0.5$ is suitable for ideal conditions, where there is no imbalance between classes in the training database considering equal costs for misclassification errors.

Furthermore, decision trees can handle high-dimensional feature spaces, as irrelevant (low information value) features are excluded automatically in the tree learning process. A decision tree structure for binary classification is illustrated in Fig. 4.17.

As large scale power system security problems require high-dimensional operation space, decision trees are suitable to address these. Many applications of decision trees for power system security problems can be found in the literature, such as the prediction of stability limits [41], the identification of preventive and corrective actions against contingencies [42], or security assessment [43]. System operators can obtain additional information from the tree path of prediction, because the decision tree is an interpretable white box model [44]. An other advantage of decision trees and other ML models in decision making processes is that these do not raise computational costs (between $O(n)$–$O(1)$). As a DT solves high-order nonlinear differential algebraic equations, the computation effort here is expected to be higher. Thus both applications can be executed on the same computation platform.

Online DSA can have critical impact on secure power system operation. While false alarms (false negative) of the DSA can trigger unnecessary or undesired control actions, any missed alarm (false positive) can jeopardize the secure power system operation. These two types of errors in the assessment process impose different follow-up costs for system operators, e.g., cost of missed alarms C_{FP} and cost of false alarms C_{FN}. Undetected insecure cases can cause a power outage that have severe effects on society, industry, and service facilities, hence the alarm costs must comply to $C_{FP} >> C_{FN}$. However, during the training stage of ML models, the main objective is minimizing the misclassification error or the other commonly used classification metrics that were introduced in the section entitled "Training the ML model from data." However, without considering the error costs, ML models cannot prioritize a specific class over the others. To approach this problem the receiver operating characteristics (ROC) curve is used; the latter is a visual metric that shows how true positive rate $TPR = \frac{TP}{\pi_+}$ and false positive rate $FPR = \frac{FP}{\pi_+}$ evolves with respect to Z, where π_+ corresponds to positive samples in the dataset. Fig. 4.18 shows the sensitivity of Z on the ROC of a decision tree model.

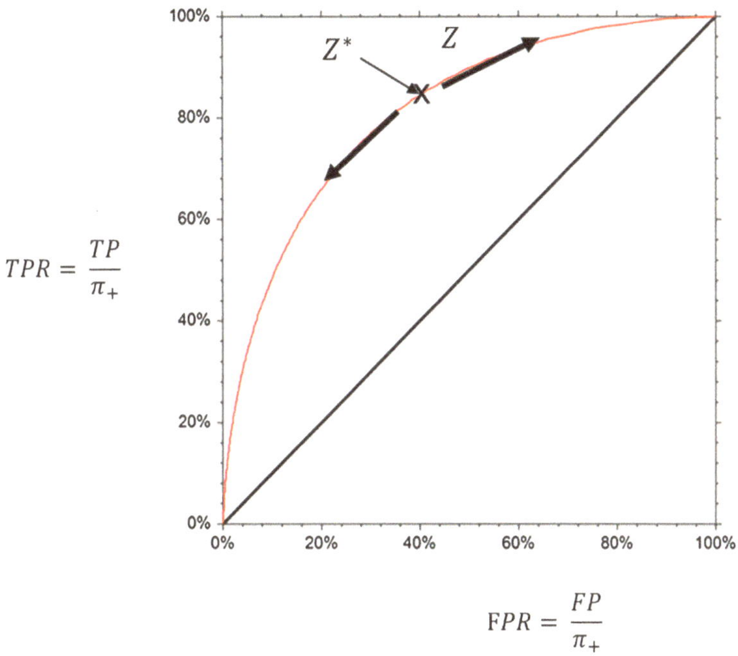

Figure 4.18 ROC and decision boundary.

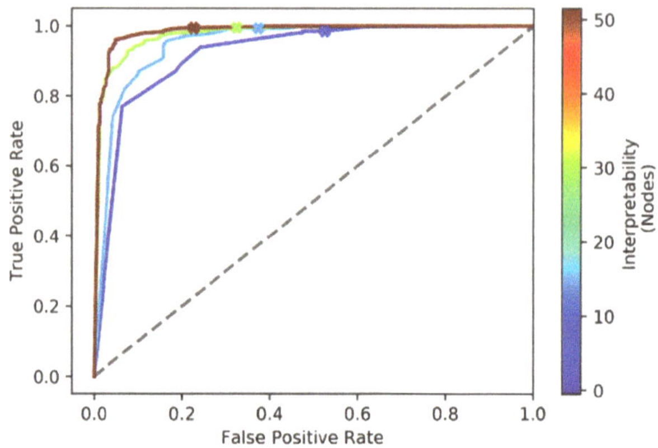

Figure 4.19 Selection with the ROC curve for IEEE 68 bus system [43].

Model predictions can be altered by changing the value of Z. For example, selecting a higher threshold reduces the number of secure predictions, which can reduce the overall model performance, but reduces the number of missed alarms. Also, the cost-optimum threshold value Z^* on the ROC curve is determined by the ratio of insecure cases π_- and the predefined error costs C_{FP} and C_{FN}. This cost optimal point is calculated by

$$Z^* = \frac{\pi_- C_{FP}}{\pi_- C_{FP} + \pi_+ C_{FN}}.$$

Power systems have a dynamic structure, where cost of the contingency (cost of load loss) and probability of the contingency are parameters that can change in real-time [45]. Fortunately, these parameters can be obtained and updated online to calculate the new optimal threshold value, without retraining the ML model.

The ROC curve can be further used to analyze several decision tree models with the optimal threshold. Although decision trees are trained with the same training data, their interpretabilities are different, because parameters such as maximum depth have been varied between models.

The ROC curves of four separate decision tree models for a case study conducted with the IEEE 68 bus test system are illustrated in Fig. 4.19. Here, the dots on the curve show the optimal cost point Z^*. As the tree grows interpretability, the capabilities are lower. The power system nodes are represented by numbers within the color bar. Based on such visual graphs, system operators can select the best classifier that suits the operation [43].

4.3.3 Probabilistic and risk analysis with DT and ML

Most power system disturbances are unforeseeable and highly unpredictable, but their impact can be modeled with scenario-based probabilistic risk analysis. A probabilistic approach to security assessment is illustrated in Fig. 4.20, where the approach

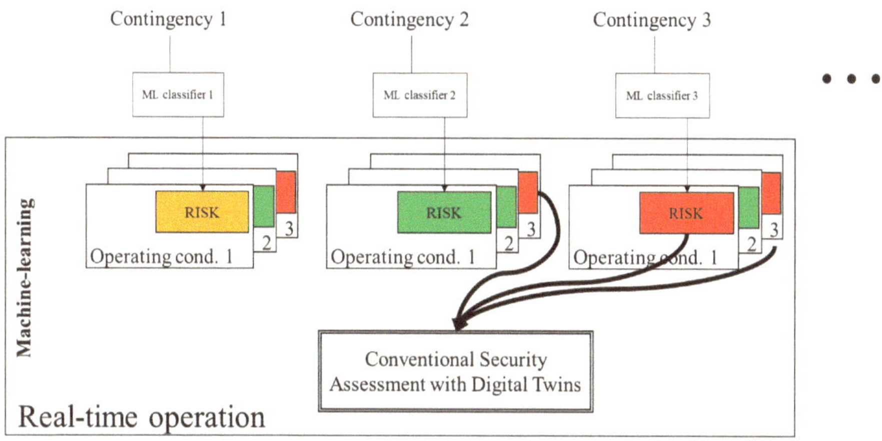

Figure 4.20 Use of ML for security assessment.

evaluates the risk in short-term operation range for multiple contingency scenarios simultaneously by the corresponding ML classifiers [45]. Estimated high-risk scenarios are further investigated by DT simulations to reveal overall system response with high accuracy.

The ML classifier models compute the individual risks of all scenarios $s \in \Omega^S$, which is equivalent to the Cartesian product of operating conditions Ω^I and contingencies Ω^C, $\Omega^S = \Omega^C \times \Omega^I$. The probability of a scenario is $p_s^S = p_c^C\ p_i^I$, or the joint probability of conditions p_i^I and contingencies p_c^C. Then, the ML classifiers compute the scenario risk $R_s = R_c(x_i)$ based on the associated security rules.

Subsequently, the scenario set Ω^S is divided into high risk $\Omega^{S,H}$ and low risk $\Omega^{S,L}$, where $\Omega^S = \Omega^{S,H} \cup \Omega^{S,L}$. Low-risk samples are excluded from DT simulations, since these samples are relatively less costly and DT simulations are computationally more expensive than ML. The total risk (low risk) computed by ML is

$$RISK^{ML} = \sum_{s \in \Omega^{S,L}} p_s^S R_s.$$

The size of the high-risk subset is limited by the DT computational capacity C, which must be greater than the size of high-risk subset $C \geq \left|\Omega^{S,H}\right|$. Thus, the short-term computational capacity of the DT simulations is determined by the amount of time domain security assessments. The DT conducts the conventional security assessment for the high-risk scenario samples only. The DT-based risk assessment is computed as follows

$$RISK^{DT} = \sum_{s \in \Omega^{S,H}} p_s^S \gamma_s.$$

The parameter γ_s represents the severity level of the scenario s security impact, which is computed by time domain simulations from the DT. The final estimated residual risk is computed with the combination of both DT and ML.

$$RISK^{TOT} = RISK^{SA} + RISK^{ML}.$$

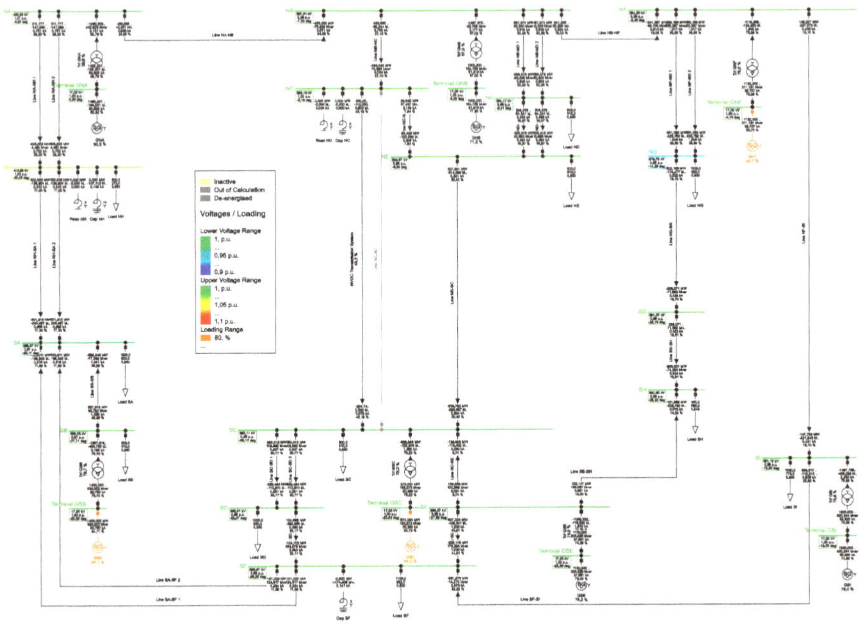

Figure 4.21 Didactic example: base scenario. The time-domain model is adapted from Cigré Technical Brochure 536 [46].

4.4 Conclusion

Power system operation is a promising example that DT simulations and ML can complement one another. Combined, the reliability of DT modeling can be met, while increasing solution speed significantly.

In conclusion, this chapter has revisited concepts from DT and ML and showcased the opportunity in combining them for power system operation. The example of dynamic security assessment has the promise to combine probabilistically the two approaches, ML and DT. This presents the opportunity to investigate ML and DTs for other promising applications, such as emergency control, corrective and preventive control, or congestion management.

4.5 Didactic example

This chapter aimed at providing a strong high-level overview for the reader with references to deepen the learnings. Therefore a didactic example is provided that comprises a selection of case studies from real-world research. The learning objective of the didactic example is to analyze the simulation data (digital twin results) and the training of a machine learning model. Therefore the risk assessment example contains data from a power system model, which is illustrated in Fig. 4.21. The power system model is adapted from Cigré Technical Brochure 536 [46]. It comprises 17 400-kV buses connected by several AC transmission lines, 7 synchronous generators, 11 loads, several shunts, and a VSC-HVDC transmission system connecting bus NC and SC. Details on the didactic example are outlined in the corresponding Jupyter notebook "Dynamic_Risk_Assessment.ipynb" and the related files "FeatureMatrix.csv," "Label.npy" and "SimulationExample.csv" (see Appendix 4.A).

Appendix 4.A Supplementary material

Supplementary material related to this chapter can be found online at https://doi.org/10.1016/B978-0-32-399904-5.00010-7.

References

[1] B.F. Wollenberg, T. Sakaguchi, Artificial intelligence in power system operations, Proceedings of the IEEE 75 (12) (1987) 1678–1685, https://doi.org/10.1109/PROC.1987.13935.
[2] R.P. Schulte, S.L. Larsen, G.B. Sheble, J.N. Wrubel, B.F. Wollenberg, Artificial intelligence solutions to power system operating problems, IEEE Transactions on Power Systems 2 (4) (1987) 920–922, https://doi.org/10.1109/TPWRS.1987.4335274.
[3] K.P. Wong, Applications of artificial intelligence and expert systems in power engineering, in: S. Tzafestas (Ed.), Expert Systems in Engineering Applications, Springer Berlin, Heidelberg, Berlin, Heidelberg, 1993, pp. 160–178.
[4] T.E. Dy-Liacco, Enhancing power system security control, IEEE Computer Applications in Power 10 (3) (1997) 38–41, https://doi.org/10.1109/67.595291.
[5] D. Niebur, Expert systems for power system control in Western Europe, IEEE Control Systems 11 (4) (1991) 34–40, https://doi.org/10.1109/37.88588.
[6] C.-C. Liu, T. Dillon, State-of-the-art of expert system applications to power systems, International Journal of Electrical Power & Energy Systems 14 (2–3) (1992) 86–96, https://doi.org/10.1016/0142-0615(92)90031-4.
[7] Tharam S. Dillon, Michael Laughton (Eds.), Expert System Applications in Power Systems, Prentice Hall International (UK) Ltd., 1990.
[8] M. Nazari-Heris, S. Asadi, B. Mohammadi-Ivatloo, M. Abdar, H. Jebelli, M. Sadat-Mohammadi, Application of Machine Learning and Deep Learning Methods to Power System Problems, Springer International Publishing, Cham, 2021.
[9] F. Bellizio, J.L. Cremer, G. Strbac, Machine-learned security assessment for changing system topologies, International Journal of Electrical Power & Energy Systems 134 (5) (2022) 107380, https://doi.org/10.1016/j.ijepes.2021.107380.
[10] C. Brosinsky, A. Kummerow, M. Richter, A. Naumann, P. Wiest, S. Nicolai, D. Westermann, The role of digital twins in power system automation and control: necessity,

requirements, and benefits, in: ETG-Fachberichte, ETG-Kongress: Das Gesamtsystem Im Fokus der Energiewende, 18–19 Mai 2021, Online-Veranstaltung, 1st ed., VDE Verlag, Berlin, 2021.

[11] A.M. Prostejovsky, C. Brosinsky, K. Heussen, D. Westermann, J. Kreusel, M. Marinelli, The future role of human operators in highly automated electric power systems, Electric Power Systems Research 175 (2019), https://doi.org/10.1016/j.epsr.2019.105883.

[12] C. Brosinsky, D. Westermann, R. Krebs, Recent and prospective developments in power system control centers: Adapting the digital twin technology for application in power system control centers, in: 2018 IEEE International Energy Conference (ENERGYCON), Limassol, Cyprus, 2018, pp. 1–6.

[13] M. Mazhar Rathore, S.A. Shah, D. Shukla, E. Bentafat, S. Bakiras, The role of AI, machine learning, and big data in digital twinning: a systematic literature review, challenges, and opportunities, IEEE Access (2021) 1, https://doi.org/10.1109/ACCESS.2021.3060863.

[14] L. Wright, S. Davidson, How to tell the difference between a model and a digital twin, Advanced Modeling and Simulation in Engineering Sciences 7 (1) (2020) A6, https://doi.org/10.1186/s40323-020-00147-4.

[15] ARUP, Digital twin: towards a meaningful framework, [Online]. Available www.arup.com/digitaltwinreport. (Accessed 18 December 2020).

[16] A. Kummerow, S. Nicolai, C. Brosinsky, D. Westermann, A. Naumann, M. Richter, Digital-twin based services for advanced monitoring and control of future power systems, in: 2020 IEEE Power & Energy Society General Meeting (PESGM), Montreal, QC, Canada, 2020, pp. 1–5.

[17] W. Kritzinger, M. Karner, G. Traar, J. Henjes, W. Sihn, Digital twin in manufacturing: a categorical literature review and classification, IFAC-Papers OnLine 51 (11) (2018) 1016–1022, https://doi.org/10.1016/j.ifacol.2018.08.474.

[18] B. Tekinerdogan, C. Verdouw, Systems architecture design pattern catalog for developing digital twins, Sensors (Basel, Switzerland) 20 (18) (2020), https://doi.org/10.3390/s20185103.

[19] Christoph Brosinsky, Rainer Krebs, Dirk Westermann, Embedded digital twins in future energy management systems: paving the way for automated grid control, Automatisierungstechnik 68 (9) (2020) 1–15.

[20] W. Kang, A.J. Krener, M. Xiao, L. Xu, A Survey of Observers for Nonlinear Dynamical Systems, Data Assimilation for Atmospheric, Oceanic and Hydrologic Applications, vol. II, Springer, Berlin Heidelberg, 2013.

[21] A. Agarwal, A. Pandey, L. Pileggi, Robust event-driven dynamic simulation using power flow, Electric Power Systems Research 189 (2020) 106752.

[22] IEEE Power System Operation, Planning and Economics (PSOPE) Committee: Task Force on Power System Dynamic State and Parameter Estimation, Power System Dynamic State and Parameter Estimation-Transition to Power Electronics-Dominated Clean Energy Systems, Technical Report PES-TR88, 2021.

[23] J.B. Rawlings, D.Q. Mayne, M.M. Diehl, Model Predictive Control: Theory, Computation, and Design, 2nd ed., Nob Hill Publishing, Madison, Wisconsin, 2017.

[24] M. Diehl, H.J. Ferreau, N. Haverbeke, Efficient numerical methods for nonlinear MPC and moving horizon estimation, Lecture Notes in Control and Information Sciences (2009) 391–417.

[25] G. James, D. Witten, T. Hastie, R. Tibshirani, An Introduction to Statistical Learning: With Applications in R, 8th ed., Springer, New York, Heidelberg, Dordrecht, London, 2017.

[26] S. Ruder, An overview of gradient descent optimization algorithms, 2016.

[27] I. Konstantelos, M. Sun, S.H. Tindemans, S. Issad, P. Panciatici, G. Strbac, Using vine copulas to generate representative system states for machine learning, IEEE Transactions on Power Systems 34 (1) (2019) 225–235, https://doi.org/10.1109/TPWRS.2018.2859367.
[28] D.P. Kroese, T. Taimre, Z.I. Botev, Handbook of Monte Carlo Methods, John Wiley & Sons, Hoboken, New Jersey, 2011.
[29] C. Liu, F. Tang, C. Leth Bak, An accurate online dynamic security assessment scheme based on random forest, Energies 11 (7) (2018) 1914, https://doi.org/10.3390/en11071914.
[30] C. Liu, Z.H. Rather, Z. Chen, C.L. Bak, P. Thogersen, Importance sampling based decision trees for security assessment and the corresponding preventive control schemes: the Danish case study, in: 2013 IEEE Grenoble Conference, Grenoble, 2013, pp. 1–6.
[31] IEEE/CIGRE Joint Task Force on Stability Terms and Definitions, Definition and classification of power system stability IEEE/CIGRE joint task force on stability terms and definitions, IEEE Transactions on Power Systems 19 (3) (2004) 1387–1401, https://doi.org/10.1109/TPWRS.2004.825981.
[32] N. Hatziargyriou, et al., Definition and classification of power system stability revisited & extended, IEEE Transactions on Power Systems (2020) 1, https://doi.org/10.1109/TPWRS.2020.3041774.
[33] U. Kerin, C. Heyde, R. Krebs, E. Lerch, Real-time dynamic security assessment of power grids, The European Physical Journal Special Topics 223 (12) (2014) 2503–2516, https://doi.org/10.1140/epjst/e2014-02272-1.
[34] C.O. Heyde, Dynamic Voltage Security Assessment for on-Line Control Room Application (Dynamische Spannungsstabilitätsrechnungen als online Entscheidungsgrundlage für die Leitwarte), Diss., 1st ed., Magdeburg, Univ., Fak. für Elektrotechnik und Informationstechnik, Magdeburg, 2010.
[35] O. Ruhle, E. Lerch, Ranking of system contingencies in DSA systems – first experiences, in: IEEE PES General Meeting, Minneapolis, MN, 2010, pp. 1–6.
[36] J.M. Gimenez Alvarez, P.E. Mercado, Online inference of the dynamic security level of power systems using fuzzy techniques, IEEE Transactions on Power Systems 22 (2) (2007) 717–726, https://doi.org/10.1109/TPWRS.2007.895161.
[37] Cigré WG C4.601, Review of on-line dynamic security assessment tools and techniques, Technical Brochure No. 325, 2007.
[38] ENTSO-E, Dynamic security assessment (DSA), SPD DSA Task Force, RG-CE System Protection & Dynamics Sub Group [Online]. Available https://docstore.entsoe.eu/Documents/SOC%20documents/Regional_Groups_Continental_Europe/2017/DSA_REPORT_Public.pdf, Apr. 2017.
[39] Commission Regulation (EU) 2017/1485 of 2 August 2017 establishing a guideline on electricity transmission system operation: OJ L 220 Official Journal of the European Union (2017), [Online]. Available http://data.europa.eu/eli/reg/2017/1485/oj.
[40] S.C. Savulescu, Real-Time Stability in Power Systems: Techniques for Early Detection of the Risk of Blackout, 2nd ed., Springer International Publishing, Cham, 2014.
[41] L. Wehenkel, M. Pavella, Decision trees and transient stability of electric power systems, Automatica 27 (1) (1991) 115–134, https://doi.org/10.1016/0005-1098(91)90010-Y.
[42] E.S. Karapidakis, N.D. Hatziargyriou, On-line preventive dynamic security of isolated power systems using decision trees, IEEE Power Engineering Review 22 (4) (2002) 75, https://doi.org/10.1109/MPER.2002.4312129.
[43] A.-A.B. Bugaje, J.L. Cremer, M. Sun, G. Strbac, Selecting decision trees for power system security assessment, Energy and AI 6 (5) (2021) 100110, https://doi.org/10.1016/j.egyai.2021.100110.

[44] J.L. Cremer, I. Konstantelos, G. Strbac, From optimization-based machine learning to interpretable security rules for operation, IEEE Transactions on Power Systems 34 (5) (2019) 3826–3836, https://doi.org/10.1109/TPWRS.2019.2911598.

[45] J.L. Cremer, G. Strbac, A machine-learning based probabilistic perspective on dynamic security assessment, International Journal of Electrical Power & Energy Systems 128 (5) (2021) 106571, https://doi.org/10.1016/j.ijepes.2020.106571.

[46] Cigré Joint Working Group C4/B4/C1.604, Influence of Embedded HVDC Transmission on System Security and AC Network Performance, Technical Brochure 536, 2013.

Synchrophasor applications in distribution systems: real-life experience

5

Juan Sebastián Rocha-Doria, José Guadalupe Fuentes-Velázquez, and César Angeles-Camacho
National Autonomous University of Mexico, Mexico City, Mexico

Acronyms

ABC Artificial bee colony
API Application programming interface
BCBV Branch current to bus voltage
BIBC Bus injection to branch current
CFE *Comisión Federal de Energía*
CSV Comma-separated values
DFL Direct flow load
DG Distributed generation
DGPL Distributed generation penetration level
EMS Energy management system
FACTS Flexible alternative current transmission system
GPS Global Positioning System
HC Hosting capacity
HIL Hardware- in-the-loop
HMI Human machine interface
IEC International Electrotechnical Commission
IED Intelligent electronic device
IEEE Institute of Electrical and Electronic Engineers
LIREI *Laboratorio de investigación en redes eléctricas inteligentes*
NAS Network attached storage
NASPI North American SynchroPhasor initiative
PDC Phasor data concentrator
PDC-L Local phasor data concentrator
PDC-LIREI Phasor data concentrator installed in LIREI
PDC-RTS-HIL Phasor data concentrator real-time simulation hardware-in-the-loop
PF Power factor
PMU Phasor measurement unit
PVS Photovoltaic system
RMS Root mean square
RMT CU-UNAM *Red de Media Tensión de la Ciudad Universitaria de la Universidad Nacional Autónoma de México*
ROCOF Rate of change of frequency
ROCOV Rate of change of voltage
SCADA Supervisory control and data acquisition
SPMS Synchrophasor measurement system
SQL Structure query language
TCP/IP Transmission control protocol / Internet protocol

Monitoring and Control of Electrical Power Systems Using Machine Learning Techniques
https://doi.org/10.1016/B978-0-32-399904-5.00011-9

THD	Total harmonic distortion
TVE	Total vector error
UDP	User datagram protocol
UNAM	*Universidad Nacional Autónoma de México*
WAMS	Wide-area measurement systems

5.1 Introduction

In recent years, the emergence of distributed and aggregated renewable energy sources has caused challenges, particularly in modern distribution systems. The integration of distributed energy resources includes not only renewable generation, but also electric vehicles and demand response plans. Both the former are uncontrollable and intermittent, which causes distribution systems more dynamics, and which affects the quality of the power supply, the reliability of the system operation, and the cost of electricity [1–3]. The incorporation of these models of generations has led to the evolution of distribution systems using smart grid technologies, such as big data analysis, communication technologies, real-time monitoring, stability control, and so on. The WAMS (wide-area measurement systems) are presented as a monitoring system that could capture a full-scale snapshot of system-wide dynamic behavior in real time and could identify and mitigate risky conditions through on/off-line monitoring and post-mortem analysis. Numerous universities and private/public entities have promoted the development of WAMS application, particularly to enhance the modern power systems in four areas: automation, reliability, planning, and operation.

The RMT CU-UNAM (*Red de Media Tensión de la Ciudad Universitaria de la Universidad Nacional Autónoma de México* for its Spanish abbreviation) has experimented with two punctual issues in recent years: the gradual increase of DG (distributed generation) units causing changes in power consumption patterns and the lack of appropriated information and software solutions for analyzing power failures of external or internal origin. Therefore, it has become necessary to implement an advanced monitoring system that improves the capacity of visualization and observation of phenomena, which can be observed with conventional SCADA (supervisory control and data acquisition) systems. Recent research mentioned the potential of synchrophasor applications for power distribution networks in post-mortem analysis and DG monitoring. This chapter addresses synchrophasor applications for medium-voltage systems and is organized as follows: First, a brief overview about architecture, standardization, and application of synchrophasor monitoring systems is mentioned. After that, the architecture and structure of the synchrophasor monitoring system implemented in the UNAM is detailed. Then, the post-mortem analysis of some events recorded in the medium-voltage network inside the RMT CU-UNAM are explained. Next, preliminary results of real-time monitoring simulated with an HIL (hardware- in-the-loop) testbed of PDC (phasor data concentrator) are presented. Afterwards, a brief description of methodology proposed to analyze hosting capacity. Finally, conclusions and future works are drawn.

5.1.1 Synchrophasor monitoring system

Real-time monitoring systems based on synchrophasor technology have become a suitable opportunity for improving the modern power system's capability to improve on-line the grid situational awareness/analytics and protection /control actions against disturbance [4–6]. The heart of these platforms, PMU (phasor measurement unit), is a device that can provide voltage/current phasor and frequency with a precise time stamping to compare phase angle with different nodes in a power system. This technology was used almost exclusively at the transmission level. In last few years, however, some notable benefits were observed after PMU implementation at the distribution level. In contrast to the transmission level, the angle differences on the distribution level are smaller because of smaller power flows and shorter distances. Moreover, distribution systems contain a higher level of harmonic distortion. Therefore, some laboratories, such as the Power Standards Lab, have developed a special PMU model that offers high-resolution angle measurements and power quality measurements of harmonics and transients, which is called the micro-PMU [7].

The synchrophasor measurement systems are shaped mainly according to the diagram illustrated in Fig. 5.1, which is described below.

a. The PMU is a device designed to provide measurements of RMS (root mean square) value of voltage and current, ROCOF (rate of change of frequency). This measurement contains precise time stamping offered by the time source with GPS (Global Positioning System) time synchronization.
b. The PDC is the device or communication node of WAMS architecture performed in data collection, according to time stamp and routing. There are different hierarchy levels of PDC (local, corporative and regional) according to system position and available function. This element connects PMU data with databases, HMIs (human machine interfaces), APIs (application programming interfaces), and other PDCs.
c. Communication system corresponds to synchrophasor data transmitting using serial or Ethernet communication. The trend, however, is via Ethernet through TCP and UDP (user datagram protocol) protocols.
d. Data repository corresponds to the database design for storage, according to the dimension of synchrophasor data. This repository allows accessing data for post-mortem application.
e. APIs: set of definitions and protocols for building and integrating on/off-line or real-time application into WAMS.
f. HMI: interface developed to visualize all variables contained in synchrophasor data.

5.1.2 Standardization for synchrophasor systems

The PMU definition and the first prototypes were publicized by the research group of Arun Phadke from Virginia Tech University in the1980s. The first commercial prototype PMU was manufactured by Virginia Tech University and Macrodyne in 1991

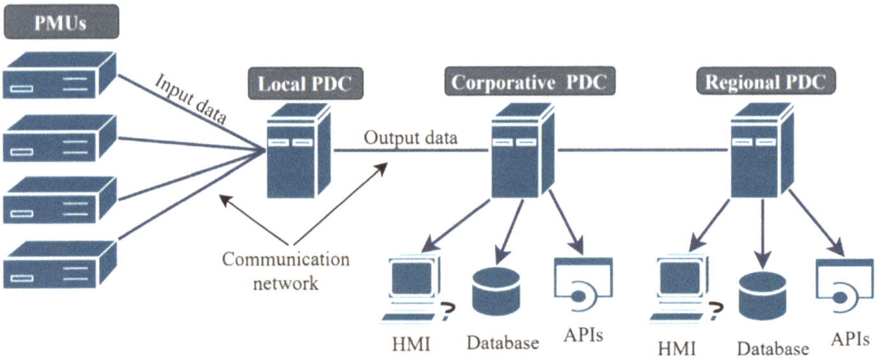

Figure 5.1 Synchrophasor data collection network.

[8]. Several experimental projects related to the application in transmission systems were developed. In fact, different companies manufactured IED (intelligent electronic device) with a PMU function.

The emergence of synchrophasor technology brought the need for standardizing general aspects, such as data transmission, steady and dynamic state estimation, and test conditions, dataframe structure, PDC requirements, and so on. Some organizations, such as NASPI (North American SynchroPhasor Initiative), IEEE (Institute of Electrical and Electronic Engineers), IEC (International Electrotechnical Commission) and others, have worked on the standardization of synchrophasor technology. Below, a brief description of the most relevant standards in the last years is presented.

a. IEEE 1344-1995 [9] is a standard, the first related to synchrophasor, focused on defining the message format, time stamping guidelines, PMU synchronization, and a recommendation for low latency.

b. IEEE C37.118-2005 [10] is a revision of standard IEEE 1344-1995, which implements the following: considerable improvements related to the definition of time stamping, useful requirements for the message format for real-time communication, a method for checking accuracy requirement by TVE (total vector error).

c. IEEE C37.118.1-2011 [11] is the standard that performed important enhancement related to PMU interoperability under dynamic conditions. It focused on the definition and requirements of measurement, frequency ramp tests, time response, and latency tests.

d. IEEE C37.118.2-2011 [12] defines a method for synchrophasor data transmission, the message format requirement for communication among PMUs, PDCs, and applications.

e. IEEE C37.118.1a-2014 [13] is an amendment of IEEE C37.118-2011, which resolves some inconsistencies and limitations related to performance requirements.

f. IEC/IEEE 60255-118-1-2018 [14] combines IEC and IEEE criteria to clarify, simplify, and remove ambiguities about certain definition and procedures. This standard is considered as amendment IEEE C37.118a-2014.

g. IEEE C37.244-2013 [15] is a guide that describes the need for performance, functionality, and communication required by a PDC for protection, control, and monitoring application in power systems.

h. Synchrophasor monitoring for distribution systems (NASPI (2018)) [16] developed by the NASPI, which is a working group comprised of the United States of America Department of Energy, North American Electric Reliability Corporation, and other entities (manufactures, consultants, researchers, and so on). In this guide, published in 2018, synchrophasor and other synchronized measurement applications for planning, operation, reliability, and automation in distribution systems were described.

5.1.3 Synchrophasor monitoring for transmission and distribution systems

A current tendency in power systems consists of integrated SPMS (synchrophasor measurement system) or WAMS with SCADA and/or EMS (energy management system), which can be applied as transmission, such as distribution systems. The SPMS looks forward to interact with SCADA/EMS by providing concise and clear information to control center operators. The SPMS has a higher message rate from PMUs (1 to 120 messages per second), whereas SCADA/EMS receives data from RTUs and IEDs with a lower rate (1 to 5 message per second). Therefore, to integrate these platforms, filtering, down-sampling, and analysis data are required. In Fig. 5.2, a typical structure of an integrated SPMS and EMS system is presented, emphasizing some differences between transmission and distribution levels. The synchrophasor applications are mentioned according to their field of application in transmission or distribution networks. The main differences for transmission and distribution networks are mentioned below.

a. Transmission systems are characterized by having highly inductive line impedances and by demonstrating a balanced behavior in voltage and current. Transmission PMUs are widely used for applications to characterize the dynamic behavior of phenomena, such as power oscillations and islanding/resynchronizing, to guarantee the power reliability.

b. Distribution systems are comprised of short lines that involve impedances with a highly resistive component, a voltage angular difference in the order of 0.010°, and a high level of current imbalance. D-PMUs were developed to monitor dynamic power flows with fidelity requirements and to incorporate some applications, such as DG characterization, microgrid operation, load modeling, topology detection, etc., as described in Section 5.1.3. Thus it is proposed that measurement accuracy of D-PMUs must be around 0.0001 p.u. and 0.01° for the magnitude and phase angle, respectively. These types of high-precision models are mentioned in [7] and [16] as micro-PMU.

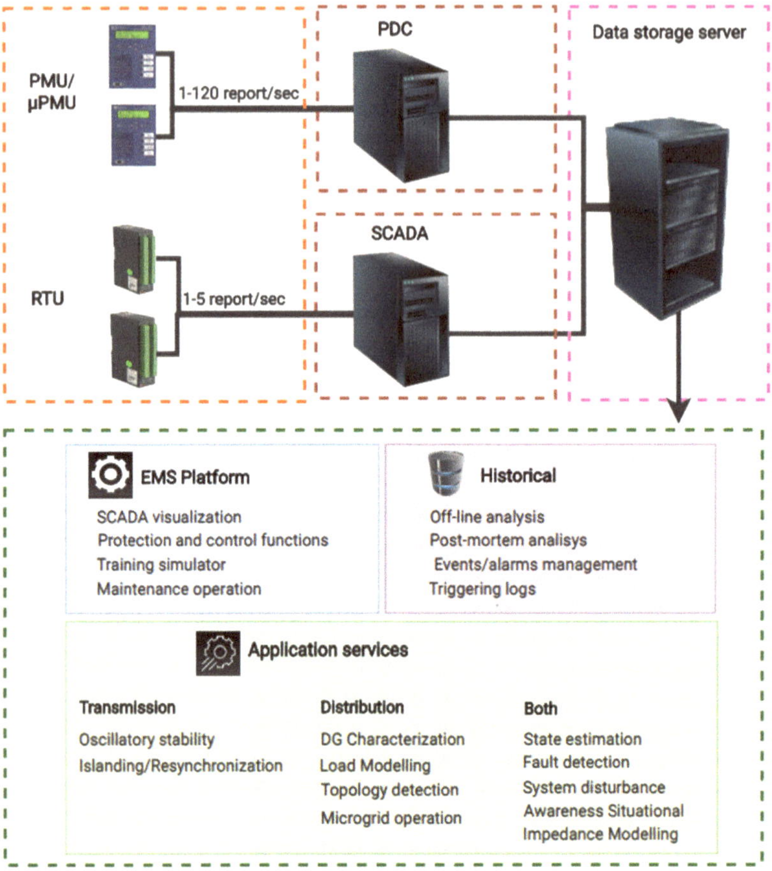

Figure 5.2 Integrated SPSM and SCADA/EMS infrastructure for transmission and distribution level.

5.1.4 Applications for distribution level

In contrast to transmission level, distribution systems are characterized by a higher level of current imbalance and voltage/current fluctuation, as well as a smaller angle difference. Therefore, synchrophasor monitoring on the distribution level requires measurements with higher resolution and more accuracy. Synchrophasor applications for distribution systems must consider these operation condition differences with transmission systems. NASPI developed and overviewed a taxonomy of applications for the distribution level, as mentioned in [17], classified according to features of accuracy, availability/reliability, latency, time alignment, and message rate, as shown in Fig. 5.3. Below, these classes are described briefly.

a. Class A (Automation): These applications seek to improve the reliability and security of the distribution system. This class is composed of automated control and protection applications, such as microgrid operation, fault detection and location, and operation using FACTS (flexible alternative current transmission system). To

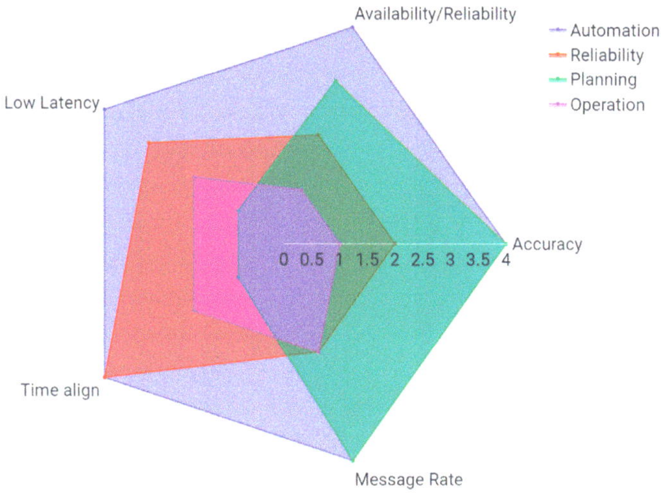

Figure 5.3 Synchrophasor applications' classification for distribution systems.

guarantee an effective communication with control and protection devices, a high level of availability, accuracy, and message rate (more than 60 frames per second), and low latency are required.

b. Class B (Reliability): This class is grouped by applications related to situational awareness because of low resolution and the message rate of SCADA systems. Therefore, the priority requirements are low latency and time alignment. To handle these challenges, there are applications such as the real-time load situational awareness dashboard and anomaly characterization and alarms.

c. Class C (Planning): This cluster of applications seeks a better power system understanding and enhances modeling techniques. The priority requirements are high accuracy and message report and low latency. Some examples of applications are state estimation, voltage stability, power quality analysis, DG characterization, and model validation.

d. Class D (Operation): This class is mainly composed of applications for monitoring and visualization of direct and derivate measurement to help with daily tasks to control center operators. With respect to the requirements, this class presents the less-demanding levels. Some examples of applications are topology and disturbance detection, phase identification, outage management, real-time monitoring with reliability standards, and thermal monitoring (overload).

5.2 Structure of SPMS proposed

A SPMS was installed in a real distribution system, known as the RMT CU-UNAM, and is located on the UNAM campus. This distribution system operates at 23 kV,

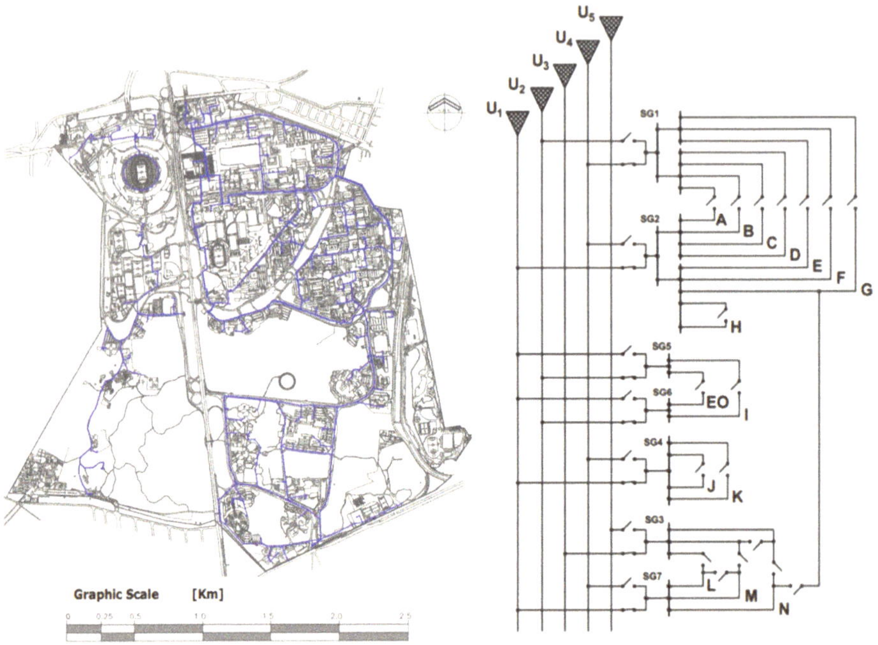

Figure 5.4 RMT CU-UNAM: a) the map, showing the branches, and b) the single-line diagram.

supplies energy through seven general substations, and has a fiber optic network for monitoring, control, and protection functions. The RMT CU-UNAM distributes energy by underground power lines through 15 rings to 156 derived substations, as mentioned in [18]. In addition, the RMT CU-UNAM are illustrated in a geographical map (Fig. 5.4.a) and, similarly, in an electrical single-diagram (Fig. 5.4.b). The campus of UNAM has several types of loads, including the following: museums, laboratories, classrooms, libraries, air conditioning, pumping systems, incinerators, offices, and sports arenas, among others.

Currently, the SPMS proposed has installed 39 relays with a PMU function; however, there are 11 of them with enabled PMUs' function. PMUs' transmit synchrophasor data using a fiber optic network to local and remote PDC. The PDC collects synchrophasor and configuration data to store it in a data repository; these components are described at Section 5.2.1. An HMI requests and interprets PDC buffer data to visualize synchrophasor data in real time, which is summarized in Section 5.2.2. The synchrophasor data are handled for common issues data, such as data loss, duplicated data, and a signal with a high level of noise, described in Section 5.3.1. Finally, the platform includes applications for post-mortem, off-line, and hosting capacity analysis, mentioned in Sections 5.3 and 5.5. All components of the proposed SPSM are illustrated in Fig. 5.5, which gives a general view of its architecture.

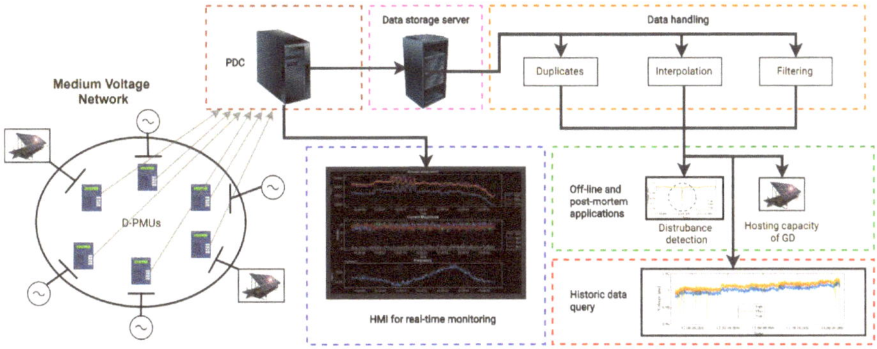

Figure 5.5 Architecture of the SPSM implemented in RMT CU-UNAM.

5.2.1 Phasor data concentrator

The phasor data concentration was developed to achieve different functions, according to requirements of the RMT CU-UNAM and technical features of PMUs installed. The PDC functions are established in the standard IEEE C37.244-2012 [15] "Guide for PDC Requirements for Power System Protection, Control, and Monitoring." These functions allow testing the performance of transmission data, the interface with PMUs for real-time access, and data validation. Below, the PDC functions implemented are briefly mentioned.

- **a.** Data aggregation
- **b.** Data communication
- **c.** Data validation
- **d.** Data format and coordinate conversion
- **e.** Data latency calculation
- **f.** Output data buffering
- **g.** Configuration
- **h.** Phase and magnitude adjustment
- **i.** Performance monitoring
- **j.** Redundant data handling

The RMT CU-UNAM requires a SPSM with a high level of data reliability/availability. Therefore, there are several options for backing up synchrophasor data, and one of these methods consists of establishing data redundance using two types of PDC. Both of them are classified regarding the hierarchical level and the location in the RMT CU-UNAM: PDC-L and PDC-LIREI. This software application was developed in Python 3 programming language and MariaDB. The following briefly explains both PDCs:

- **a.** The PDC-L, local phasor data concentrator, is enabled in microcomputers, known as Raspberry Pi 4, consisting of single-board computers with basic hardware to perform the PDC functions mentioned. This type of PDC uses an operating system based on UNIX, which permits monitoring processing performance and temperature, as well as installing programs and applications, such as SQL (structure query language) manager, remote access and Python. All substations have one PDC-L per four PMUs, which has its own buffering data and redundant management.

b. The PDC-LIREI, phasor data concentrator installed in LIREI (*Laboratorio de investigación en redes eléctricas inteligentes* for its Spanish abbreviation), is a higher level of data management system, interfaces with PMUs using a fiber optic network. This PDC, running on a rack server with two 8-core Intel Xeon processors, 64 Gb of RAM, and 16 Tb of hard drives, can collect data from almost 40 PMUs located in general substations.

The synchrophasor data, obtained from PMU, are time-aligned and collected in files with a CSV (comma-separated values) format through a PDC system. After that, these files are sent to a data repository, running on an NAS (network attached storage) server. This server is used for historical request and allows transferring historical data to authorized external users, such as central control operators. The data repository interfaces with applications for off-line, post-mortem and hosting capacity analysis, and other software for real-time simulation, as mentioned in Section 5.4. In addition, the storing capacity is of 32 Tb, with an expandable capacity to 64 Tb, ensuring data prevalence.

5.2.2 Real-time monitoring

The importance of real-time monitoring of a power system's performance has constantly increased in the last years. For the SPSM proposed was implemented an HMI, which is a visualization tool of synchrophasor measurement at the distribution level and developed in Python 3 programming language.

The application accesses the PDC buffer to monitor all PMUs deployed in the distribution system. Moreover, the HMI perform the PMU setting query and control of synchrophasor streaming. This HMI could be accessed in LIREI and the control center of RMT CU-UNAM. As shown in Fig. 5.6, the interface was developed as a dashboard with several functions, such as individual PMU control, multiple PMUs control for synchrophasor streaming, and PMU setting query. Finally, the main function is the real-time monitoring interface, as shown in Fig. 5.6, which consists of line graphs with a user-defined buffer for voltage/current magnitude and angle, as well as frequency. Similarly, polar diagrams give a snapshot of voltage/current phasor to recognize phase sequence.

5.2.3 Real-life experience: RMT CU-UNAM

5.2.3.1 Requirement of data repository

Evidently, the PMUs' transmitted data volume is considerably larger than a conventional SCADA system. Based on the first year of operation of the SPSM installed in the RMT CU-UNAM, storage data were assessed for the following:

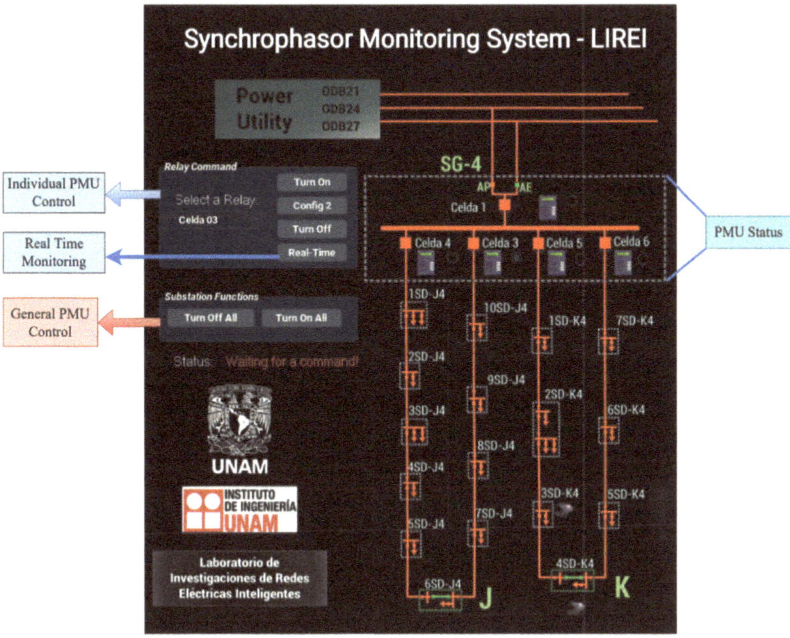

Figure 5.6 Dashboard of synchrophasor HMI. Note real-time and synchrophasor.

Table 5.1 Synchrophasor data storing volume of SMSP in RMT CU-UNAM.

Period	Individual PMU		RMT CU-UNAM (39 PMUs)	
	Digital information volume (Gb)	Data number (Million)	Digital information volume (Tb)	Data number (Million)
Day	0.105	0.864	0.004	33.696
Month	3.11	10.368	0.121	404.352
Year	37.8	315.36	1.474	12299.04

- Measurement periods: daily, monthly, annual
- Number of PMUs: performance for just one and all PMUs installed in the RMT CU-UNAM (39 PMUs)
- Message rate:10 message per second (low message rate)

Table 5.1 summarizes the storage data performance using two data volume measures: digital information (Gb or Tb) and data/records number in millions. Considering the data volume for just one PMU that occupied almost 38 Gb/year, to reach the requirement of storage capacity for the 39 PMUs installed in the RMT CU-UNAM, a data repository must have a memory capacity of around 1.5 Tb per year.

The current NAS storage server is a device dedicated to storing and backing up historical PMU data, RMS values, and detected events in a CSV format. This server is connected to the RMT CU-UNAM private communication network, and it has its own protection and backup algorithms to guarantee the prevalence of data. The stor-

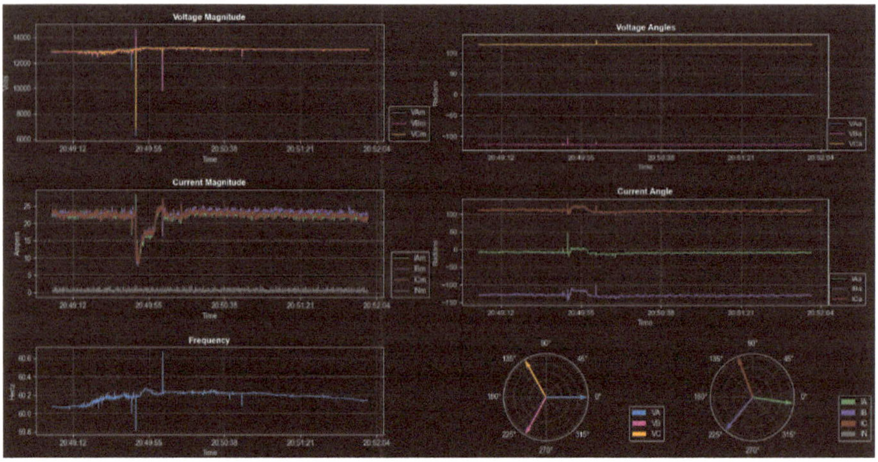

Figure 5.7 Screenshot of real-time monitoring interface.

age capacity server of 32 Tb allows storing almost 20 years of synchrophasor data, according to the digital information volume of 39 PMUs per year, as established in Table 5.1.

5.2.3.2 Visualization of events using real-time monitoring

By using the HMI for real-time monitoring developed in Python 3, a disturbance could be identified, because of an earthquake that occurred on September 07, 2020, in Mexico, as illustrated in Fig. 5.7. The imbalance in current and voltage is not significant (less than 5% in magnitude). In contrast, a series of voltage overshoots were recorded beginning at 20:49:15.8, with a descending peak that reached 0.47 p.u. Similarly, the line current suffered a drop that lasted about 12 s. Furthermore, the frequency had a stable performance close to 60 Hz, with some peaks related to sudden changes in the voltage profile. The current/voltage angle performance is tied to the system frequency; therefore if the frequency slightly rises to 60 [Hz], the current and voltage angle increases constantly, as seen in Fig. 5.7. Finally, the phasor diagram is helpful for operators and technicians in identifying phasor sequence and angular deviations.

5.2.4 PDC real-time simulation HIL testbed

HIL simulation is a type of real-time simulation linking a virtual environment with physical devices or subsystems to enhance the fidelity and consistency of results [19], [20]. For power system modeling, it is valuable when there are complex system components to model. In this section, the PDC performance validation in the RMT CU-UNAM is described using HIL simulation, known as the PDC-RTS-HIL (phasor data concentrator real-time simulation hardware-in-the-loop) testbed. In this HIL simulation, physical PMUs stream synchrophasor data through TCP/IP (transmission

Table 5.2 OP5700 hardware and software features.

Feature	Description
Operating system	Linux-based real-time
Processor	Intel Xeon E5 CPU with 8 processor cores, up to 3.2 GHz
Motherboard	ATX
Cache Memory	20 Mb
RAM Memory	16 GB of DRAM
Memory	512 GB SSD disk
PCIs slots	6 PCIe slots used to connect the internal FPGA board and PCIe or PCI third-party I/O and communication cards

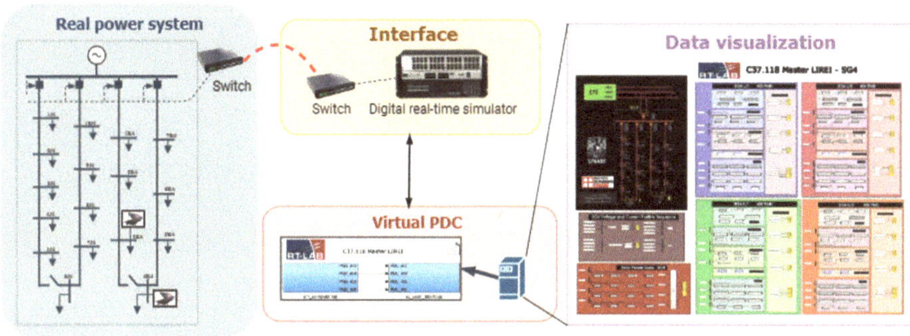

Figure 5.8 PDC-RTS-HIL testbed architecture.

control protocol/Internet protocol) by using a digital real-time simulator as a communication interface with a virtual PDC. In this application, the simulator implemented and the modeling compilator were OP5700 and RT-LAB from OPAL-RT, respectively. The OP5700 hardware/software features are given in Table 5.2. In addition, the virtual PDC was modeled to collect and interpret synchrophasor data from real PMUs deployed in RMT CU-UNAM. Synchrophasor data are interpreted and visualized in an interface for data visualization using MATLAB® and Simulink. This HIL simulation explained above is summarized in the diagram in Fig. 5.8.

The PDC's SPSM performance validation was tested through the visualization of direct and derivate variables obtained from PMU streamers. Firstly, a distribution system's general view is assessed through positive sequence voltage and current line plots as state indicators for a control center operator, as shown in Fig. 5.9. Whereas the RMT CU-UNAM has a rated voltage phase of 13.28 kV, the voltage profile has a normal performance with fluctuations less than 5% of the nominal value, and the current profile indicated a low power consumption between 2–8 A per feeder.

It is worth mentioning that before the SPSM installation, a power quality study was performed in the general substation to estimate the THD (total harmonic distortion) per feeder. The study lasted seven days and concluded that there is a low level of THD, less than 5% in all feeders, which suggests that voltage and current phasors from synchrophasor data (RMS values) could be used to calculate power variables as apparent

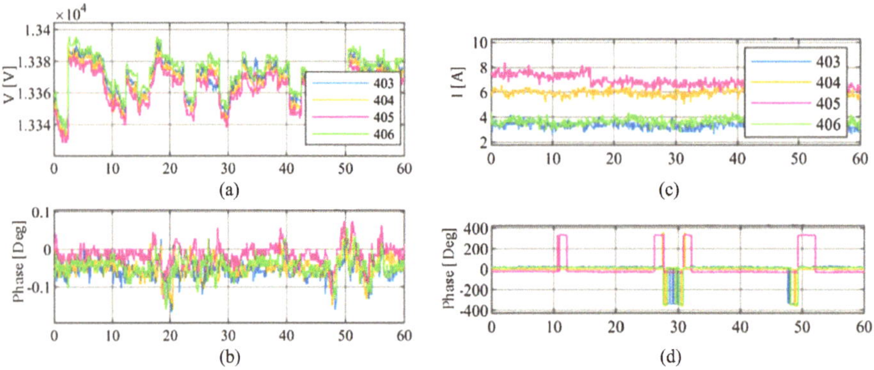

Figure 5.9 PDC-RTS-HIL testbed real-time monitoring for positive sequence voltage: (a) magnitude and (b) phase, and positive sequence current (c) magnitude and (d) phase.

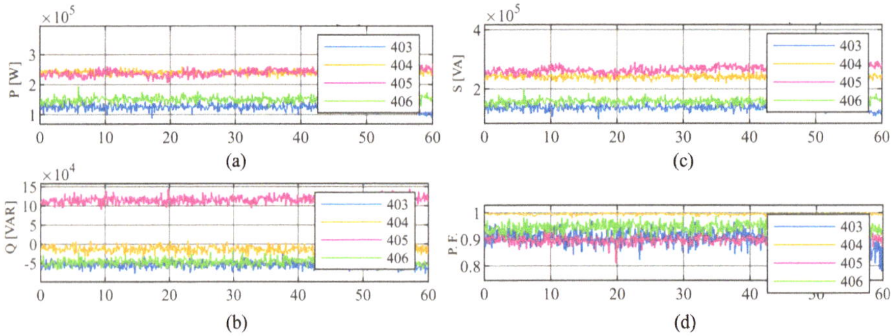

Figure 5.10 Real-time monitoring of PDC-RTS-HIL testbed for the power measurements of (a) active, (b) reactive, (c) apparent, and (d) the PF.

power, reactive power, active power, and PF (power factor), as mentioned in [21]. In the PDC-RTS-HIL testbed, an algorithm was implemented to calculate and visualize power variables in real time, as shown in Fig. 5.10. Similar to current monitoring, apparent and active power monitoring imply a low power consumption in the range of 10–30 kVA per feeder. On the other hand, the PF is in the range of 0.8–1, with an average of about 0.9, indicating a normal performance.

5.3 Post-mortem analysis

SPMS has become part of the measurement system which operates as real-time, wide-area, accurate and time-synchronized. Specifically, this platform simplifies the methodology for collecting, correlating, and comparing records related to disturbances, most well-known as post-mortem analysis. The analysis of disturbance recording with a synchrophasor could be on-line monitoring, as well as off-line by using trigger logic or an algorithm for certain conditions. The on-line monitoring of disturbance demands is constrained for communication network bandwidth and data repository facility. Thus the method based on trigger logic was implemented in the SPSM for post-mortem analysis and is described in this section as preliminary experimental results. The main challenge for the control center operator is to steadily improve situational awareness using tools such as open-source applications. The analysis of a large-scale disturbance into a distribution system could be exhausting and time-demanding process, as mentioned in [22–24]. Thus the proposed method for post-mortem analysis consists of using an alarm notification from the SCADA system for analyzing disturbances or events through historical data requests to the data repository, PDC-L, and PDC-LIREI. The historical synchrophasor data provide mainly voltage/current root mean square (RMS) and frequency through CSV files. Synchrophasor data must be preprocessed before applying the trigger logic, as described in Section 5.3.1. This method for post-mortem analysis was developed to accomplish the objectives below:

- Acquisition of recorded disturbance data to identify causes, origin, and duration.
- Analysis of the sequence of events.
- Analysis of the predisturbance system operating condition and normal behavior.
- Planning of appropriate sequence of restoration.

5.3.1 Preprocessing data

The signal preprocessing stage seeks to condition data for enhancing the event identification algorithm. This stage is composed of several functions, and not all of them are strictly necessary: for instance, filtering is applied if there is a high noise level. The following describes this conditioning data function:

a. Duplicate data: data remove by time stamping

b. Linear interpolation for data loss: using a first-degree polynomial (Eq. (5.1)) to estimate data in data loss intervals.

$$f(x|x_1; x_2) = f(x_1) + \frac{f(x_2) - f(x_1)}{(x_2 - x_1)}(x - x_1). \tag{5.1}$$

c. Filtering data: use a Savitzky–Golay filter as a technique for parameter fitting to reduce high frequency noise. For adequate filtering, it is necessary to set an appropriate width length and degree polynomial, as mentioned in [25].

d. Down-sampling: This function consists of reducing the synchrophasor sample rate to minimize the data volume and ease data handling for post-mortem analysis. Overall, it is important to maintain information quality by obtaining average, maximum, and minimum values per interval of down-sampling.

Fig. 5.7 condenses the process preprocessing stage described above. This practical example includes a signal data set of phase A voltage magnitude measured on June 19, 2020. The Savitzky–Golay filter was applied with a window width of 101 and a polynomial order of 7. As shown in Fig. 5.11, the output signal preserves the fidelity of information with less data volume.

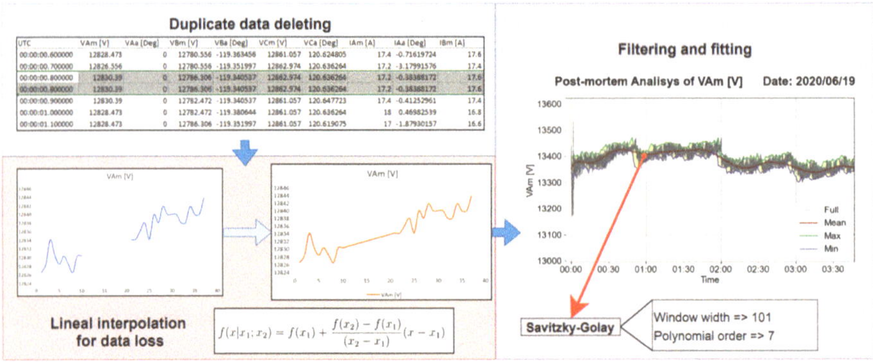

Figure 5.11 Synchrophasor data preprocessing.

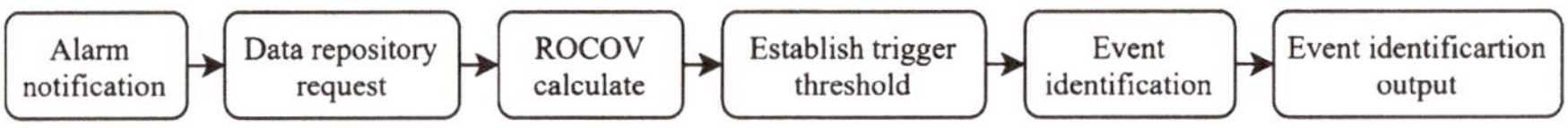

Figure 5.12 Proposed methodology for post-mortem analysis.

5.3.2 Post-mortem methodology

The proposed methodology for post-mortem analysis provides a consistent picture of disturbance and other transient events, such as overload, load shedding, undervoltage, and so on. The SCADA system provides alarms indicating time and equipment related to the event under analysis. This alarm system is used to request for a specific time interval of synchrophasor data in data repository. For the initiating values associated with the fault itself, the ROCOV (rate of change of voltage) and ROCOF were used. The ROCOF is obtained directly with synchrophasor data, whereas the ROCOV must be calculated from the line voltage through Eq. (5.2).

$$ROCOV = \frac{dV(t)}{dt}. \tag{5.2}$$

Previously, a trigger threshold had to be established to provide the lower and upper limit of the voltage and frequency to identify the outliers' data. An algorithm was developed that precisely correlates and processes data from all PMUs and shows outliers' tables and line plot with voltage and frequency signals and respective identified outliers. This kind of report yields valuable information that brings a series of panoramic snapshots of disturbance sequence to identify origin, duration, and consequence. This methodology is explained step-by-step in Fig. 5.12.

5.3.3 Case study: blackout and earthquake

In this section, the post-mortem analysis of two examples of power disturbance recorded by SPMS in the RMT CU-UNAM is introduced. These events occurred at the

Table 5.3 Frequency and ROCOF outliers caused by the 2020 blackout.

UTC	ROCOF [Hz/s]	Frequency [Hz]
14:28:40.9	−0.6	59.79
14:28:41.1	−1.22	59.625
14:28:41.3	−1.33	59.396
14:28:42.0	−0.73	59.152
14:28:42.2	−0.57	59.079
14:28:42.4	−1.11	58.926
14:28:42.9	−0.76	58.804

Table 5.4 Voltage and ROCOV outliers caused by the 2020 blackout.

UTC	ROCOV [pu/s]	VAm (pu)
14:28:41.4	−0.11556837	0.97400687
14:28:41.6	0.14735012	0.98021349

transmission level; therefore they were external phenomena to the RMT CU-UNAM, and they were described as follows.

5.3.3.1 Blackout in 2020

The first event is related to a blackout that occurred in the national transmission system on December 28, 2020, which affected 30% of the total national demand (equivalent to 10.3 million users.) Several distribution systems, such as RMT CU-UNAM, were affected by this blackout. The event started at 14:27 and lasted about 2 h; however, the SPMS detected the disturbance in the RMT CU-UNAM at 14:28:40.9, as seen in Table 5.3 and Fig. 5.12.a. Furthermore, the frequency/ROCOF and voltage/ROCOV are compared, as shown in Fig. 5.13, and some ROCOF and ROCOV outliers were detected and listed in Table 5.3 and 5.4. The trigger threshold for outlier detection was adjusted to 0.5 Hz/s and 0.1 p.u./s for ROCOF and ROCOV, respectively. The frequency was the most fluctuating variable and lasted around 2 s; the voltage only fluctuated for 1 s. The blackout's impact on the RMT CU-UNAM indicated a drastic change of frequency because of the ROCOF's fluctuation reaching values of 1.33 Hz/s. At the transmission level, a power oscillation was implied that could shed the RMT CU-UNAM. Therefore, it is necessary to implement controlled islanding-based PMUs that could improve a distribution system's safety and reliability, although a load shedding control could be implemented to diminish the blackout's impact on the distribution system.

5.3.3.2 Earthquake in 2021

The second event corresponds to a disturbance provoked by the 7.1-magnitude earthquake that occurred on September 7, 2021, on the west coast of Mexico. The event did

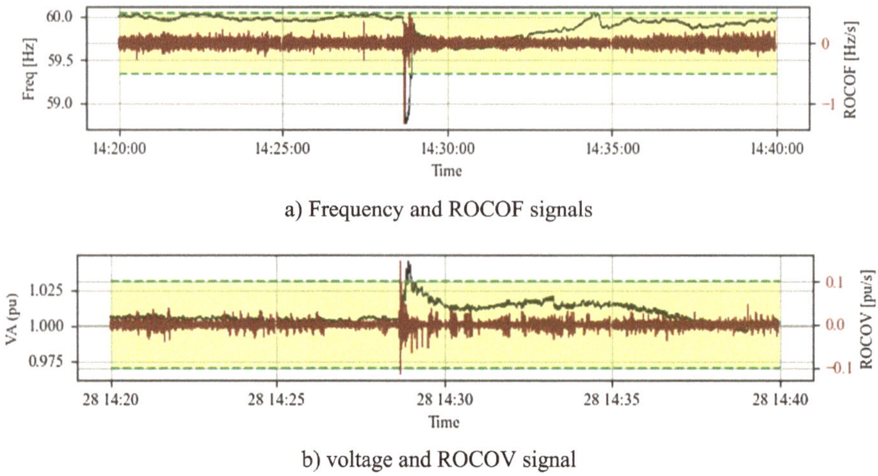

Figure 5.13 Power disturbance recorded on December 28, 2020.

not leave casualties, but it affected the energy supply of 1.6 million users. The earthquake started at 20:47:57, and minutes later some aftershocks were witnessed. The SPMS detected the first outlier data in the voltage profile at 20:49:23.2, 26.2 s after the earthquake began, as shown in Fig. 5.14.b and Table 5.6. The ROCOF and ROCOV line plots show the first disturbance that lasted around 1 min and 25 s, and four pronounced aftershocks minutes later, as shown in Fig. 5.14. As seen in Tables 5.5 and 5.6, this event has stronger variations than the 2020 blackout with maximum absolute values of 14.96 Hz/s and 6.29 p.u./s. Probably, the multiple outages were caused by falling objects on top of distribution lines, and consequently that triggered preventive line disconnections.

Countries such as Mexico that are constantly affected by natural phenomena, such as earthquakes and hurricanes, could improve the reliability and safety of national distribution systems by implementing a monitoring and control operating system that uses a national SPSM through a distribution and forecasting institute, such as National Institute of Seismology. This cooperative system could prepare strategies against natural phenomena, such as load shedding and controlled islanding to diminish or even avoid the impact on highly important users, such as hospitals, research institutes, and others.

5.4 Distributed generation hosting capacity in distribution systems

The integration of excessive distributed generation could negatively affect distribution systems by causing problems related to overvoltage, thermal overload in networks assets, the increment of short-circuit levels and the incorrect operation of protection

Table 5.5 Frequency and ROCOF outliers caused by the 2021 earthquake.

UTC	ROCOF [Hz/s]	Frequency [Hz]
20:49:24.8	0.5	60.152
20:49:25.6	−0.5	60.12
20:49:26.3	−0.63	60.051
20:49:26.4	1.08	60.159
20:49:31.4	0.52	60.225
20:49:35.5	−0.62	60.168
20:49:43.6	−0.65	60.158
20:49:45.5	−1.88	59.999
20:49:45.6	1.38	60.137
20:49:46.5	0.55	60.23
20:49:46.6	−0.72	60.158
20:49:47.9	−3.65	59.813
20:49:48.0	3.88	60.201
20:50:03.3	2.92	60.513
20:50:03.5	−4.69	60.21
20:50:41.6	−0.54	60.158
20:50:41.7	0.61	60.219
20:50:48.8	−1.34	60.083
20:50:48.9	1.3	60.213
20:53:29.1	−1.94	59.845
20:53:29.2	2.61	60.106
20:53:29.3	−0.82	60.024
20:53:35.8	1.28	60.185
20:53:35.9	−1.31	60.054
20:54:27.4	−3.08	59.747
20:54:27.5	3.02	60.049
20:55:36.3	−5.83	59.411
20:55:36.4	14.96	60.907
20:55:36.5	−8.83	60.024

system [26], [27]. Therefore, it is important to estimate the DG capacity that could be integrated to the distribution system without exceeding operational limits, such as voltage variations, the level of power loss, and power quality. This concept is known as HC (hosting capacity) for DG, as mentioned in [2]. The main objective of HC consists of determining the optimal location and sizing of the DG. There are numerous methodologies to estimate the HC of DG, such as heuristic methods based on genetic algorithms [28]; the meta-heurism method, such as ABC (artificial bee colony) [29], and stochastic methods [30].

Table 5.6 Voltage and ROCOV outliers caused by the earthquake in 2021.

UTC	ROCOV [pu/s]	VAm (pu)
20:49:23.2	−0.14450063	0.950768098
20:49:23.3	0.239870901	0.974728536
20:49:25.6	−0.14450063	0.959139777
20:49:28.4	−0.18640559	0.961304916
20:49:29.8	−0.22397561	0.951345473
20:49:30.3	0.135830747	0.977615337
20:49:32.3	−0.11993545	0.970542696
20:49:36.1	−0.10548539	0.968666283
20:49:36.2	0.105485387	0.979203101
20:49:39.7	−0.1314958	0.969099296
20:49:43.4	−0.19652147	0.965490756
20:49:44.0	0.10693062	0.982667278
20:49:45.5	−1.15311461	0.862287638
20:49:45.6	1.22970062	0.985121067
20:49:46.1	−0.11849022	0.969532308
20:49:46.5	0.157505465	0.978481363
20:49:47.9	1.157450303	1.101459154
20:49:48.0	−6.29589269	0.472569429
20:49:48.1	4.996831464	0.971697372
20:49:50.5	−0.11271004	0.977326687
20:49:50.7	0.11704574	0.993059733
20:50:03.3	−0.40893672	0.94946906
20:50:03.4	0.439281324	0.993348383
20:50:26.3	0.119935451	0.993348383
20:50:26.4	−0.10115044	0.983244578
20:50:34.3	0.104040155	0.991038957
20:50:48.8	−0.3020061	0.962026654
20:50:48.9	0.148835577	0.976893675
20:53:29.0	−1.62563175	0.815665801
20:53:29.1	1.674762119	0.982955928
20:53:35.7	−1.5215916	0.826635631
20:53:35.8	1.562052077	0.982667278
20:54:27.3	−0.35836112	0.940375643
20:54:27.4	0.497081577	0.990028569
20:54:27.5	−0.15750546	0.974295523
20:55:36.3	−2.37414534	0.737433433
20:55:36.4	2.306230718	0.967800257

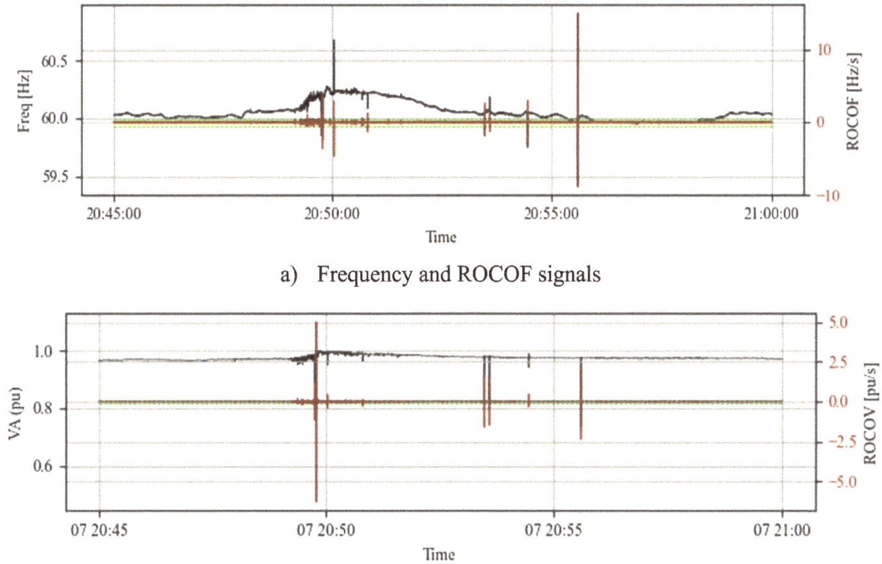

a) Frequency and ROCOF signals

b) voltage and ROCOV signals

Figure 5.14 Power disturbance recorded on September 7, 2021.

5.4.1 Proposed methodology of HC analysis

On the one hand, DG integration could unleash an increase of voltage magnitude, especially when the DG produces more energy than the power consumed in the DG installation bus. On the other hand, an optimal DG sizing and location could minimize system losses and diminish the risk of overload in some network assets, such as feeders and power transformers, enhancing the efficient operation and extending useful life. Therefore, in this section, a methodology is presented to analyze the impact of DG in radial distribution systems, according to the limits of voltage variation and total system losses, as explained in the flow diagram of Fig. 5.15.

To estimate power flow with the DG in radial distribution systems, a power flow algorithm was implemented based on a fast decoupled method documented in [31], [32] and described below.

1. Request synchrophasor data from feeders and SCADA measurements as initial input data for the power flow algorithm. Measurement data used in the present case study are shown in Fig. 5.16.
2. Create a single-line diagram from distribution network numbering nodes and branch currents (BC), as illustrated in Fig. 5.16.
3. The algorithm implemented uses the relationship between injected current and branch current, called the BIBC (bus injection to branch current) matrix, obtained by Kirchhoff's current law. For example, this relationship is applied to RMT-CU

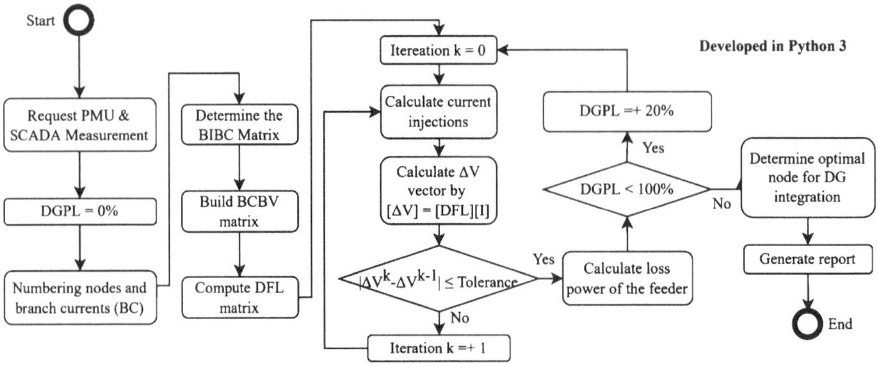

Figure 5.15 Flow diagram of methodology for HC analysis.

UNAM as in (5.3):

$$\begin{aligned} BC1 &= I2 + I3 + I4 + I5 + I6 \\ BC2 &= I3 + I4 + I5 + I6 \\ BC3 &= I4 + I5 + I6 \\ BC4 &= I5 + I6 \\ BC5 &= I6. \end{aligned} \tag{5.3}$$

Thus the BIBC matrix is defined as follows in (5.4):

$$[BC]_{mx1} = [BICB]_{mxn}\,[I]_{nx1}\,, \tag{5.4}$$

where m is the total number of branches and n is the total number of nodes

4. Similarly, the relationship between branch current and bus voltage is obtained for the RMT-CU UNAM as follows:

$$\begin{aligned} V_2 &= V_1 - Z_{12}BC_1 \\ V_3 &= V_2 - Z_{23}BC_2 \\ V_4 &= V_3 - Z_{34}BC_3 \\ V_5 &= V_4 - Z_{45}BC_4 \\ V_6 &= V_5 - Z_{56}BC_5. \end{aligned} \tag{5.5}$$

where Z_{ik} is the impedance between buses i and k.
As shown (5.6), (5.5) can be rewritten using a substitution:

$$\begin{bmatrix} V_1 - V_2 \\ V_1 - V_3 \\ V_1 - V_4 \\ V_1 - V_5 \\ V_1 - V_6 \end{bmatrix} = \begin{bmatrix} Z_{12} & 0 & 0 & 0 & 0 \\ Z_{12} & Z_{23} & 0 & 0 & 0 \\ Z_{12} & Z_{23} & Z_{34} & 0 & 0 \\ Z_{12} & Z_{23} & Z_{34} & Z_{45} & 0 \\ Z_{12} & Z_{23} & Z_{34} & Z_{45} & Z_{46} \end{bmatrix} \begin{bmatrix} BC_1 \\ BC_2 \\ BC_3 \\ BC_4 \\ BC_5 \end{bmatrix} \tag{5.6}$$

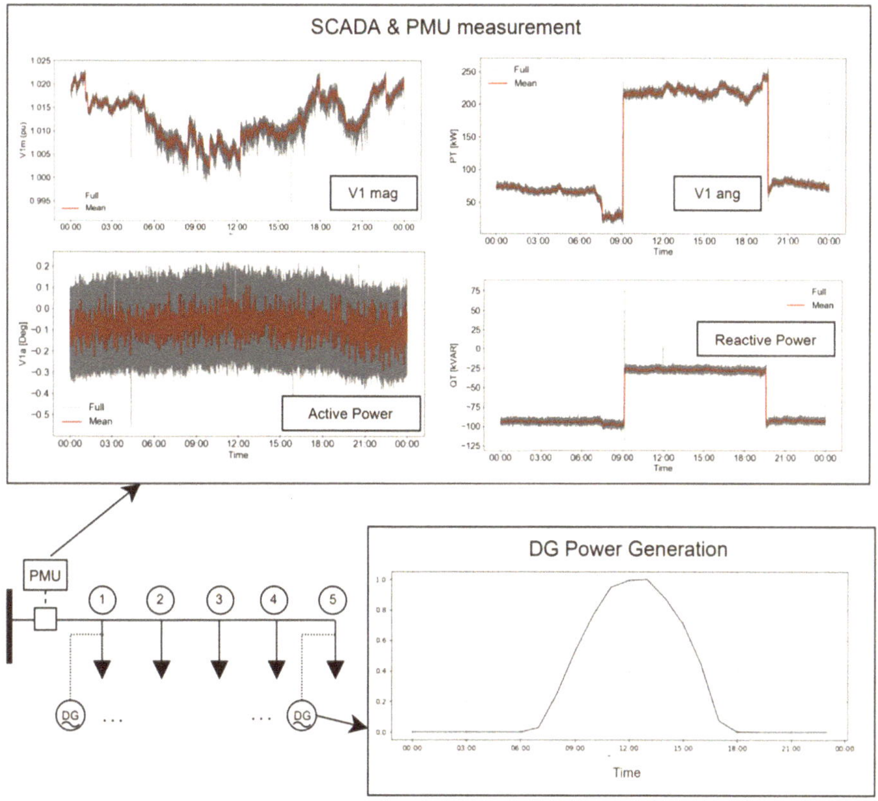

Figure 5.16 Distribution network under analysis.

This relationship is called the BCBV (branch current to bus voltage) matrix and is expressed as follows:

$$[\Delta V]_{nx1} = [BCBV]_{nxm} [BC_i]_{mx1} . \tag{5.7}$$

5. Calculate the DFL (direct flow load) matrix that indicates the relation between injection current and the difference of bus voltages with slack node as explained in (5.8):

$$[\Delta V]_{nx1} = [BCBV]_{nxm} [BICB]_{mxn} [BC]_{mx1} ,$$
$$[\Delta V]_{nx1} = [DFL]_{nxn} [I]_n . \tag{5.8}$$

6. This formulation consists of an iterative process to determine the power flow, where the injection currents and the new values of ΔV are calculated through

Eqs. (5.9) and (5.10):

$$I_i^k = \left(\frac{P_i + jQ_i}{V_i^k}\right)^*, \tag{5.9}$$

$$\left[\Delta V^{k+1}\right]_{nx1} = [DFL]_{nxn}\left[I^k\right]_{nx1}, \tag{5.10}$$

where $i = 1, 2, \ldots n$, n: total number of buses, k: the iteration number.

7. Thus the voltage values are updated using (5.11):

$$\left[V^{k+1}\right]_{nx1} = V_1 - \left[\Delta V^{k+1}\right]_{nx1}. \tag{5.11}$$

8. If an absolute voltage error reaches a tolerance of 10^{-6}, as shown in (5.12), the iteration process gets finished and calculates the total power loss. If not, it advances into the next iteration with updated values of voltage and currents:

$$\left|\left[V^{k+1}\right]_{nx1} - \left[V^k\right]_{nx1}\right| < Tolerance \tag{5.12}$$

9. To calculate the line power loss, which connects buses i and i+1, (5.13) is used:

$$P_{loss}(i, i+1) = R_{i,i+1}\left(\frac{P_i^2 + Q_i^2}{|V_i|^2}\right). \tag{5.13}$$

Therefore, the total power losses of the feeder (PF_{loss}) are calculated by summing up the line power losses through (5.14):

$$PF_{loss} = \sum_{i=1}^{N-1} P_{loss}(i, i+1). \tag{5.14}$$

To determine the optimal placement, the total power loss and voltage node are estimated, where the DG was installed for several scenarios with different DG penetration levels for each bus. The DGPL (distributed generation penetration level) is evaluated from 0% to 100%, where 100% of the LP corresponds to a maximum level of load of the DG installed node.

5.4.2 Case study: HC in the RMT CU-UNAM

Over the past years, several PVSs (photovoltaic systems) have been installed in RMT CU-UNAM, which are the main DG source on the campus. Therefore, the PVS integration impact was analyzed for different DGPL (0 to 100%) in a radial feeder of the RMT CU-UNAM, illustrated in Fig. 5.15. This distribution system operates with a rated line voltage of 23 kV and base apparent power of 100 kVA. This case study corresponds to one day of measurement data, and a power generation input measured

Table 5.7 SFV integration impact in the radial brach of RMT CU-UNAM.

Node	**2**		**3**		**4**		**5**		**6**	
LP [%]	**P_loss**	**V_DG**	**P_loss**	**V_DG**	**P_loss**	**V_DG**	**P_loss**	**V_DG**	**P_loss**	**V_DG**
0	2.7884	1.0002	2.2677	1.0013	2.2593	1.0022	2.2325	1.0029	2.2085	1.0036
20	2.6009	1.0003	2.2647	1.0014	2.2497	1.0023	2.2239	1.0030	2.2085	1.0037
40	2.3877	1.0004	2.2613	1.0016	2.2387	1.0024	2.2140	1.0030	2.2085	1.0037
60	2.2869	1.0005	2.2597	1.0017	2.2335	1.0025	2.2094	1.0031	2.2085	1.0037
80	2.2685	1.0006	2.2594	1.0018	2.2326	1.0026	**2.2085**	**1.0031**	2.2085	1.0037
100	2.2677	1.0007	2.2593	1.0019	2.2325	1.0027	2.2085	1.0032	2.2085	1.0037

from the PVS installed in the RMT CU-UNAM was used. Afterwards, the methodology described above was applied, and the obtained results were collected in Table 5.7, where the maximum values of total power losses and the respectively installed DG bus voltage were obtained. In Table 5.7, it can be seen that the optimal placement and sizing of the PVS were in node 5 with a DGPL of 80%, because these conditions present the lower level of total power loss and the installed DG bus voltage closest to 1 p.u.

After the optimal placement in node 5, the DG integration impact was evaluated using the line plots of total power loss and voltage variation at bus DG, as shown in Figs. 5.17 and 5.18, respectively. In this analysis is observed the fluctuation for a period of one day and different LP of DG. Additionally, these line plots include the PVS power generation used for the power flow study. Regarding total power losses, there is a low level of losses because of power line robustness of the RMT CU-UNAM, and the progressive losses' reduction could be valued according to the increase of DG penetration. Moreover, the level of total power losses would be more considerable if the grid demanded higher energy consumption and more robust power lines.

Similarly, installed DG voltage variation for node 5 do not present a considerable fluctuation during the generation peak hours (11 to 14 h), as shown in Fig. 5.18. Therefore, it is suggested to implement other methods of analysis of the HC of DG to analyze the excess of electrical energy generated to the supply network and the effects on the protection system, as mentioned in [27], [33].

These kinds of studies are useful for distribution planning; however, the implemented algorithm is limited because of the low number of D-PMUs in the distribution system. In addition, the RMT CU-UNAM has a robust infrastructure in distribution lines. Furthermore, the power transformers have a low utilization factor, which allows accommodating distributed generation blocks higher than the load demanded by the UNAM campus facilities. Therefore, in the present study, the capacity of GD was limited to the maximum load value demanded on the UNAM campus, since otherwise the study would have to be extended according to the regulatory guidelines of the energy supply company, CFE (*Comisión Federal de Energía* for its Spanish abbreviation), for the purposes of export and the sale of energy.

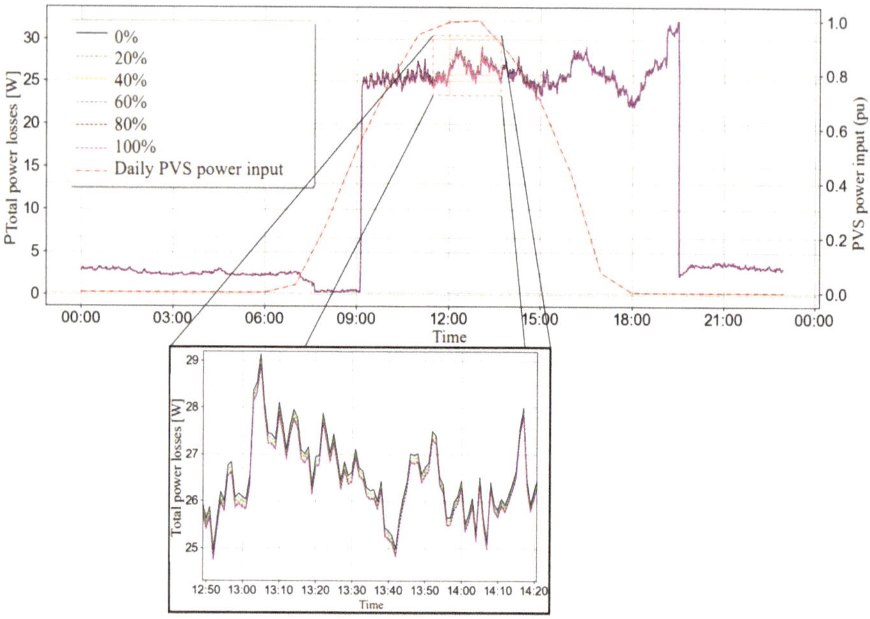

Figure 5.17 Total power losses for different DG penetration levels.

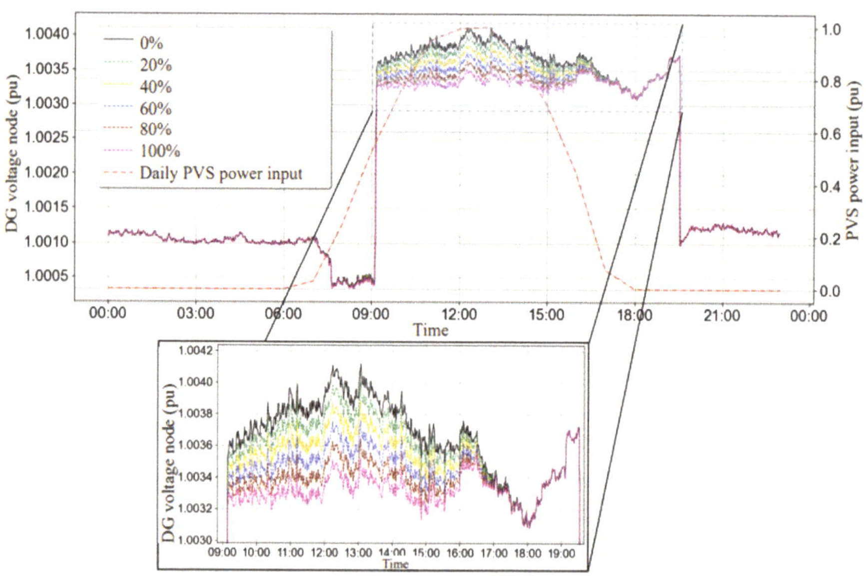

Figure 5.18 Voltage variation for different DG penetration levels.

5.5 Conclusions

This paper presented a synchrophasor measurement system for distribution systems that faced several challenges: the provision of adequate hardware and configuration for existing IEDs, which could be used as PMU; the implementation of PDC-L, as an auxiliary PDC, in distribution substations using low-cost computers; and the development of the PDC-LIREI as the main concentration system. The development of an open-source synchrophasor measurement system by LIREI, according to IEEE C37.118-2011 and IEEE C37.244-2013, encourages future academic projects, thus software license limitations could be avoided. In addition, these kinds of efforts stimulate not only university exchanges of resources, but also cooperation with the industrial sector.

The PDC-RT-HIL testbed simulated a real-time monitoring of a distribution network using physical PMUs. Synchrophasor handling allowed visualizing not only voltage and current, but also power measurements and positive sequence components. This model of HIL simulation could be replicated to communicate multiple PMUs of different manufacturers. Moreover, this initiative might open opportunities for future analyses in developing simulation scenarios of real field resemblance to propose and validate control and protection algorithms for wide-area system operation in distribution systems using synchronized measurements.

The SPMS performs not only data concentration, but also the processing and storage of synchrophasor data from PMUs deployed in distribution systems. The HMI developed offers a suitable opportunity to enhance SCADA systems' service through the assessment of situational awareness and synchrophasor monitoring technology at the distribution level. Post-mortem analysis correlates and compares synchronized measurements from several deployed PMUs. Moreover, it was demonstrated that ROCOV and ROCOF are appropriate indicators for identifying power disturbances, which reduces data processing time.

The excessive integration of DG units leads to performance problems in the distribution network related to overvoltage, increased short-circuit level, the poor operation of protective devices, overload by the reverse flow of power, and power quality, among others. Therefore, it is important to conduct studies to determine the optimal location and sizing of the DG units, which will improve the feeder voltage profile and reduce total losses in the system, according to the level of penetration of the DG. The synchrophasor data could help to analyze the DG integration in distribution systems by the power element modeling, such as lines, transformers, loads and generators, without incurring additional field studies. Therefore, an analysis of DG integration for radial distribution systems was presented using an algorithm for decoupled power flows and SCADA and synchrophasor data.

5.6 Future works

The evolution of distribution systems through DG integration has led to research about synchrophasor applications to improve the operation and planning in distribution networks. Therefore, the following research is proposed to widen the area of synchrophasors at the distribution level:

a. Control based on D-FACTS: integration of synchrophasor measurement system with FACTS for distribution networks or D-FACTS to provided fast control responses to improve the voltage profile, reduce power losses, and balance loads;
b. State estimation in distribution system: the use of synchrophasor data for the estimation of the state of the power system that allows observing parameters unknown to SCADA systems directly, which implies a drastic reduction in processing time;
c. Load modeling: uses synchrophasor data repositories to estimate ZIP and exponential load models with fidelity and accuracy margins required for power system simulations;
d. Hosting capacity: implementation of heuristic methodologies for the calculation of optimal location and sizing, considering voltage constraints, short-circuit level increment, protection coordination, power quality, and power losses.

Acknowledgments

This paper highlights the contribution made by the Survalent company, providing a platform of control, measurement, and the protection of power systems, which was installed in the RMT CU-UNAM. Juan Sebstián Rocha-Doria and José Guadalupe Fuentes-Velázquez (with codes 928337 and 486627) thank the Consejo Nacional de Ciencia y Tecnología (CONACyT) from Mexico for the support provided through the graduate scholarship. In addition, César Angeles-Camacho are grateful for the financial aid received from the project CEMIE-REDES/PE-A-03/2019 and the 246949 to conduct this research.

References

[1] A. Keyhani, Design of Smart Power Grid Renewable Energy System, John Wiley & Sons, 2019.
[2] T. Funabashi, Integration of Distributed Energy Resources in Power Systems: Implementation, Operation and Control, Academic Press, ISBN 9780128032121, 2016.
[3] R. Bansal, Handbook of Distributed Generation. Electric Power Technologies, Economics and Environmental Impacts, Springer, ISBN 9783319513423, 2017.
[4] A. von Meier, R. Arghandeh, Renewable Energy Integration, Every Moment Counts: Synchrophasors for Distribution Networks with Variable Resources, second edition, Academic Press, ISBN 9780128095928, 2017.
[5] A.G. Phadke, J.S. Thorp, Synchronized Phasor Measurements and Their Applications, Springer, ISBN 9783319505848, 2017.
[6] Z. Dong, P. Zhang, Emerging Techniques in Power System Analysis, Springer, ISBN 9783642042829, 2010.
[7] A. von Meier, E. Stewart, A. McEachern, M. Andersen, L. Mehrmanesh, Precision micro-synchrophasors for distribution systems: a summary of applications, IEEE Transactions on Smart Grid 8 (6) (2017) 2926–2936, https://doi.org/10.1109/TSG.2017.2720543.

[8] A.G. Phadke, Synchronized phasor measurements – a historical overview, in: Proceedings of the IEEE Power Engineering Society Transmission and Distribution Conference, vol. 1, Asia Pacific, 2002, pp. 476–479.
[9] K.E. Martin, et al., IEEE standard for synchrophasors for power systems, IEEE Transactions on Power Delivery 13 (1) (1998), https://doi.org/10.1109/61.660853.
[10] K.E. Martin, et al., C37.118-2005-IEEE standard for synchrophasors for power systems, in: IEEE Std C37.118-2005 (Revision of IEEE Std 1344-1995), 2005.
[11] C37.118.1-2011 – IEEE Standard for Synchrophasor Measurements for Power Systems, IEEE Std C37.118.1-2011 (Revision of IEEE Std C37.118-2005), 2011.
[12] C37.118.2-2011 – IEEE Standard for Synchrophasor Data Transfer for Power Systems, IEEE Std C37.118.2-2011 (Revision of IEEE Std C37.118-2005), 2011.
[13] IEEE Standard for Synchrophasor Measurements for Power Systems – Amendment 1: Modification of Selected Performance Requirements, IEEE Std C37.118.1a-2014 (Amendment to IEEE Std C37.118.1-2011), 2014.
[14] IEEE/IEC International Standard – Measuring relays and protection equipment – Part 118-1: Synchrophasor for power systems – Measurements, IEC/IEEE 60255-118-1:2018, 2018.
[15] IEEE Guide for Phasor Data Concentrator Requirements for Power System Protection, Control, and Monitoring, Power System Relaying Committee IEEE Power and Energy Society, 2013.
[16] A. von Meier, et al., Synchrophasor Monitoring for Distribution Systems: Technical Foundations and Applications, White Paper, NASPI Distribution Task Team, 2018.
[17] M. Hojabri, U. Dersch, A. Papaemmanouil, P. Bosshart, A comprehensive survey on phasor measurement unit applications in distribution systems, Energies 12 (23) (2019) 1–23, https://doi.org/10.3390/en12234552.
[18] J.G. Deras Campos, Modelado de la Red Eléctrica en Media Tensión de Ciudad Universitaria y Diseño de un sistema de monitoreo sincrofasorial, M.Ing. Tesis, Universidad Nacional Autónoma de México, México, 2017.
[19] M.O. Faruque, et al., Real-time simulation technologies for power systems design, testing, and analysis, IEEE Power and Energy Technology Systems Journal 2 (2) (2015) 63–73, https://doi.org/10.1109/JPETS.2015.2427370.
[20] M. Stifter, J. Cordova, J. Kazmi, R. Arghandeh, Real-time simulation and hardware-in-the-loop testbed for distribution synchrophasor applications, Energies 11 (4) (2018) 1–21, https://doi.org/10.3390/en11040876.
[21] IEEE Power & Energy Society, Definitions for the measurement of electric power quantities under sinusoidal, nonsinusoidal, balanced, or unbalanced conditions, https://doi.org/10.1109/IEEESTD.2010.5953405, 2010.
[22] R. Bansal, Power System Protection in Smart Grid Environment, CRC Press, ISBN 9780429401756, 2019.
[23] H. Gharavi, B. Hu, Synchrophasor sensor networks for grid communication and protection, Proceedings of the IEEE 105 (7) (Jul. 2017) 1408–1428, https://doi.org/10.1109/JPROC.2017.2696881.
[24] V. Roy, S.S. Noureen, S.B. Bayne, A. Bilbao, M. Giesselmann, Event detection from PMU generated big data using R programming, in: 2018 IEEE Conference on Technologies for Sustainability, SusTech 2018, 2019.
[25] R. Schafer, What is a Savitzky-Golay filter? [lecture notes], IEEE Signal Processing Magazine 28 (4) (Jul. 2011) 111–117, https://doi.org/10.1109/MSP.2011.941097.
[26] M. Bollen, F. Hassan, Integration of Distributed Generation in the Power System, John Wiley & Sons, Inc., ISBN 9781118029039, 2011.

[27] S.M. Ismael, S.H.E. Abdel Aleem, A.Y. Abdelaziz, A.F. Zobaa, State-of-the-art of hosting capacity in modern power systems with distributed generation, Renewable Energy 130 (2019) 1002–1020, https://doi.org/10.1016/j.renene.2018.07.008.
[28] A. Alhamali, M.E. Farrag, G. Bevan, D.M. Hepburn, Determination of optimal site and capacity of DG systems in distribution network based on genetic algorithm, in: 2017 52nd International Universities Power Engineering Conference, UPEC 2017, vol. 2017-Janua, 2017, pp. 1–6.
[29] F.S. Abu-Mouti, M.E. El-Hawary, Modified artificial bee colony algorithm for optimal distributed generation sizing and allocation in distribution systems, IEEE Transactions on Power Delivery 26 (4) (Oct. 2011) 2090–2101, https://doi.org/10.1109/TPWRD.2011.2158246.
[30] X. Chen, W. Wu, B. Zhang, C. Lin, Data-driven DG capacity assessment method for active distribution networks, IEEE Transactions on Power Systems 32 (5) (2017) 3946–3957, https://doi.org/10.1109/TPWRS.2016.2633299.
[31] I.T. Papaioannou, A. Purvins, A methodology to calculate maximum generation capacity in low voltage distribution feeders, International Journal of Electrical Power and Energy Systems 57 (2014) 141–147, https://doi.org/10.1016/j.ijepes.2013.11.047.
[32] S. Dahal, A.K. Jha, N. Bhattarai, A.K. Panjiyar, Determining the hosting capacity of solar photovoltaic in a radial distribution network using an analytical approach, Journal of Advanced College of Engineering and Management 5 (2019) 15–25, https://doi.org/10.3126/jacem.v5i0.26673.
[33] M. Delfanti, M. Merlo, G. Monfredini, V. Olivieri, M. Pozzi, A. Silvestri, Hosting dispersed generation on Italian MV networks: towards smart grids, in: Proceedings of 14th International Conference on Harmonics and Quality of Power – ICHQP 2010, Sep. 2010, pp. 1–6, Synchrophasorial monitoring system for Medium Voltage Networks.

A graph mapping based supervised machine learning strategy for PMU voltage anomalies' detection and classification in distribution networks

6

Ernesto Beltran, Emilio Barocio Espejo, and Dunstano del Puerto Flores
University of Guadalajara, Guadalajara, Mexico

6.1 Introduction

The electrical grid is a dynamic system due to the continuous interaction between generation and load. To ensure the safety and reliability of the electrical grid, as well as allow monitoring and controlling in real time, the grid operator is assisted by simulators based on mathematical models, analysis of historical data, and monitoring systems, such as SCADA systems. However, the granularity of the information is not sufficient to observe or predict disturbances associated with stochastic variations of the load (such as incipient faults or electromechanical oscillations), topological changes in the network, or anomalies of unintentional (starting engines, smelting furnaces and switched sources) and intentional (cyber-attacks) origins [1].

To overcome these limitations, more advanced monitoring and control systems have been implemented, among the most notorious are the phasorial measuring units (PMUs). The PMUs are able to capture the random behavior that involves multiple operating scenarios and load conditions in the electrical grid at a processing rate of 10 to 60 samples per second [2]; whereas, that of the μPMUs is 120 samples per second [3]. The higher processing rate allows the development of new algorithms and a better understanding of dynamic behaviors in the power grid, where PMU voltage signals contain relevant information regarding the operational state of the grid. Moreover, it allows to take preventive actions against dangerous conditions through the early detection and classification of anomalies in the voltage data. Voltage anomalies can be grouped into two classes: within and outside the normal operating range.

According to the grid code [4], at distribution level, the voltage anomalies related to the power quality are located outside the normal operating range of the voltage (0.9–1.1 pu). Among the main causes of these events are the start-up of certain equipment, such as compressors and pumps, sudden increases in demand, loads far from the interconnection point, and the recycling of protection devices within the first second after detection of a failure [5]. The other type of anomalies, within the normal operating range, are mainly caused by slight changes in generation-load. These types of

Monitoring and Control of Electrical Power Systems Using Machine Learning Techniques
https://doi.org/10.1016/B978-0-32-399904-5.00012-0

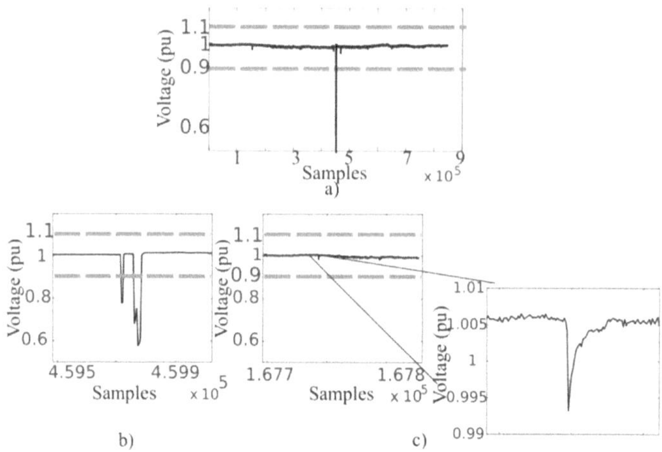

Figure 6.1 a) Voltage PMU records at distribution level voltage, b) anomaly outside normal operating limits, and c) anomaly within normal operating conditions.

anomalies have no direct impact on the power quality at short-term; however, if not resolved in time, they can trigger failures in the continuity of service or permanent damage to sensitive equipment [6]. Both scenarios can be seen in Fig. 6.1.

To avoid damage or loss of strategic equipment, as well as partial or total power outages, due to anomalies such as those shown in Fig. 6.1, timely anomaly detection and classification in voltage signals opens the opportunity to the design of preventive and corrective control actions. A brief review of main papers with similar approaches based on data-driven techniques for the detection and classification of anomalies is presented below.

In [7], two approaches are proposed to classify events such as capacitor bank switching or tap changes. The first is based on principal component analysis (PCA) and support vector machines (SVM); the second employees an autoencoder along with a softmax classifier. Nevertheless, all of the presented examples are entirely composed by synthetic PMU signals.

In [8], the detection of sags is carried out by independent component analysis (ICA) and teaching-learning-based optimization (TLBO), where later a higher-order statistics (HOS), TLBO, recurrent neural network (RNN), and SVM are employed for the classification stage. However, it focuses on other specific problems, such as failures, transformer saturation, and induction motor starting, based on the analysis of both simulated and real signals.

In [9], an autoencoder as well as a softmax classifier are employed twice in a row to classify malfunctions in the switching of capacitor banks, changes on transformer taps, abrupt power load connections, and reconfiguration maneuvers. However, just as in [7], this work only presents results from simulated networks.

In [10], the size and granularity of the data from PMU anomalies is higher than those presented in this chapter. Additionally, they present results for a binary-SVM, a multi-SVM, and a *k*-NN and decision-tree classifiers. Nevertheless, the detection stage is carried by a simple statistical-based method which, unlike the proposed method, makes it more susceptible to the effects of the noise.

In this chapter, a supervised machine learning-based strategy is presented; the strategy combines graph theory with recurrence quantification analysis for feature extraction that allows anomaly detection and subsequent classification through a multiclass support vector machine algorithm. The proposed methodology is robust to noise, is free of a filtering stage, with low computational burden and easy interpretation; therefore its application on online monitoring of distribution systems is viable. To validate the proposed methodology, case studies are presented to analyze scenarios with voltage anomalies related to power quality events and small-magnitude voltage anomalies that occur during normal operation conditions. Thus synthetic records generated through a Monte Carlo model and PMU records obtained from a distribution system are processed.

6.2 Monte Carlo model for anomaly simulation

The voltage in distribution systems is characterized by a random dynamic behavior during the ambient operational conditions. The PMUs are able to capture the random framework that involves multiple operating scenarios and load conditions, but the identification of PMU records for the training process in machine learning methods is a complex task due to the lack of registered anomalies in available data. Thus a Monte Carlo model is employed to generate multiple scenarios associated with anomaly behavior and to model the dynamic characteristics.

The Monte Carlo model uses a series of random variables to determine the value of a deterministic variable or expression. Hence, it permits training and guarantees the effectiveness of the proposed machine learning algorithm [11].

To model each scenario, let the sample space Ω be a set containing different scenarios related to random variables involved into

$$\Omega = \{(\beta, T) \mid \beta \in (\beta_{min}, \beta_{max}), T \in (T_{min}, T_{max})\}, \tag{6.1}$$

where $\beta \in \mathbb{R}$ is a continuous random variable with uniform distribution for the magnitude of the anomaly given in pu; $T \in \mathbb{Z}^+$ is a discrete random variable with uniform distribution, which represents the temporality of the event given in samples; $(\beta_{min}, \beta_{max})$ denotes the minimum and maximum limits for β, respectively, and $(T_{min}, T_{\max})$ corresponds to the minimum and maximum limits for T, respectively.

The PMU voltage signal can be represented as a time series, where the numerical proposed models for the syntethized anomalies are described as follows.

The step anomaly model x_i^s is described by its magnitude β_i^s, and time as follows:

$$x_i^s[n] = \beta_i^s u[n - \tau_1], \tag{6.2}$$

where $u[n]$ denotes the unitary step, and τ_1 is the start time of the step.

The pulse anomaly model x_i^p represents a short abrupt change in the signal:

$$x_i^p[n] = \beta_i^p (u[n-\tau_1] - u[n-\tau_2]); \tau_2 > \tau_1, \tag{6.3}$$

where $u[n]$ denotes the unitary step, and τ_1 is the start time of step, β_i^p is the magnitude, and $\tau_2 - \tau_1$ denotes the pulse width.

The ramp model x_i^r represents a smooth short abrupt change and is modeled as follows:

$$x_i^r[n] = \beta_i^r \left[\frac{n-\tau_1}{\mathrm{T}} (u[n-\tau_1] - u[n-\tau_2]) + u[n-\tau_2]\right]; \mathrm{T} = \tau_2 - \tau_1, \tau_2 > \tau_1, \tag{6.4}$$

where τ_1 is the start time of ramp, β_i^r/T denotes the ramp slope, T is the transition time.

Finally, the resulting time series for the proposed numerical model of a dynamic PMU data is described as follows:

$$x[n] = v_{rms} + \sum_{i=1}^{K_s} x_i^s[n] + \sum_{i=1}^{K_p} x_i^p[n] + \sum_{i=1}^{K_r} x_i^r[n] + \zeta[n], \tag{6.5}$$

where the mathematical model is composed by the sum of the RMS voltage (steady state phasor magnitude) and anomaly functions, plus Gaussian noise denoted by $\zeta[n]$, where the terms K_s, K_p, and K_r are the total numbers of model functions for the step, pulse, and ramp anomalies, respectively.

The density function for Monte Carlo model is defined as follows:

$$f(\beta, T) = \begin{cases} \frac{1}{(\beta_{max}-\beta_{min})(T_{max}-T_{min})} & \text{if } \beta \in (\beta_{min}, \beta_{max}), T \in \{T_{min}, T_{min+1}, \ldots, T_{max}\} \\ 0 & \text{otherwise} \end{cases}. \tag{6.6}$$

Such density distribution function is denoted as

$$\begin{aligned} F(\beta, T) &= \sum_{T_i \le T} f_T(T_i) \int_{-\infty}^{\infty} f_\beta(\boldsymbol{\beta})\, d\boldsymbol{\beta} \\ f_T(T) &= \begin{cases} \frac{1}{T_{max}-T_{min}} & \text{if } T \in \{T_{min}, T_{min+1}, \ldots, T_{max}\} \\ 0 & \text{otherwise} \end{cases} \\ f_\beta(\beta) &= \begin{cases} \frac{1}{\beta_{max}-\beta_{min}} & \text{if } \beta \in (\beta_{min}, \beta_{max}) \\ 0 & \text{otherwise} \end{cases} \end{aligned} \tag{6.7}$$

where $f_T(T)$ and $f_\beta(\beta)$ correspond to the marginal distribution functions for the discrete random variable T and the continuous random variable β, respectively. The

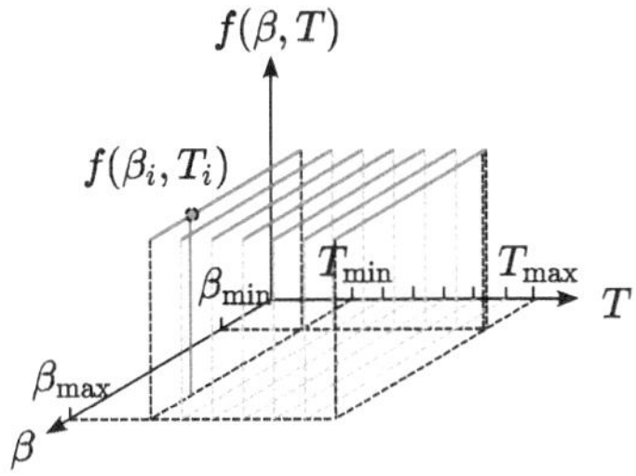

Figure 6.2 Bivariate density function for anomaly generation.

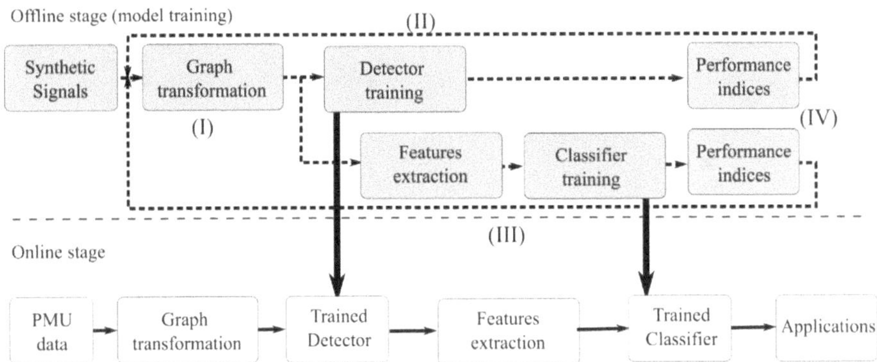

Figure 6.3 Proposed methodology.

sample space Ω for the Monte Carlo scheme is observed through its density function in Fig. 6.2.

6.3 Methodology for the supervised machine learning strategy

The proposed methodology for the supervised machine learning strategy is shown in Fig. 6.3. The strategy is carried out in two main stages: offline and online. The steps of both stages are described next.

- **Training set procedure (offline stage).** Composed mainly of a training and test stage for both detector and classifier.
  - **I.** Graph transformation. To facilitate the anomaly detection, the data window is mapped to a graph that highlights the abrupt changes or anomalies present in the signal.
  - **II.** Detector training. Detector parameters are configured to fit the model and adequately capture the anomaly windows.

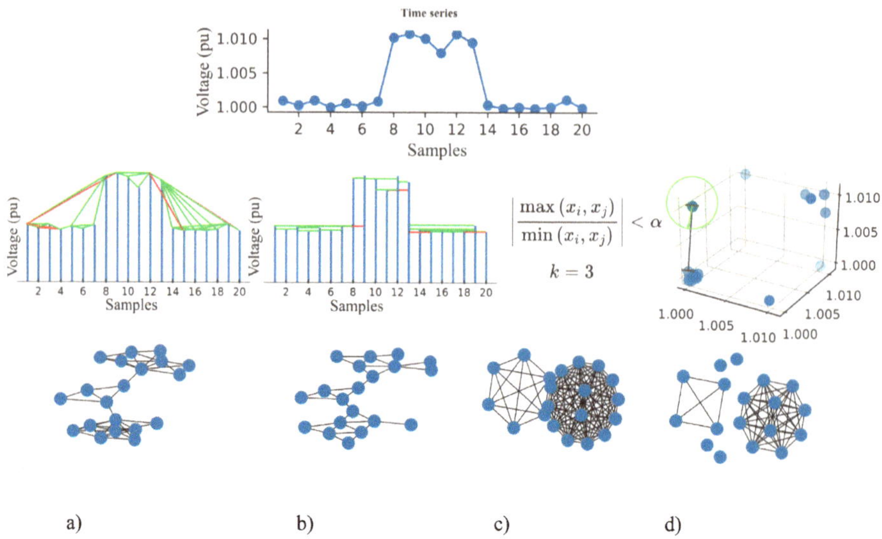

Figure 6.4 Time series to graph mapping methods: a) NVG, b) HVH, c) Similarity, d) Recurrence.

III. Classifier training. Anomaly windows are sent to the classifier, where the RQA feature extraction procedure is applied to the recurrence matrix. Next, computed features are sent to the SVM classifier.

IV. Validation. For both detector and classifier models, performance indices are obtained form a synthetic signal to evaluate them.

- **Online stage**. The PMU voltage data window is mapped to the same graph as the one for the offline stage. Detection is performed by the trained model obtained in the offline stage. The same occurs with the trained classifier for the captured windows from the detection stage.

6.3.1 Graph transformation

The methods for obtaining the adjacency matrix from a univariate time series presented in this chapter can be grouped into three categories: visibility, similarity, and recurrence. Fig. 6.4 shows the mapping process for the different methods described below.

6.3.1.1 Natural visibility graph (NVG)

This graph is in the group of visibility graphs. In the natural visibility graph (NVG), each node v_i in the graph corresponds to a sample x_i from the time series. For this graph, the construction of the adjacency matrix $\boldsymbol{A}$ is made according to the following condition for each element a_{ij}:

$$a_{ij} = \begin{cases} 1, & \text{if } x_c < x_j + (x_i - x_j)\frac{j-c}{j-i}, \ \forall c \in (i, j) \\ 0, & \text{otherwise} \end{cases}, \tag{6.8}$$

where x_c corresponds to an intermediate sample between x_i and x_j.

The connectivity search is done by tracing a line between both nodes and verifying that none of the intermediate nodes intercept it [12].

This method allows to examine linear properties of the time series. The NVG presents a topology capable of capturing local maxima in the corresponding time series. The main advantage for this algorithm is that its implementation is simple [13].

6.3.1.2 Horizontal visibility graph (HVG)

This graph is also grouped within the visibility graphs. The horizontal visibility graph (HVG) is a simplified version of the NVG, resulting in a subgraph of the graph obtained from the NVG [14].

For HVG, each node v_i in the graph also corresponds to a sample x_i of the time series. The visibility conditions for each element a_{ij} of its adjacency matrix $\boldsymbol{A}$ are obtained by the expression:

$$a_{ij} = \begin{cases} 1, & \text{if } x_i, x_j > x_c, \ \forall c \in (i, j) \\ 0, & \text{otherwise} \end{cases} \tag{6.9}$$

where x_c corresponds to an intermediate sample between x_i and x_j. This graph presents an alternative version of visibility.

6.3.1.3 Similarity graph

The similarity graph is a heuristic algorithm focused on the discrete properties of a time series. As in visibility graphs, each node of the resulting graph represents a sample of the time series. The algorithm has been employed in [15] to analyze patients with different types of mental disorders. The primary property used is intrinsic to the definition of graph: if two vertices are adjacent, then there is some kind of relationship (similarity) between them.

The formulation for the similarity graph is as follows: for a time series $X = \{x_1, x_2, \ldots, x_n\}$, each member x_i is associated with a graph node v_i. Each element a_{ij} of the adjacency matrix $\boldsymbol{A}$ is defined by the condition of similarity:

$$a_{ij} = \begin{cases} 1, & \text{if } \left| \frac{\max\ (x_i, x_j)}{min(x_i, x_j)} \right| < \alpha \\ 0, & \text{otherwise} \end{cases}, \tag{6.10}$$

where α is a fixed threshold for check similarity between nodes.

The search for similarity is carried out within the limits set by $k \in \mathbb{Z}^+$. This parameter sets the maximum search distance between an item in the set and its k nearest neighbors and limits the maximum number of neighbors to $2k$.

6.3.1.4 Recurrence graph

The conventional recurrence analysis, developed by Eckmann et al. [16], aims to obtain useful and easy-to-interpret information from dynamic systems. To construct the recurrence-based graph, first calculate a phase space trajectory equivalent to the original one using the time delay method.

Reconstruction of phase space using the time delay method

Let $X = \{x_1, x_2, \ldots, x_{N_X}\}$ be a time series of length N_X, corresponding to the observation of one of the system's state variables: the reconstruction of its phase space, for an embedding dimensión m and a time delay τ, is given by

$$\boldsymbol{x}_i = \sum_{j=1}^{m} x_{i+(j-1)\tau} \hat{\boldsymbol{u}}_j, i = 1, 2, \ldots, N_X - (m-1)\tau \tag{6.11}$$

$$\left\| \hat{\boldsymbol{u}}_j \right\| = 1, \left\langle \hat{\boldsymbol{u}}_i, \hat{\boldsymbol{u}}_j \right\rangle = \delta_{ij}$$

where δ_{ij} denotes delta Kronecker function ($\delta_{ij} = 1$ if $i = j$, $\delta_{ij} = 0$ if $i \neq j$), and $\|\cdot\|$ represents the Euclidean norm for its argument.

From the reconstructed phase space, features of the time series can be obtained by comparing the phase space vectors to each other; an example of this is by using recurrence diagrams.

A recurrence diagram (RP) is a tool that quantifies the recurrences of a trajectory in phase space.

A state is considered recurring if the system path approaches the state $\boldsymbol{x}_i$ in phase space at a distance not greater than a certain recurrence threshold [17]. To calculate the RP, the recurrence of a state at a time i relative to a different time j is mapped to a square matrix [18], according to the expression

$$r_{ij}(\varepsilon) = \Theta\left(\varepsilon - \left\|\boldsymbol{x}_i - \boldsymbol{x}_j\right\|\right), \tag{6.12}$$

where Θ represent the Heaviside function ($\Theta(\mathrm{x}) = 1$ if $x \geq 0$, $\Theta(x) = 0$ if $x < 0$), and ε is the recurrence threshold.

$\boldsymbol{R}$ is a binary matrix, equivalent to a two-dimensional coordinate system with both axes being time axes. By convention, the color black is used for the elements $r_{ij} = 1$, and the color white for the elements $r_{ij} = 0$; in addition, both axes are built in an increasing right (abscissa) and upward (ordinate) direction [19]. By definition $r_{ii} = 1$, so the main diagonal is composed of black elements and is known as line of identity (LOI).

The structure of the adjacency matrix $\boldsymbol{A}$ coincides with that of the recurrence matrix, the only exception being its LOI, since the graph does not allow edges containing the same vertex at both ends. Thus the value of each element $a_{ij} \in \boldsymbol{A}$ is constructed from the element r_{ij} of its RP by the following expression:

$$a_{ij} = r_{ij} - \delta_{ij}, \tag{6.13}$$

where δ_{ij} denotes delta Kronecker function ($\delta_{ij} = 1$ if $i = j$, $\delta_{ij} = 0$ if $i \neq j$).

The matrix obtained in Eq. (6.13) contains qualitative information about the observed window, from which additional information can be obtained using techniques for the features extraction. Recurrence quantification analysis is an example. Furthermore, the fundamental properties of a dynamic system are preserved in its recurrence diagram [17,20].

6.3.1.5 Density-based recurrence graph

The conventional recurrence plot has been widely employed in various works [16–20]. However, it presents problems in the small magnitude detection stage due to the noise content. To mitigate the effects of noise at the detection stage, alternating construction method for adjacency matrix is proposed. The proposal has a density-based approach using the k-distance, rather than the classical Euclidean distance-based approach.

k-distance based recurrence graph

Let $\boldsymbol{X}_{phsp}$ be the reconstructed phase space obtained from Eq. (6.11), then k-distance for $\boldsymbol{x}_i \in \boldsymbol{X}_{phsp}$, denoted by kd_i, is the one that meets the following conditions for a $k \in \mathbb{Z}^+$:

- $\left\|\boldsymbol{x}_i - \boldsymbol{x}_j\right\| \leq kd_j$ for at least k elements $\boldsymbol{x}_j \in \boldsymbol{X}_{phsp}, \boldsymbol{x}_j \neq \boldsymbol{x}_i$
- $\left\|\boldsymbol{x}_i - \boldsymbol{x}_j\right\| < kd_j$ at most to $k-1$ elements $\boldsymbol{x}_j \in \boldsymbol{X}_{phsp}, \boldsymbol{x}_j \neq \boldsymbol{x}_i$

The k-distance represents the radius of an n-sphere with center $\boldsymbol{x}_i$. It is a density-based detection strategy, where a large radius means a bigger separation between $\boldsymbol{x}_i$ and its k-nearest neighbor [21]. Replacing the distance-based approach used in Eq. (6.12) with a density-based approach involves the calculation of a new matrix $\mathbf{R}_k$, where each element rk_{ij} is determined by

$$rk_{ij}(\varepsilon) = \Theta\left(\varepsilon - \max\left(kd_i, kd_j\right)\right). \tag{6.14}$$

This matrix, just as that of Eq. (6.12), is symmetrical, and is characterized by emphasizing significant changes in the signal (relative to the value ε). The threshold ε must be set proportionally to the smallest anomaly magnitude. In a noisy signal, the k-distance provides a better resolution than that obtained from the Euclidean distance; moreover, it emphasizes the separation of the farthest points in the phase space (suspected anomalies), as shown in Fig. 6.5.

6.3.1.6 Comparison of graph mapping methods for anomaly characterization

The capacity of the graph mapping methods presented for detecting anomalies (pulses, ramps, and steps) and for extracting qualitative information from them in the presence of noise is now evaluated. The operating principle of the detector is to detect abnormal changes in the voltage magnitude from the signal. Thus the detector is expected to be able to detect changes in the voltage signal of small magnitude in a noisy environment, so at this stage the robustness to the noise is evaluated for all the algorithms. Fig. 6.6 shows the graphs obtained for a noisy window observed without the presence of anomalies. The results obtained from it are discussed below.

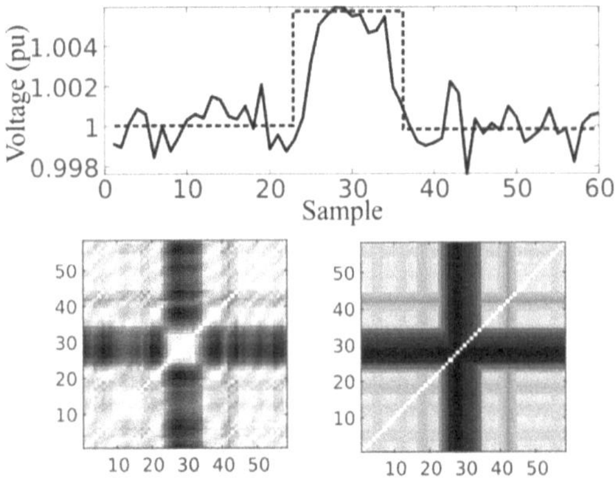

Figure 6.5 Signal (top) with a comparison of Euclidean distance (left) and k-distance (right) approaches.

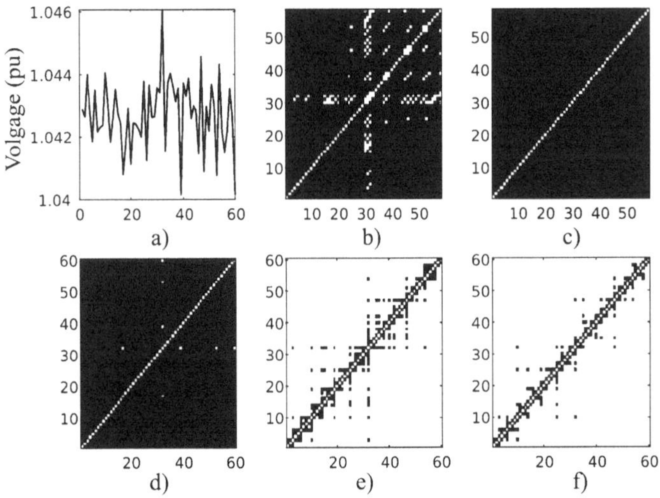

Figure 6.6 Comparison of detection capabilities: a) window without anomalies, b) recurrence graph, c) density-based recurrence graph, d) similarity graph, e) natural visibility graph, and f) horizontal visibility graph.

Due to the visibility graphs, construction is made by looking for intersections between sample pairs; the presence of noise in the signal causes the adjacency matrix to consist essentially of zeros (sparse matrix). A sample with a magnitude considerably greater than the rest is characterized by greater connectivity for the case of the NVG (columns with several elements of value equal to the unit). Meanwhile, in the case of HVG, samples tend to connect only with their nearest neighbors (relative to their indices in the time series, not to their magnitudes). There is no proposal for the detection scheme using this kind of algorithm, but this visibility approach can be employed for the detection scheme using metrics related to vertical lines.

Since the detection capability of the algorithm is being evaluated and not the computational burden, the value of k is adjusted so that each sample is compared with all the others ($2k = N_W$, where N_W is the size of the window) to verify the similarity between each pair of samples for the observed window. Samples not similar to each other indicate the possible presence of some anomaly. The graph obtained using this method is slightly more affected by the noise than that obtained with the k-distance, as seen in Fig. 6.6, for windows without the presence of anomalies.

The voltage magnitude given by the PMU is constant under normal conditions; therefore in the absence of changes (anomalies) in its dynamics, the resulting graph must be complete. For the classical recurrence graph approach, the noise in the signal causes some points to move away from the rest in the phase space. Despite the signal being in normal conditions, the distance between some points exceeds the established limit, which generates the presence of some zeros in the columns (rows) of the resulting adjacency matrix, as observed in Fig. 6.6. In contrast, the density-based k-distance recurrence graph derivates in a complete graph, even in the presence of noise. This is because k-distance mitigates the effects of noise by comparing sample radius (densities), rather than distances relative to neighbors.

For the feature extraction capabilities comparison, consider the defined anomalies, which are shown in Fig. 6.7. There are two cases: for the first case, noise-free (ideal); for the second case; Gaussian noise (white noise) is added to the anomalies to observe the robustness of the algorithms.

Table 6.1 summarizes the general observations made to the different mapping methods and their feature extraction capabilities.

Recurrence-based graph methods are distinguished from the rest due to the resultant graph has a lower number of nodes than the number of samples in the original window as result from the phase space reconstruction. Nevertheless, this reconstruction gets the signal embedded to a higher dimension (delayed copies of the signal), which makes it easy to compare the window with itself, as well it allows to obtain extra features from the data, through RQA feature extraction.

Based on the above observations, the density-based recurrence graph is chosen for the detection stage due to its robustness to noise. For the feature extraction stage, the conventional recurrence graph has a higher performance than the rest, so these are the options used in the training and testing of the detection and classification algorithms presented in what follows. The rest of the algorithms have the potential to detect and extract features by using different combinations; however, more research is required, which is outside the scope of this research, so they won't be boarded at the moment.

6.3.2 Laplacian-based anomaly detection algorithm

Let $\deg v_i$ denote the degree of the vertex v_i from graph G, computed as

$$\deg v_i = \sum_{v_j \in V, j \neq i} a_{ij}, \tag{6.15}$$

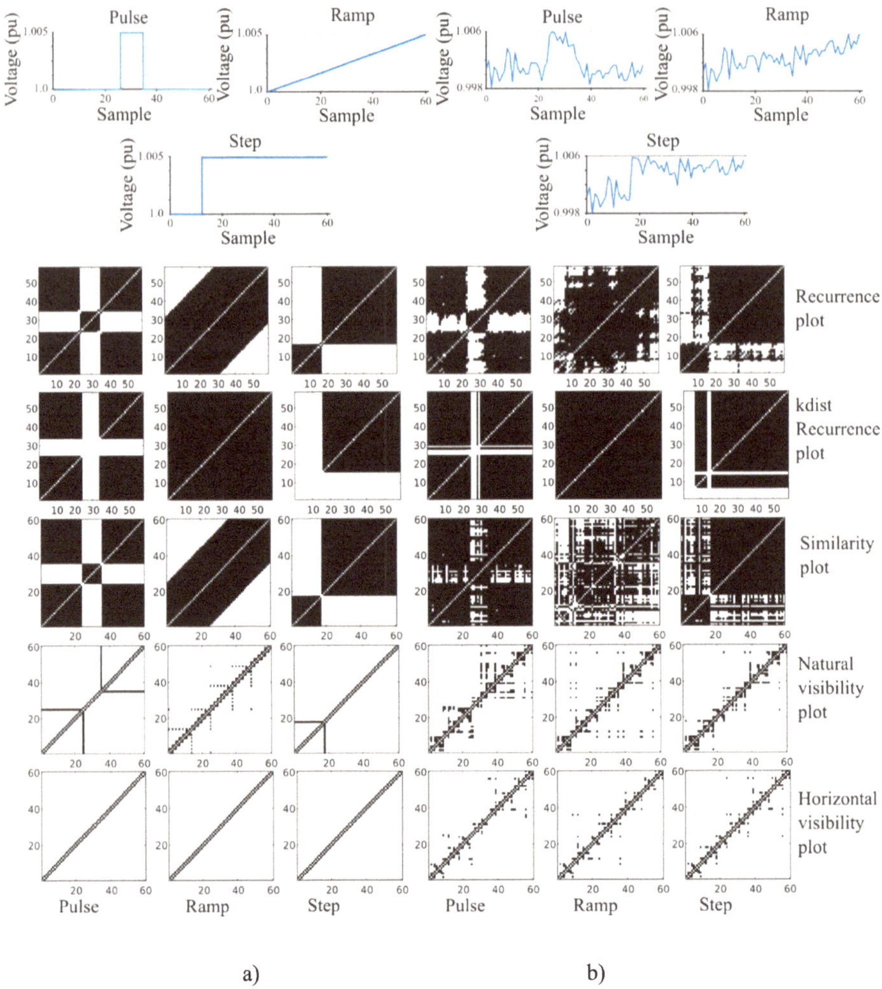

Figure 6.7 Color diagram of the adjacency matrix for graph mapping methods: a) without noise and b) with white noise added.

where V denotes the set of vertex from G. Then, the Laplacian matrix $\boldsymbol{L}(G) = (l_{ij}) = \boldsymbol{L}$ is defined by

$$\boldsymbol{L}(G) = \boldsymbol{D}(G) - \boldsymbol{A}(G), \tag{6.16}$$

where $\boldsymbol{D}(G)$ is the degree diagonal matrix.

Let X be the time series obtained from Eq. (6.5), and let $\boldsymbol{W}$ be the set of all N_W length sliding windows W_i from X. The sliding window, denoted by W_i, is defined for X as follows:

$$W_i : [x_j, x_{j+N_W}), \ j = 1 + (i-1)\, w_S, \tag{6.17}$$

Table 6.1 General remarks for different mapping methods.

Graph	**Capabilities to distinguish between anomalies**	**Remarks**
Natural Visibility Graph	Weak	Presents a precarious distinction between classes. Along with similarity, it presents a less computational complexity than the case of recurrence, but does not compensate for its low performance in the distinction of anomalies.
Horizontal Visibility Graph	Null	Being a subgraph of the NVG, a similar performance is expected; however, for the case study, it is unable to distinguish between any of the classes.
Similarity Graph	Appropriate	It is able to distinguish between classes, although the effects of noise are greater than those for the case of recurrence. The k parameter reduces computational complexity, making it the fastest algorithm. However, it is not a notorious algorithm, and lacks from well-defined metrics.
Recurrence Graph	Optima	It is the method for which a better distinction between classes is observed. It has a whole theory for its analysis and has been successfully tested in several disciplines.
Recurrence Graph (Density-based)	Weak	This method has the best detection performance. However, in feature extraction it is severely affected by the presence of noise in the signal.

where w_S is the window step.

Let $\boldsymbol{L}_i$ be the Laplacian matrix for the recurrence graph G_i from observed window $W_i \in \boldsymbol{W}$. It can be shown that $\boldsymbol{L}_i$ is always symmetric; furthermore, $l_{ij} \in \mathbb{R}$, thus $\boldsymbol{L}_i$ is a positive semidefinite matrix. Therefore it is characterized by N_{phsp} real eigenvalues $\lambda_i(\boldsymbol{L}_i) = \lambda_i \in \mathbb{R}$, which can be organized in non-decreasing order $(0 = \lambda_1 \leq \lambda_2 \leq \cdots \leq \lambda_{N_{\text{phsp}}})$. These eigenvalues define the spectrum of the Laplacian matrix of the graph. For the detection stage, the energy from Laplacian matrix [22] for window W_i is computed as

$$\sum_{i=1}^{N_{\text{phsp}}} \lambda_i^2 = \sum_{i=1}^{N_{\text{phsp}}} ((\deg v_i)^2 + \deg v_i). \tag{6.18}$$

For a steady state system, vectors (points) in $\boldsymbol{X}_{\text{phsp}}$ form an extremely dense cluster $(\boldsymbol{x}_i \approx \boldsymbol{x}_j; \forall \boldsymbol{x}_i, \boldsymbol{x}_j \in \boldsymbol{X}_{phsp})$. Thus quantifying the separation between these vectors makes it possible to detect changes in the signal.

For a steady state signal $\deg v_i \rightarrow N_{phsp} - 1,\ \forall i$. Thus

$$LE(G_i) = N_{\text{phsp}}\left[\left(N_{\text{phsp}} - 1\right)^2 + \left(N_{\text{phsp}} - 1\right)\right] = N_{\text{phsp}}^2\left(N_{\text{phsp}} - 1\right). \tag{6.19}$$

The objective of the detection model is to identify adjacent windows with abnormal values of $LE(G_i)$. Therefore changes in the fixed value $LE(G_i) \neq N_{\text{phsp}}^2(N_{\text{phsp}} - 1)$ indicate the presence of anomalies in the signal.

For the window W_i, the detection flag $D(W_i)$ is described by

$$D(W_i) = \begin{cases} 1, & \text{if} LE(G_i) < LE_{\text{lim}} \\ 0, & \text{otherwise} \end{cases}. \tag{6.20}$$

$LE_{\text{lim}} = N_{\text{phsp}}^2(N_{\text{phsp}} - 1)$ correspond to Laplacian energy for a complete graph (steady state window).

6.3.3 Anomaly classification with DAGSVM

A support vector machine (SVM) corresponds to a binary classifier that finds an hyperplane in a D-dimensional space, where D denote the features number, which maximizes separation margin between classes [23]. Decision function for SVM is given by

$$f(\boldsymbol{x}) = \boldsymbol{w}\boldsymbol{x} + b \tag{6.21}$$

where $\boldsymbol{w}$ correspond to a vector normal to hyperplane, $\boldsymbol{x}$ is the features vector sample and b denote the bias.

SVM binary concept can be extended to multiclass classification for N classes through the directed acyclic graph (DAG). The DAG scheme makes $N - 1$ comparisons by employing $N(N-1)/2$ binary classifiers to determine the class of an object among the N possible classes [24].

6.3.3.1 Feature extraction and selection

Feature extraction and selection is a crucial part for the machine learning training stage. In this regard, it is possible to classify the different anomalies employing specific RQA features from the recurrence graphs in a hybrid qualitative-quantitative analysis of the data. Furthermore, the chosen features allow obtaining a low error linear model, avoiding overfitting and reducing computational burden, as well as extracting a visual representation of the classification process. The chosen RQA metrics are described below [17,18].

The recurrence rate is defined by

$$RR = \frac{1}{N_{\text{phsp}}(N_{\text{phsp}} - 1)} \sum_{i \neq j} rk_{ij} \tag{6.22}$$

which is the arithmetic mean of ones contained in adjacency matrix.

The average diagonal line length measures the proximity between two segments of the trajectory in the phase space at different times and is determined by

$$L_D = \frac{\sum_{l_d=d_{min}}^{N} l_d P(l_d)}{\sum_{l_d=d_{min}}^{N} P(l_d)} \tag{6.23}$$

where $P(l)$ correspond to the histogram.

The longest diagonal line is given by

$$L_{Dmax} = \max\left(\{l_{di} \mid P(l_{di}) > 0, i = d_{min}, \ldots, N_{\text{phsp}}\}\right) \tag{6.24}$$

The average vertical line length its calculated as follows:

$$L_V = \frac{\sum_{l_v=v_{min}}^{N} l_v P(l_v)}{\sum_{l_v=v_{min}}^{N} P(l_v)} \tag{6.25}$$

which is an estimate of the average time the system will remain in a certain state.

To calculate the longest vertical line, the following equation is used:

$$L_{Vmax} = \max\left(\{l_{vi} \mid P(l_{vi}) > 0, i = v_{min}, \ldots, N_{\text{phsp}}\}\right) \tag{6.26}$$

Both the average and the longest diagonal are useful for identifying transitions in the phase space.

The RQA compares matrices of the same order; nevertheless, anomalies of different temporalities generate matrices of different orders. To overcome these limitations, feature normalization is required.

Normalization is performed by dividing the feature under consideration by $N_{\text{phsp}} - 1$, which corresponds to the maximum theoretical value for both the diagonal and vertical lengths, limiting the value of the features to the closed interval [0, 1], regardless of its temporality.

Unless otherwise specified, to simplify nomenclature, all features are considered to be normalized. The only exception is the recurrence rate, which is already calculated based on the matrix dimensions.

6.3.3.2 Anomaly classification with DAGSVM

Let $\boldsymbol{\alpha}$ be the Euclidean space with axes that correspond to the RQA features:

$$\boldsymbol{\alpha} = [RR, L_D, L_{Dmax}, L_V, L_{Vmax}]. \tag{6.27}$$

Thus a subspace $\boldsymbol{\alpha}_{ij}$ containing the most meaningful features for classes i, j is assigned to each of the $N(N-1)/2$ binary blocks from the DAG classification model.

Let $\boldsymbol{\alpha}_k \in \boldsymbol{\alpha}$ be the features vector for the kth sample. The subvector $\boldsymbol{\alpha}_{kij}$ is then extracted from $\boldsymbol{\alpha}_k$, according to subspace $\boldsymbol{\alpha}_{ij}$ for the pairwise class comparison. The proposed DAGSVM classification scheme consists of a three-classes multiclass classification problem, where each binary classification model $\boldsymbol{\alpha}_{ij}$ is modeled according

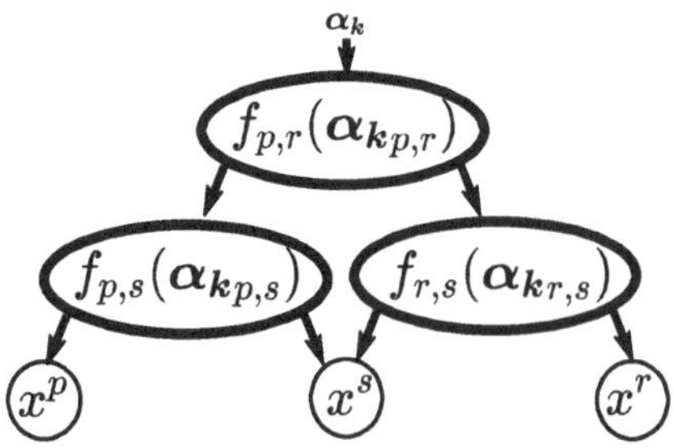

Figure 6.8 DAGSVM.

to the best linear fit subspace:

$$
\begin{aligned}
\alpha_{p,r} &= [RR, L_{Dmax}, L_V] \\
\alpha_{p,s} &= [RR, L_D, L_{Vmax}] \\
\alpha_{r,s} &= [RR, L_D, L_V]
\end{aligned}
\tag{6.28}
$$

where p, r, s subscripts indicate the pulse, ramp, and step classes, respectively. Proposed DAG classifier blocks are shown in Fig. 6.8.

Decision functions for each block correspond to a SVM model with a linear kernel in a 3-D features subspace, according to Eqs. (6.21) and (6.28).

6.3.4 Performance indices for the assessment of the algorithm

Let D_1 and D_0 be the training sets of positive detection flags (anomaly) and negative detection flags, respectively. Let D_1' and D_0' be the sets obtained from the tested detector analogous to D_1 and D_0, respectively. Then, the indices described below can be used to determine the detector and classifier performance [25].

The true positive rate (TPR), also known as sensitivity, measures the ratio between detected anomalies and the overall anomalies present in the signal, calculated as follows:

$$
TPR = \frac{|TP|}{|TP| + |FN|} = \frac{|D_1 \cap D_1'|}{|D_1|} \tag{6.29}
$$

where $|\cdot|$ denotes set cardinality, TP corresponds to True Positives and FN is for False Negatives.

True negative rate (TNR), or specificity, is the rate of negative flag detections with respect to negative flags from training data:

$$
TNR = \frac{|TN|}{|TN| + |FP|} = \frac{|D_0 \cap D_0'|}{|D_0|}. \tag{6.30}
$$

Both TPR and TNR are useful for measuring algorithm accuracy.

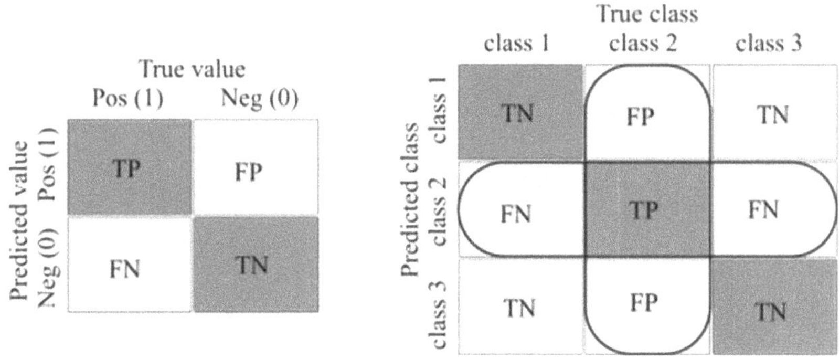

Figure 6.9 Confusion matrix for a binary (left) and multiclass (right) classifiers.

False positive rate (FPR) indicates algorithm false alarms rate:

$$FPR = 1 - TNR, \tag{6.31}$$

and it is a quality algorithm measure.

Precision evaluates the algorithm capabilities for detecting only anomalies:

$$prec = \frac{|TP|}{|TP| + |FF|} = \frac{|D_1 \cap D_1'|}{|D_1'|}. \tag{6.32}$$

High precision means that the algorithm has a low false alarm rate.

The $F1$-score is an index to measure algorithm performance given by

$$F_1 = \frac{2|TP|}{2|TP| + |FN| + |FP|} \tag{6.33}$$

The Jaccard index compares the similarity or diversity between pairs of sets. Furthermore, for anomaly detection, it compares training and test sets similarity:

$$Jaccard = \frac{|TP|}{|TP| + |FP| + |FN|} = \frac{|D_1 \cap D_1'|}{|D_1 \cup D_1'|} \tag{6.34}$$

Accuracy refers to the arithmetic mean of true detections in respect to the total samples:

$$Acc = \frac{|TP| + |TN|}{|TP| + |TN| + |FP| + |FN|} = \frac{|D_1 \cap D_1'| + |D_0 \cap D_0'|}{|D_1 \cup D_0|} \tag{6.35}$$

The detection stage corresponds to a case of binary classifier. For a multiclass classifier, the same indices can be employed according to Fig. 6.9

Table 6.2 Monte Carlo model parameters.

Magnitude (pu)	Temporality (samples)	Anomalies per class
0.005–0.01	30–120	10,000

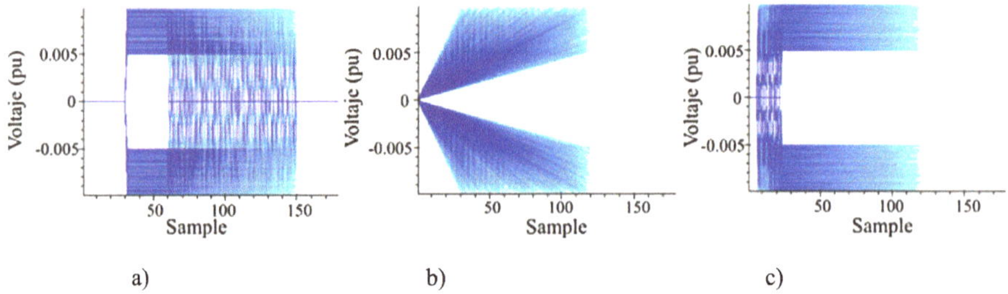

Figure 6.10 Anomalies generated through Monte Carlo model: a) pulse, b) ramp, c) step.

6.4 Example: anomaly detection and classification in a synthetic voltage signal

The following example shows the code for the training and testing stages employed in the proposed algorithm for the synthetic voltage signal, and can be accessed for its reproducibility through the public github repository available at the following link: https://nbviewer.org/github/NetoIE/ML_PMU_Example/blob/main/Example_detector_classifier.ipynb.

6.4.1 Monte Carlo model for PMU synthetic voltage signal

To carry out the simulation of anomaly data, a great amount of anomalies are generated employing a Monte Carlo model with a uniform distribution random sample space to cover as many values as possible within the closed intervals $[\beta_{min}, \beta_{max}]$ and $[T_{min}, T_{max}]$. Simulation parameters employed for anomaly generation are summarized in Table 6.2.

To maintain homogeneity in the number of data samples per anomaly class, Monte Carlo model generates the same number of anomalies for each class. Anomalies resulting from Monte Carlo model are illustrated in Fig. 6.10

Next step consists of creating a synthetic signal, employing the generated anomalies in the previous step and incorporating them into the steady state signal v_{rms} as in Eq. (6.5). In this step, Gaussian noise is also included. For distribution system, the SNR values typically ranges between 40 to 80 dB; for this case a signal to noise ratio (SNR) of 60 dB is considered. The resulting signal corresponds to a time series denoted by X and can be observed in Fig. 6.11.

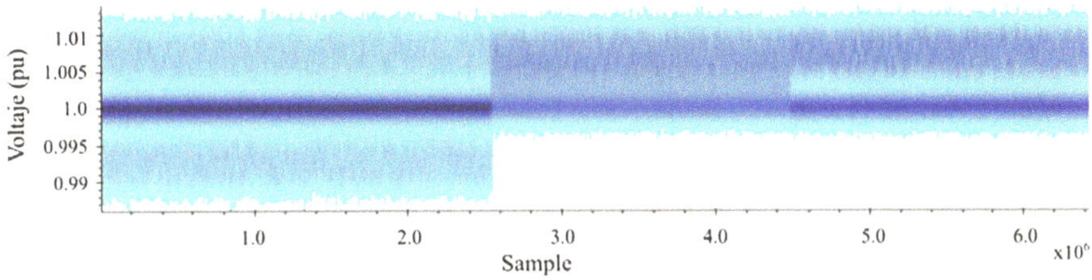

Figure 6.11 PMU voltage synthetic signal with anomalies within normal operational condition.

6.4.2 Supervised machine learning training stage

With the synthetic PMU data obtained from previous step, now it is possible to train a machine learning detections and classification model. For this, the first step consists of splitting data for training and test stage.

6.4.2.1 Split training/test data

Let $\boldsymbol{W_A}$ be the set of N_A windows for each of the $N_A = K_p + K_r + K_s$ generated anomalies. For the learning stage, a typically used 80/20 rate is chosen to split training and test data. Thus set $\boldsymbol{W_A}$ is splitted into X_{TRAIN} and X_{TEST} subsets. Splitering is done in such way that both the training and test subsets contains the same number of anomalies for each of the anomaly classes (pulse, ramp, and step).

Detector split data

Detection stage must distinguish between normal windows (noisy windows) and anomaly windows; thus it's a binary classifier. To test the detection stage, the test set X_{TEST} from anomaly windows is merged with a second set with the same cardinality, composed by windows without anomalies (noisy windows). New set is denoted by XD_{TEST}.

For the detection stage, there are two flags (tags), 1 for positive flag (detected anomaly) and 0 for negative flag (normal window). Anomaly window W_{Ai} size differs from that of the sliding window. Therefore to test the detector, every anomaly window from subset XD_{TEST} is carried by a sliding window that determines if it contains an anomaly. This step leads to the creation of two sets, D and D', which correspond to the real flag (training) and the detected flag (test), respectively.

Classifier split data

RQA features extraction is performed for all windows (samples) from $\boldsymbol{W_A}$ $(X_{TRAIN} \cup X_{TEST})$. Tags for anomaly classes are inherited from data generation step and correspond to the set C_A, composed by classes 1, 2, and 3 for pulse, ramp, and step anomalies, respectively.

Samples and tags from X_{TRAIN} are sent to the classifier for the training stage. For the other subset, X_{TEST}, analogously to the case of the detector, this subset leads to the creation of two sets, C and C', which correspond to the trained classes and the

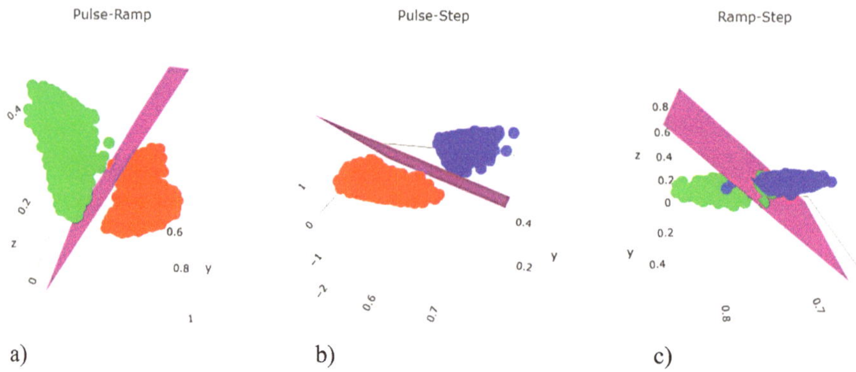

Figure 6.12 Features subspaces for DAGSVM binatu blocks: a) Pulse-ramp, b) Pulse-step, c) Ramp-step.

predicted classes, respectively. C is inherited from data generation; meanwhile, C' is constructed from test stage.

6.4.2.2 Laplacian energy detector parameters

The PMU from which voltage records are collected has a reporting rate of 10 phasors/s. The refresh time for the detection step is equal to the time it takes to the PMU report a new phasor, that is every 100 ms. The first complete window loads a buffer composed by 60 samples (6 s delay), and later the algorithm is refreshed with every new sample (100 ms). Thus the sliding window is composed by a length of $N_W = 60$ samples with a window step of 1 sample. Parameters for the k-distance recurrence graph are $k = 18$, $m = 3$, $\tau = 1$. Threshold value is set to a value close to the β_{min} ($0.8\beta_{min}$ for this case). Detector energy limit is set according to the energy for a k-distance recurrence graph from a steady state window of length N_W, $LE_{lim} = 191{,}748$, although LE_{lim} is normalized in the code.

6.4.2.3 DAGSVM classifier

RQA features were extracted for every anomaly window from X_{TRAIN} and analyzed in all possible 3-dimensional features subspaces. The DAGSVM model was trained to minimize error for a linear kernel in a 3-D features subspace for each pair of classes, taking advantage of the block composition provided by the DAG scheme. This model is constructed by three binary models, each trained with 3 features from the available RQA features, all chosen features subspaces are shown in Fig. 6.12

6.4.3 Supervised machine learning test stage and performance metrics

Next step consists of the test part for the proposed method. In this step, the test sets XD_{TEST} and X_{TEST} are employed to determine the performance of the proposed detector and classifier models. For this purpose, common performance indices employed

Table 6.3 Performance indices for Laplacian energy detector.

	TPR	TNR	FPR	PREC	F1	JACCARD	ACC
Pulse	0.989	1.000	0.000	1.000	0.994	0.989	0.994
Ramp	0.984	0.973	0.027	0.973	0.978	0.958	0.978
Step	0.983	1.000	0.000	1.000	0.991	0.983	0.992

Table 6.4 Performance indices for DAGSVM classifier.

	TPR	TNR	FPR	PREC	F1	JACCARD	ACC
Pulse	1.000	1.000	0.000	1.000	1.000	1.000	1.000
Ramp	0.990	0.996	0.004	0.992	0.991	0.982	0.994
Step	0.992	0.995	0.005	0.990	0.991	0.982	0.994

in literature for the test of the detection/classifications models are presented in the next section.

6.4.3.1 Assessment of performance for the Laplacian energy detector

Laplacian energy detector is applied to all samples from the set XD_{TEST}, and the results are incorporated into the set D'. Sets D and D' are compared and its results are summarized in Table 6.3 and discussed below.

As shown in Table 6.3, results provided by the proposed detector demonstrate its feasibility in detecting the studied anomalies with a high True Positive Rate (TPR) and Accuracy (ACC), as well as its capabilities to distinguish between normal windows and those with the presence of anomalies, with a low level of false alarms, False Positive Rate (FPR). Jaccard and Precision (PREC) indices, as well as TPR and $F1$-score, are defined by the number of correct detections; therefore they are especially useful for measuring algorithm performance and the obtained results are very promising for the proposed detection model.

Additional to the aforementioned, the density-based recurrence matrix optimizes the results for the detection stage by highlighting the abrupt changes in the observed window. Therefore Laplacian energy detector results in a promising method to carry the anomaly detection stage.

6.4.3.2 Assessment of performance for the DAGSVM classifier

The trained DAGSVM classifier model is applied to all sample windows X_{TEST} to predict the anomaly class. Predicted results correspond to the set C'. Sets C and C' results are summarized in Table 6.4 and discussed below.

The DAGSVM classification stage performs well in general and its observed that it can distinguish between the studied classes (anomalies). Nevertheless, the DAGSVM classifier has some troubles distinguishing between ramp and step-type anomalies. The main reason is that the slopes of badly classified ramps are very steep and tend behave like a step anomaly (instantaneous change).

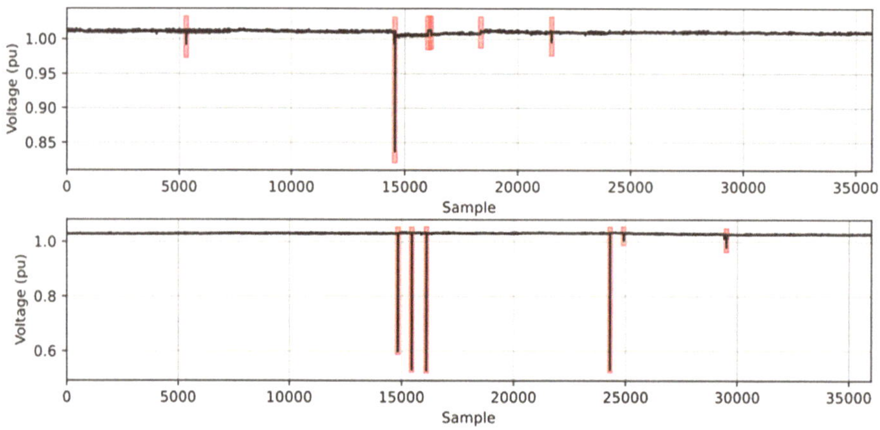

Figure 6.13 PMU voltage magnitude signal from a distribution network with the presence of anomalies.

It should be noted that the proposed DAGSVM method decomposes the multiclass classification problem into several binary classification subproblems, with selected feature spaces to maximize the separation between samples of each class. Therefore the proposed block scheme simplifies the classification stage, employing only the 3 more significant features for each pair of classes. Furthermore, the minification of the feature space allows to obtain an easy-to-interpret visual representation.

6.5 Real PMU data anomaly detection and classification for a distribution network

For the next case study, a PMU's voltage data record from a distribution network is employed. The studied distribution network is comprised of 30 branches. Each branch is monitored by a SEL-751A, which operates simultaneously as protection relay and PMU monitoring device. Each data frame obtained by the PMUs contains the GPS time stamp, in addition to the measurements of voltage and current for the three phases, the frequency, and ROCOF. Raw data storage is done at the rate of 10 phasors/s with a total of 864,000 data frames per day. The PMU data is captured by a phasor data concentrator (PDC), which save the measurements into CSV files for long-term storage. One of the aforementioned branches is chosen for the presented case study. In particular, two fragments from the signal were extracted for the analysis due to the presence of some of the anomalies studied in this chapter, both within and outside normal operational range; therefore its suitable for the test of the proposed method. The first signal contains 3 pulse and 3 step-type anomalies; the second signal contains 6 pulse-type anomalies. Ramp anomaly types were not observed in the collected data so they are omitted in this part. Studied signals are shown in Fig. 6.13, where the anomalies have been highlighted.

Table 6.5 Performance indices for Laplacian detector.

	TPR	TNR	FPR	PREC	F1	JACCARD	ACC
Anomaly	1.000	0.889	0.111	0.995	0.998	0.995	0.995

Table 6.6 Performance indices for DAGSVM classifier.

	TPR	TNR	FPR	PREC	F1	JACCARD	ACC
Pulse	0.889	1.000	0.000	1.000	0.941	0.889	0.917
Ramp	–	–	–	–	–	–	–
Step	0.667	0.889	0.111	0.667	0.667	0.500	0.833

6.5.1 Detection stage

Laplacian detector is employed in the signals from the PMU for the described network, and the obtained results are shown in Table 6.5. In this step, anomalies are considered together due to the small number of available data. Thus all are grouped into the Anomaly class from Table 6.5.

Laplacian energy detection stage does not present major drawbacks in the anomaly detection stage. In this case study, the observed anomalies are successfully detected by the proposed algorithm. In addition, windows without the presence of anomalies are not affected by the noise level in the signal, and the false alarm level is very low (FPR). Therefore implementation of the proposed detection method in real PMU data its factible.

6.5.2 Classification stage

For the classification stage, anomaly windows are sent to the classifier to determine their anomaly type. The results from the classification stage are shown in Table 6.6.

Both step and pulse class anomalies in the study case were correctly captured by the detection step. However, even when the classifier fails to classify only 1 of the 3 steps, there exist only a few step anomalies, resulting in a low TPR, PREC, F1-score, Jaccard, and Accuracy indices. This misclassification could be due to the magnitude of the step being near the fixed threshold ε for the detection stage, or due to the similarity between a ramp with a steep slope and the step type anomalies.

Since 2 from the 3 observed step and 8 from the 9 pulse anomalies were correctly classified, results show that the proposed classifier has the capabilities to distinguish between pulse-type anomaly and the other two without major complications, nevertheless, there exist some trouble for the step-type classification due to a lack of step-type anomaly samples.

6.6 Conclusions

The density-based recurrence matrix optimizes the results of the detection stage by highlighting the abrupt changes in the observed window. In the case of the classification stage, the original recurrence matrix preserves information about the waveform of the graph and performs better in the feature extraction stage as well as in the classification process in general. Furthermore, the classification stage is carried by a low dimensional featured space, which allows to obtain a fast and easy-to-interpret visual representation of the method and makes it feasible for its implementation in real-time schemes.

Meanwhile, some drawbacks are that the k-distance graph threshold for the detection stage must be calibrated for the noisy conditions in the data. It is important to ensure that the detection stage is carried out correctly; otherwise, favorable results cannot be expected in the classification stage. Nevertheless, both detection and classification stages have promising results in the presented synthetic and real cases.

Among the limitations of the algorithm is that it is restricted to a single voltage signal data, so it must be implemented in parallel for multiple PMUs, which increases the computational burden. Thus future work includes to extend the proposed algorithm to a framework based on a spatiotemporal algorithm to extract the relevant features in a single stage from the PMU data. In addition, to improve classification process, future work includes the use of ensemble-based techniques.

References

[1] M. Coutinho, G. Lambert-Torres, L. Silva, H. Lazarek, Improving detection attacks in electric power system critical infrastructure using rough classification algorithm, in: Proceedings of the Second International Conference on Forensic Computer Science, January 2007, pp. 18–23.

[2] L. Sun, K. Zhou, X. Zhang, S. Yang, Outlier data treatment methods toward smart grid applications, in: IEEE Access, July 2018, pp. 39849–39859.

[3] M. Jamei, S. Anna, C. Roberts, E. Stewart, S. Peisert, C. McParland, A. McEachern, Anomaly detection using optimally-placed micro-PMU sensors in distribution grids, IEEE Transactions on Power Systems (July 2017) 3611–3623.

[4] RES/151/2016 – Código de red, Diario Oficial de la Federación, 2016.

[5] D. Macii, D. Fontanelli, D. Petri, Performance of phasor measurement units for power quality event detection in urban distribution grids, in: 2016 IEEE International Smart Cities Conference (ISC2), 2016.

[6] N. Save, M. Popov, A. Jongepier, G. Rietveld, PMU-based power system analysis of a medium-voltage distribution grid, CIRED – Open Access Proceedings Journal 2017 (October 2017) 1927–1930.

[7] I. Niazazari, H. Livani, Disruptive event classification using PMU data in distribution networks, in: 2017 IEEE Power Energy Society General Meeting, 2017.

[8] E. Nagata, D. Ferreira, M. Bollen, B. Barbosa, E. Ribeiro, C.A. Duque, P. Ribeiro, Real-time voltage sag detection and classification for power quality diagnostics, Measurement 164 (June 2020) 108097.

[9] H.L. Iman Niazazari, A PMU-data-driven disruptive event classification in distribution systems, Electric Power Systems Research 157 (December 2018) 251–260.

[10] A. Shahsavari, M. Farajollahi, E. Stewart, E. Cortez, H. Mohsenian-Rad, Situational awareness in distribution grid using micro-PMU data: a machine learning approach, Transactions on Smart Grid (2019).

[11] J.E. Gentle, Random Number Generation and Monte Carlo Methods, 2nd ed., Springer, 2003, p. 229.
[12] L. Lacasa, B. Luque, F. Ballesteros, J. Luque, J.C. Nuño, From time series to complex networks: the visibility graph, Proceedings of the National Academy of Sciences 105 (May 2008) 4972–4985.
[13] V. Freitas, Time Series Analysis based on Complex Networks, Porto, 2018.
[14] B. Luque, L. Lacasa, F. Ballesteros, J. Luque, Horizontal visibility graphs: exact results for random time series, Physical Review. E, Statistical, Nonlinear, and Soft Matter Physics 80 (October 2009) 1–11.
[15] E. Fasmer, O.B. Fasmer, J. Berle, K. Oedegaard, E. Hauge, Graph theory applied to the analysis of motor activity in patients with schizophrenia and depression, PLoS ONE 13 (April 2018) e0194791.
[16] J.-P. Eckmann, S. Kamphorst, D. Ruelle, Recurrence plots of dynamical systems, Europhysics Letters 4 (November 1987) 973–977.
[17] R. Donner, M. Small, J. Donges, N. Marwan, Y. Zou, R. Xiang, J. Kurths, Recurrence-based time series analysis by means of complex network methods, International Journal of Bifurcation and Chaos 21 (October 2010).
[18] C. Webber, N. Marwan, Recurrence Quantification Analysis – Theory and Best Practices, Springer, 2015, pp. 3–10, 13–16.
[19] N. Marwan, M. Romano, M. Thiel, J. Kurths, Recurrence plots for the analysis of complex systems, Physics Reports 438 (January 2007) 246–250, 263–270.
[20] R. Donner, Y. Zou, J. Donges, N. Marwan, J. Kurths, Recurrence networks – a novel paradigm for nonlinear time series analysis, New Journal of Physics 12 (August 2009).
[21] M. Breunig, H.-P. Kriegel, R. Ng, J. Sander, LOF: identifying density-based local outliers, in: ACM Sigmod Record, 2000.
[22] M. Lazić, On the Laplacian energy of a graph, Czechoslovak Mathematical Journal 56 (December 2006) 1207–1213.
[23] L. Ladicky, P. Torr, Linear support vector machines, in: Proceedings of the 28th International Conference on Machine Learning, 2011.
[24] Chih-Wei Hsu, Chih-Jen Lin, A comparison of methods for multiclass support vector machines, IEEE Transactions on Neural Networks 13 (2002) 415–425.
[25] M. Gaur, S. Makonin, I. Bajic, A. Majumdar, Performance evaluation of techniques for identifying abnormal energy consumption in buildings, in: IEEE Access, May 2019, pp. 62721–62733.

Identification of source harmonics in electrical networks using spatiotemporal approaches

7

Ramón J. Betancourt[a], *Ramón Daniel Rodríguez-Soto*[b], *Antonio Concha Sánchez*[a], *and Emilio Barocio Espejo*[b]
[a]University of Colima, Colima, Mexico, [b]University of Guadalajara, Guadalajara, Mexico

7.1 Introduction

The modern electric power network as is shown in Fig. 7.1, are large-scale interconnected networks, with power plants of different primary nature, storage systems, and large loads that can include their proper generation and storage. Electrical power networks are constantly subjected to a wide variety of disturbances affecting their proper functioning. The disturbances, such as faults, can cause small or large topological changes in the network endangering the integrity of the entire electrical network. During a disturbance, transient conditions are observed and are characterized by a slow response of large generators, which in conjunction with the high-speed controller actions of distributed generation cause the trigger of interarea oscillations. These phenomena involve the nonlinear dynamic interaction between group of generators spatially far away one of each other. Frequently, these oscillations are stabilized by controller actions, but other times, the protective relay equipment causes the topology to evolve towards losing and important elements of the power system, and thereby causing collapses in the system. To prevent this kind of phenomena the speed of the generators is continuously observed by the synchronized phasor measurement units (PMU) of each main substation. The PMUs form part of the wide area monitoring system (WAMS) located at the energy control center (ECC), where the information is processed, and the decisions are taken for [1]. Massive amount of data is continuously generated, updated, and collected in the data concentrators, which is processed at the same time is generated. The frequencies of oscillations contained in the data, can be determined by fast Fourier transform methods (FFT), but due the numerous PMU sensors, this task is cumbersome when different scenarios should be characterized, hence advanced tools, such as those related to machine learning methods (ML), are needed to predict future behavior from relevant data [2].

Machine learning methods (ML), are currently the most appropriate techniques for analyzing large data set of measurements, providing feature extraction and reduction, allowing to predict dynamical behavior from several operation conditions in an optimal sense [3]. A crucial step in the overall process of ML methods constitutes the extraction of features of the data, and several approaches can be used with this pur-

Monitoring and Control of Electrical Power Systems Using Machine Learning Techniques
https://doi.org/10.1016/B978-0-32-399904-5.00013-2

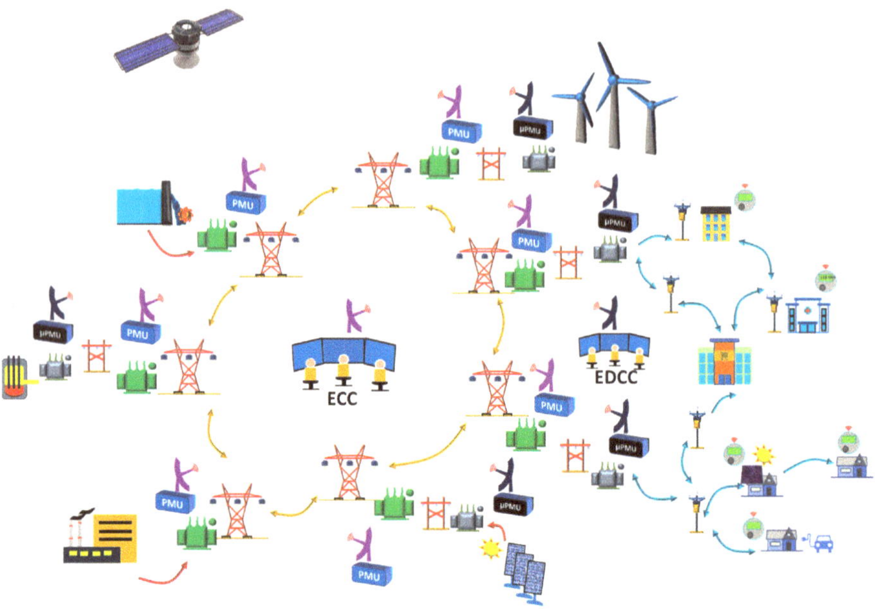

Figure 7.1 Schematic of a modern power electrical network.

pose. However, methods such as dynamic mode decomposition (DMD) provides a broader overview of how the phenomenon arises and spreads in the electrical transmission network [4,5].

Another phenomenon of major interest as the interarea oscillations is the harmonic distortion. Harmonic distortion appears in the level of subtransmission, which is mainly induced by the power electronic converters of distributed generation (DG), storage systems (SS), and large nonlinear loads, such as electric arc furnace (EAF). This phenomenon appears when DG units interact with nonlinear loads, hence high harmonic and inter-harmonic distortion are injected to the entire electrical system, and the distortion is extended to major regions of the electrical network [4–8]. The status of this harmonic distortion is monitored by higher sampling rate known as microphasor measurement units (μPMUs). The μPMU is in each substation in the level of sub-transmission and form part of the wide area power quality monitoring system (WAPQ), which is located at the energy distribution control center (EDCC) [9,10]. This phenomenon has been studied by ML methods being the main aspects to evaluate the feature extraction and data compression [3].

In this context, DMD is a powerful data-driven technique, either from numerical simulations or experimental data, to obtain reduced-order linear models for complex high-dimensional systems estimating its evolution aside from to extract coherent spatiotemporal patterns that they dominate the dynamical system. The DMD method was introduced by Schmid & Sesterhen for the analysis of dynamic fluids [11], however, DMD has been widely used for the analysis of nonlinear dynamic systems, ranging from the area of biological systems [12], phenomena climate [13], finance [14], medicine [15], even many implementations to explore. In the area of power systems, DMD is one of the most recent tools to simultaneously process spatiotemporal data from various monitoring systems, such as WAMS and SCADA. Among the various applications of DMD in the power system are the reduction modal identification analysis [16–20], coherence identification [18,19], identification of harmonic distortion [21], short-term electrical load forecasting [22], to mention a few.

The harmonic distortion in complex distribution networks, in the same way that the inter-area is a phenomenon with spatiotemporal characteristics, is needed in using methods for their proper characterization. In this sense, the DMD was applied in [21], extracting these characteristics with the operator that estimates the behavior of the system showing good performance. More recently, the Padé approach was adapted for the study of spatiotemporal harmonic phenomena, such as the source of harmonics and the process of propagation [23].

In the same way, the Padé approach becomes a global spectral decomposition method for estimating harmonic distortion in electrical systems with several levels of harmonic pollution. The Padé approach uses a rational equivalent technique and permits representing a complex oscillatory process by single-frequency linear modes, and also possesses the ability to identify and extract dominant patterns from multidimensional data sets. The simplicity of the model structure allows a direct numerical analysis of large data sets collected from software simulation or from μPMUs located strategically at key substations in the network [23]. The Padé method can be used for the treatment of multidimensional data sets without a priori knowledge of the electrical network models, extracting the spectral information and maintaining fast calculation speeds [23]. Thus it opens the possibility to develop a spatiotemporal monitoring tool in near real-time. This approach allows identifying the dominant spatial and temporal patterns, such as the harmonic frequency components, harmonic mode-shape, and participation factors of the harmonic source localities. Thus both methods are suitable tools for incorporating in unsupervised ML algorithms.

This chapter is focused on the analysis of harmonics from a space-time perspective using these two powerful methods. The chapter is structured as follows: Section 7.2 presents the formulation of the DMD and the Padé approaches, summarizing their main characteristics and emphasizing the principal similarities in the description of distorted signal. Next, in Section 7.3 an application to synthetic signals is presented, and the accuracy is evaluated facing other conventional techniques. Section 7.4 discusses the application of the proposed techniques to simulated data with multiple signals obtained from the IEEE 14-bus test system; finally, a discussion is provided.

7.2 Modeling of harmonic signals using spatiotemporal approaches

To introduce the analysis, let us assume that a set of n three-phase sensors (measures the voltage and currents at each bus) distributed across the electrical network provides the measurements. These variables are observed at N times $(0, T, \ldots, (N-1)T)$ with uniform sampling time T (sec) and satisfying the Nyquist criterion (sampling frequency in Hz, $f_s = 1/T$ up to two twice the frequency to identifying). Following the approach in [24], the system observation matrix, $\mathbf{X} \in \Re^{6nx(N-1)}$, can be defined as

$$\mathbf{X} = \begin{bmatrix} x_0^1 & x_1^1 & \cdots & x_{N-1}^1 \\ x_0^2 & x_1^2 & \cdots & x_{N-1}^2 \\ \vdots & \vdots & \ddots & \vdots \\ x_0^{6n} & x_1^{6n} & \cdots & x_{N-1}^{6n} \end{bmatrix} = \begin{bmatrix} \boldsymbol{x}_0 & \boldsymbol{x}_1 & \ldots & \boldsymbol{x}_{N-1} \end{bmatrix}, \tag{7.1}$$

where $x_k^i = \hat{x}_k^i(kT) + \varphi(kT)$, $k = 0, \cdots, N-1$, $T = 1/f_s$, $\hat{x}_k^i(kT)$, and $\varphi(kT)$ are the bus signals and their associated noise at the instant of time kT respectively and

$$\boldsymbol{x}_k = \begin{bmatrix} x_k^1 \\ x_k^2 \\ \vdots \\ x_k^{6n} \end{bmatrix} k = 0, 1, \ldots, N-1. \tag{7.2}$$

The harmonics contained in measurements (7.1) are next studied first using the dynamic mode decomposition method and later the Padé method.

7.2.1 The dynamic mode decomposition method (DMD)

The basis of the method appears in the Koopman operator theory, to describe the initial and future evolution of Hamiltonian systems in discrete time [25]. DMD is an algorithm originally developed in the fluid dynamics community by Schmidt [11], to decompose complex flows into coherent space-time patterns or structures estimating system behavior. DMD can find Koopman modes from spatiotemporal data, where each mode can be associated with a unique frequency and growth rate and interpreted as a nonlinear generalization of global eigenmodes of a linearized system [26,27]. Considering (7.2), the dynamic system in discrete time will be given in the form

$$\boldsymbol{x}_{k+1} = \boldsymbol{F}(\boldsymbol{x}_k), \tag{7.3}$$

where $\boldsymbol{F}$ represents an operator that captures and estimates the dynamics of the system in discrete time; this can be induced from the dynamics of continuous time [11]. This procedure is analogous to the Koopman operator $\boldsymbol{K}$ in discrete time; it is a linear operator of infinite dimension, which acts on the observable function g performing a linear transformation given by

$$\boldsymbol{K} g\left(\boldsymbol{x}_k\right)=g\left(F\left(\boldsymbol{x}_k\right)\right)=g\left(\boldsymbol{x}_{k+1}\right), \tag{7.4}$$

which considers a nonlinear dynamic of finite dimension to the Koopman representation, such that it considers a linear dynamic of infinite dimension [24]. The dynamics of $\boldsymbol{x}_k$ is completely determined by the operator $\boldsymbol{K}$. The dynamic evolution of the system can be completely characterized by the eigenvalues and eigenvectors given by the eigen decomposition of $\boldsymbol{K}$ [25] as

$$\boldsymbol{K} g\left(\boldsymbol{x}_k\right)=\sum_{i=1}^{\infty} \boldsymbol{K} \boldsymbol{\varphi}_i(\boldsymbol{x}) \boldsymbol{v}_i=\sum_{i=1}^{\infty} \lambda_i \boldsymbol{\varphi}_i(\boldsymbol{x}) \boldsymbol{v}_i=g\left(\boldsymbol{x}_{k+1}\right), \tag{7.5}$$

where λ_i and $\boldsymbol{\varphi}_i$ are the Koopman eigenvalues and eigenfunctions, respectively, and $\boldsymbol{v}_i$ is the kth mode associated with the Koopman eigenfunctions. This expression represents the solution of operator $\boldsymbol{K}$ by an infinite sum that can be approximated as a finite sum of modes. These modes correspond to physical components, whose temporal behavior is given by the associated eigenvalues. The phase of the eigenvalues determines their frequency, whereas their modules determine the growth rate, characterized by an exponential growth rate and/or oscillation over time, and the magnitude of $\boldsymbol{\varphi}_i(\boldsymbol{x})$, $\boldsymbol{v}_i$ is used as a measure of the relative share of a mode to modal decomposition [5]. Through this finite approximation, the DMD method was developed as an alternative to approximate the modes of the Koopman K operator. From a discrete time vector that represents the measurement of the states of the system, which is defined in (7.2), the best fit is sought to approximate it with the linear operator $\boldsymbol{F} \in \Re^{n \times \mathrm{N}}$ that advances the state of a system and predicts its behavior. This formulation can be redefined for irregularly spaced data and concatenated data from several different time series [25]. With the consideration that we have (7.1) and two sets of data given by

$$\mathbf{X_1}=\begin{bmatrix} x_0^1 & x_1^1 & \cdots & x_{N-2}^1 \\ x_0^2 & x_1^2 & \cdots & x_{N-2}^2 \\ \vdots & \vdots & \ddots & \vdots \\ x_0^{6n} & x_1^{6n} & \cdots & x_{N-2}^{6n} \end{bmatrix}, \tag{7.6}$$

$$\mathbf{X_2}=\begin{bmatrix} x_1^1 & x_2^1 & \cdots & x_{N-1}^1 \\ x_1^2 & x_2^2 & \cdots & x_{N-1}^2 \\ \vdots & \vdots & \ddots & \vdots \\ x_1^{6n} & x_2^{6n} & \cdots & x_{N-1}^{6n} \end{bmatrix}. \tag{7.7}$$

Then, the rows of $\mathbf{X_1}$ are the spatial data and the columns are discrete time measurements called the data snapshot. Therefore $\mathbf{F}$ is approximated from pairs of system snapshots. In terms of matrices, (7.3) will take the form

$$\mathbf{X_2}=\mathbf{F X_1}. \tag{7.8}$$

Mathematically, the best fit operator **F** for a linear dynamic system can be formulated as an optimization problem [5,17] which is defined as

$$\mathbf{F} = \min \|\mathbf{X_2} - \mathbf{F}\mathbf{X_1}\|_{Fr} = \mathbf{X_2}\mathbf{X_1^{\dagger}}, \tag{7.9}$$

where $\|*\|_{Fr}$ is the Frobenius norm, and † denotes the Moore–Penrose pseudo-inverse. Then it is seen that **F** in (7.9) is like the Koopman operator in (7.3) if the direct linear measures of the state are chosen, so that g(x) = x [16]. One of the advantages of the DMD algorithm to obtain the approximation of the operator **F** from an order low model using a dimensionality reduction technique to calculate the eigenvalues and the dominant eigenvectors of **F**. This technique is based on singular value decomposition (SVD) of the matrix $\mathbf{X_1} \in \mathfrak{R}^{n\times \mathrm{N}-1}$ such that

$$\mathbf{X_1} = \mathbf{U}\boldsymbol{\Sigma}\mathbf{V}^T, \tag{7.10}$$

where $\mathbf{U} \in \mathfrak{R}^{n\times \mathrm{n}}$ and $\mathbf{V}^T \in \mathfrak{R}^{\mathrm{n}\times \mathrm{N}-1}$ are orthogonal unitary matrices containing the left and right singular vectors, respectively, also known as spatial and temporal SVD modes. The matrix $\boldsymbol{\Sigma} \in \mathfrak{R}^{\mathrm{n}\times\mathrm{n}}$ is a diagonal matrix with singular values ordered in decreasing order on the diagonal. We can compute a low-rank matrix of $\mathbf{X_1}$ using the first r singular values of $\boldsymbol{\Sigma}$, such that $\tilde{\mathbf{U}} \in \mathfrak{R}^{\mathrm{n}\times r}$, $\tilde{\mathbf{V}}^{\mathrm{T}} \in \mathrm{R}^{\mathrm{r}\times \mathrm{N}-1}$, and $\tilde{\boldsymbol{\Sigma}} \in \mathfrak{R}^{r\times r}$ are reduced dimension

$$\mathbf{X_1} = \tilde{\mathbf{U}}\tilde{\boldsymbol{\Sigma}}\tilde{\mathbf{V}}^T. \tag{7.11}$$

The approximation for $\mathbf{X_2}$ is proposed by means of the operator **F** in (7.8) by means of the following expression:

$$\mathbf{X_2} \approx \mathbf{F}\tilde{\mathbf{U}}\tilde{\boldsymbol{\Sigma}}\tilde{\mathbf{V}}^T. \tag{7.12}$$

From the operator **F** in (7.12) can be obtained using $\mathbf{X_1^{\dagger}}$ obtained by SVD:

$$\mathbf{F} = \mathbf{X_2}\tilde{\mathbf{V}}\tilde{\boldsymbol{\Sigma}}^{-1}\tilde{\mathbf{U}}^T. \tag{7.13}$$

A representation of **F** is obtained from the modes of the left singular vector of **X** [5,16] by the following equation:

$$\tilde{\mathbf{F}} = \tilde{\mathbf{U}}^{\mathrm{T}}\mathbf{F}\tilde{\mathbf{U}} = \tilde{\mathbf{U}}^{\mathrm{T}}\mathbf{X_2}\tilde{\mathbf{V}}\tilde{\boldsymbol{\Sigma}}^{-1}, \tag{7.14}$$

where $\tilde{\mathbf{F}}$ is the reduced matrix, and with this approach a linear model is defined that seeks a reduced representation based on r dominant modes that capture the highest energy content in the dynamics of the system, such that

$$\tilde{x}_{k+1} = \tilde{\mathbf{F}}(\tilde{x}_k). \tag{7.15}$$

The spectral decomposition of the operator $\tilde{F}$ approximates the spectral decomposition of F, which is obtained through the eigen decomposition

$$\tilde{\mathbf{F}}\mathbf{W} = \mathbf{W}\boldsymbol{\Lambda}, \tag{7.16}$$

where $\mathbf{W}$ are eigenvectors and $\mathbf{\Lambda}$ is a diagonal matrix containing the corresponding eigenvalues λ_k. From the eigenvectors of $\mathbf{W}$ and the matrix $\mathbf{X_2}$, it is possible to reconstruct the DMD modes $\mathbf{\Phi}$ [28] by means of the following expression:

$$\mathbf{\Phi} = \mathbf{X}_2\tilde{\mathbf{V}}\tilde{\mathbf{\Sigma}}^{-1}\mathbf{W}. \tag{7.17}$$

By obtaining the DMD modes and their respective eigenvalues, it is possible to represent the state of the system in terms of the DMD expansion

$$\boldsymbol{x}_k = \mathbf{\Phi}\mathbf{\Lambda}\mathbf{b} = \sum_{i=1}^{r} \lambda_i \phi_i (\boldsymbol{x})\, b_i, \tag{7.18}$$

where ϕ_i are eigenvectors of $\boldsymbol{F}$ (DMD modes), λ_i are eigenvalues of F (DMD eigenvalues), and b_i are the mode amplitudes. The amplitudes $\mathbf{b}$ are given by

$$\mathbf{b} = \mathbf{\Phi}\boldsymbol{x}_0, \tag{7.19}$$

where $\boldsymbol{x}_0$ is the measurement of the system at initial time.

The DMD expansion (7.18) is directly analogous to the Koopman mode decomposition (7.3). The DMD modes ϕ_i approximate the Koopman modes v_i; the DMD eigenvalues λ_i approximate the corresponding Koopman eigenvalues, and the mode amplitudes λ_i approximate the Koopman eigenfunctions evaluated at the initial condition $\varphi_i\,(\boldsymbol{x}_0)$.

With the spectral expansion in (7.18), it is possible to rewrite it in continuous time from the eigenvalues such that $\omega = \frac{\log \lambda}{\Delta t}$. Then we obtain

$$x\,(t) = \sum_{i=1}^{r} \phi_i \exp(\omega_i t)\, b_i = \mathbf{\Phi} \exp(\mathbf{\Omega} t)\,\mathbf{b}, \tag{7.20}$$

where $\mathbf{\Omega}$ is a diagonal matrix containing the eigenvalues ω_i.

From the work in [26], interpreting the dynamic behavior of a system through a modal decomposition associated with the operator $\tilde{\mathbf{F}}$, we obtain a space-time structure associated with the dynamic evolution of the system, where Φ represents the spatial term, and $\exp(\mathbf{\Omega} t)\,\mathbf{b}$, the associated features of the spatiotemporal structure of the system dynamics can be identified. Each mode ϕ_i has an associated damping and frequency given by λ_i, which can be represented by the following expression [5]:

$$\sigma_{DMDi} = \frac{\Re\{\lambda_i\}}{\Delta t}, \tag{7.21}$$

$$f_{DMDi} = \frac{\Im\{\lambda_i\}}{\frac{\Delta t}{2\pi}}. \tag{7.22}$$

Additionally, a participation factor related to the modes ϕ_i can be calculated. Since the spatial part of the system dynamics is associated, it can be calculated at time t_0

using the normalized magnitude of each mode $||\phi_i||$, and with the phase $\angle\phi_i$, the existing groups with a similar dynamic behavior can be identified.

One way to visualize the participation factors for each state of the system is through the temporal structure presented in [21]; with the b_i component it can be represented as follows:

$$A_{ji} = \left|\phi_{ji}\lambda_i \, \|b_i\|\right|, \tag{7.23}$$

where the term A_{ji} is a measure of the participation factor of the mode ϕ_i in the states of the system.

7.2.2 The Padé method

An alternative way to estimate the spatiotemporal characteristics of (7.1) is by computing the z-transform of the samples of each signal. In the discrete domain, the sampled data (7.1) can be transformed to the z-domain and expressed as a continuous function of order $N-1$ of the form

$$X(z) = \begin{bmatrix} \sum_{k=0}^{N-1} x_k^1 z^{-k} \\ \sum_{k=0}^{N-1} x_k^2 z^{-k} \\ \vdots \\ \sum_{k=0}^{N-1} x_k^{6n} z^{-k} \end{bmatrix}. \tag{7.24}$$

Building upon the above model, the set of signals in (7.24) can be represented as

$$X(z) = \begin{bmatrix} \sum_{k=0}^{N-1} x_k^1 z^{-k} \\ \sum_{k=0}^{N-1} x_k^2 z^{-k} \\ \vdots \\ \sum_{k=0}^{N-1} x_k^{6n} z^{-k} \end{bmatrix} = \begin{bmatrix} \frac{P^1(z)}{Q(z)} \\ \frac{P^2(z)}{Q(z)} \\ \vdots \\ \frac{P^{6n}(z)}{Q(z)} \end{bmatrix}, \tag{7.25}$$

where $\frac{P^i(z)}{Q(z)}$ is known as the Padé approximant of the ith signal, $P^i(z) = \sum_{l=0}^{L} p_l^i z^{-l}$ and $Q(z) = \sum_{m=0}^{M} q_m z^{-m}$ are two polynomials of order L and M, respectively. The coefficients p_l^i and q_m are unknown, and they are referred to as the Padé's coefficients [29]. Note that here, in the proposed Padé approximant, is assumed that $Q\left(z^{-1}\right)$ is the same for all signal representations [23]. In this approximation, it is assumed that $M > L$ and satisfying that $L + M + 1 = N$ ($M = N/2$, $L = N/2$-1). To determining the Padé approximant, (7.25) is expressed in alternative form as

$$\begin{bmatrix} \sum_{m=0}^{M} q_m z^{-m} \sum_{k=0}^{N-1} x_k^1 z^{-k} \\ \sum_{m=0}^{M} q_m z^{-m} \sum_{k=0}^{N-1} x_k^2 z^{-k} \\ \vdots \\ \sum_{m=0}^{M} q_m z^{-m} \sum_{k=0}^{N-1} x_k^{6n} z^{-k} \end{bmatrix} = \begin{bmatrix} \sum_{l=0}^{L} p_l^1 z^{-l} \\ \sum_{l=0}^{L} p_l^2 z^{-l} \\ \vdots \\ \sum_{l=0}^{L} p_l^{6n} z^{-l} \end{bmatrix}. \tag{7.26}$$

The underlying idea behind the method of Padé is to find the coefficients that satisfy the equality (7.26). The Padé´s coefficients for the ith signal are determined by taking the product of the two polynomials ($\sum_{m=0}^{M} q_m z^{-m} \sum_{k=0}^{N-1} x_k^i z^{-k}$) [30]. Then equaling the terms of same order in z on both sides and neglecting the terms up to $z^{-(L+M+1)}$, this results in two linear systems of equations, i.e.,

$$\begin{bmatrix} x_L^i & \cdots & x_{L+M-2}^i & x_{L+M-1}^i & x_{L+M}^i \\ \vdots & \ddots & \vdots & \vdots & \vdots \\ x_1^i & \cdots & x_L^i & x_{L+1}^i & x_{L+2}^i \\ x_0^i & \cdots & x_{L-1}^i & x_L^i & x_{L+1}^i \end{bmatrix} \begin{bmatrix} q_M \\ \vdots \\ q_1 \\ q_0 \end{bmatrix} = \begin{bmatrix} 0 \\ \vdots \\ 0 \\ 0 \end{bmatrix} \tag{7.27}$$

and

$$\begin{bmatrix} p_L^i \\ \vdots \\ p_1^i \\ p_0^i \end{bmatrix} = \begin{bmatrix} x_0^j & \cdots & x_{L-1}^i & x_L^i \\ \vdots & \ddots & \vdots & \vdots \\ 0 & \cdots & x_0^i & x_1^i \\ 0 & \cdots & 0 & x_0^i \end{bmatrix} \begin{bmatrix} q_L \\ \vdots \\ q_1 \\ q_0 \end{bmatrix}. \tag{7.28}$$

Eq. (7.27), is a homogeneous $(L+1)x(M+1)$ undetermined linear system with a Hankel matrix structure and constructed with the samples of the ith signal, whereas (7.28) is a square $(L+1)$-order upper triangular linear system with a Toeplitz structure, whose solution depends of (7.27). Solution of (7.27) cannot be obtained since all the signals have in common the same set of q's coefficients [23].

To obtaining solution of the above equations, let us assume that a system such as (7.27) appears for each signal, then the next equation needs to be solved for obtaining the q's coefficients:

$$\begin{bmatrix} \mathbf{H}^1\mathbf{q} \\ \mathbf{H}^2\mathbf{q} \\ \vdots \\ \mathbf{H}^{6n}\mathbf{q} \end{bmatrix} = \overline{\mathbf{H}}\mathbf{q} = \begin{bmatrix} \mathbf{0} \\ \mathbf{0} \\ \vdots \\ \mathbf{0} \end{bmatrix}, \overline{\mathbf{H}} \in \Re^{6n(L+1)x(M+1)}, \tag{7.29}$$

where

$$\mathbf{H}^i = \begin{bmatrix} x_L^i & \cdots & x_{L+M-2}^i & x_{L+M-1}^i & x_{L+M}^i \\ \vdots & \ddots & \vdots & \vdots & \vdots \\ x & \cdots & x_L^i & x_{L+1}^i & x_{L+2}^i \\ x_0^i & \cdots & x_{L-1}^i & x_L^i & x_{L+1}^i \end{bmatrix} \in \Re^{(L+1)x(M+1)},$$

$$\mathbf{q} = \begin{bmatrix} q_M \\ \vdots \\ q_1 \\ q_0 \end{bmatrix} \in \Re^{(M+1)}, \ \mathbf{0} = \begin{bmatrix} 0 \\ \vdots \\ 0 \\ 0 \end{bmatrix} \in \Re^{(M+1)}.$$

Due to the high dimensionality of (7.29), the analysis of measured data from multiple sensors becomes intractable. This limitation is circumvented by using the invariant property of the singular values under transposition of $\overline{\mathbf{H}}$ [31]. Using this idea, Eq. (7.29) can be rearranged in the form

$$\mathbf{q}^T\overline{\mathbf{H}}^T = \begin{bmatrix} \mathbf{0}^T & \mathbf{0}^T & \cdots & \mathbf{0}^T \end{bmatrix}. \tag{7.30}$$

Singular value decomposition of $\overline{\mathbf{H}}^T$ gives

$$\overline{\mathbf{H}}^T = \begin{bmatrix} \mathbf{U}_1 & \mathbf{U}_2 \end{bmatrix} \begin{bmatrix} \boldsymbol{\Sigma} & \mathbf{0} \\ \mathbf{0} & \mathbf{0} \end{bmatrix} \begin{bmatrix} \mathbf{V}_1 \\ \mathbf{V}_2 \end{bmatrix}, \tag{7.31}$$

where it is noticeable that $\mathbf{q}^T$ are the columns of $\mathbf{U}_2$, i.e.,

$$\mathbf{q}^T \in span\{\mathbf{U}_2\}. \tag{7.32}$$

In this case, the left singular vector associated with the minimum singular value is used. This formulation can be implemented when large sets of sensors are considered since the singular vectors are computed from $\overline{\mathbf{H}}^T\overline{\mathbf{H}} = \sum_{i=1}^{6n} \mathbf{H}^{iT}\mathbf{H}^i$, thus avoiding the storage of the matrix $\overline{\mathbf{H}}^T$ [23]. Once again that $\mathbf{q}$ is obtained, the associated p´s coefficients of each signal are obtained from (7.28), i.e.,

$$\begin{bmatrix} p_L^i \\ \vdots \\ p_1^i \\ p_0^i \end{bmatrix} = \begin{bmatrix} x_0^j & \cdots & x_{L-1}^i & x_L^i \\ \vdots & \ddots & \vdots & \vdots \\ 0 & \cdots & x_0^i & x_1^i \\ 0 & \cdots & 0 & x_0^i \end{bmatrix} \begin{bmatrix} q_L \\ \vdots \\ q_1 \\ q_0 \end{bmatrix} \quad i = 1, 2, ..., 6n. \tag{7.33}$$

The Padé approximant of signals gives an approximation to transfer function, and hence the harmonics can be directly estimated from their pole-residue representation. These realizations result in high-dimensional models appearing nonphysical poles or spurious poles, and causing noise or instabilities when the signals are recovered [32]. Thus, to determine the best reconstruction of the signals, an approximate pole-residue representation of the signals is obtained as

$$X^i(z) = \frac{P^i(z)}{Q(z)} \approx \sum_{m=1}^{M} \frac{r_m^i}{1 - \alpha_m z^{-1}}, \tag{7.34}$$

where r_m^i is the mth residue, and α_m is the mth pole of the ith transfer function. A complex pole m is discarded when its residue is smaller than the criterion $2\left\|r_m^i\right\| \le \varepsilon_1$. This value is related with the individual harmonic distortion, and in this work $\varepsilon_1 = 0.01$ [23,33]. Then is obtained a low-order transfer function of the form

$$X^i(z) \approx \sum_{m=1}^{M_i} \frac{r_m^i}{1 - \alpha_m z^{-1}} M_i < M, \tag{7.35}$$

and the discrete-time signals can be recovered using the inverse z-transform of (7.35) [23].

From (7.35), the kth sample of the ith signal of (7.1) is obtained by applying the inverse z-transform of (7.35) that yields

$$x_k^i \approx \sum_{m=1}^{M_i} r_m^i (\alpha_m)^k . \tag{7.36}$$

In the more general case, Eq. (7.36) represents a mixture of a-pairs of complex-conjugate poles, b real poles, and a constant term, thus an alternative representation of (7.36), becomes

$$x_k^i \approx \sum_{m=1}^{a} A_m^i e^{k\sigma_m^i T} \cos\left(2\pi k f_m T + \theta_m^i\right) + \sum_{m=2a+1}^{b} r_m^i e^{k\sigma_m^i T} + r_1^i, 2a+b+1 = M_i, \tag{37a}$$

which is a superposition of damped sinusoidal terms, decaying terms, and the constant term (dc-offset). The terms A_m^i, σ_m^i, f_m, and θ_m^i are the amplitude, damping factor, frequency, and phase corresponding to the complex-conjugate pairs of poles, respectively. These terms are related with the poles and residues through

$$A_m^i = 2\left\|r_m^i\right\|, \quad \sigma_m^i = f_s log\{\|\alpha_m\|\} sec^{-1}, \quad f_m = f_s \frac{\measuredangle\alpha_m}{2\pi} Hz,$$
$$\theta_m^i = \measuredangle r_m^i rad, \quad m = 2, 4, \cdots, 2a. \tag{37b}$$

From (37b) are defined the participation factors and the mode shapes. For instance, A_m^i is the participation factor of the mth harmonic onto the ith signal, whereas A_m^i along with θ_m^i give the mode shape of the mth harmonic. The participation factors allow to understand how the harmonics are spatially distributed into the electrical network. Algorithm for the Padé method and complexity details are fully described in [23]. The proposed approach can be applied to large datasets, window of observation of up to three cycles of the nominal frequency, and a sampling rate up to 5 kHz.

7.3 Preliminary results with synthetic data

Both methods were tested using five synthetic signals given in (7.38a)–(7.38e) [34]. The first signal $x_1(kT)$ given in (7.38a) is a three-component harmonic signal contaminated with 25 dB of noise (i.e., a noise $\varphi(kT)$ is added to $x_1(kT)$). Second signal (7.38b) is the signal $x_1(kT)$ plus a damped sinusoid and with DC offset. The third signal (7.38c) is the signal $x_1(kT)$ plus interharmonics, an intermodulated signal (7.38d) with a bandwidth of +/−10 Hz over the fundamental frequency, and the case of off-nominal frequency (7.38e).

$$x_1(k) = A_1 \sin(k\omega T) + A_3 \sin(3k\omega T) + A_5 \sin(5k\omega T), \tag{7.38a}$$

Table 7.1 Comparison of modal estimates for $x_1 + \varphi$.

Component	MPM		FFT		DMD		Padé	
	A	*f* *(Hz)*	A	*f* *(Hz)*	A	*f* *(Hz)*	A	*f* *(Hz)*
1st	2.952	50.01	2.821	49.80	3.158	50	**3.027**	**50.094**
3rd	0.964	150	0.770	150.39	1.043	150	**1.037**	**149.85**
5th	0.635	250	0.599	250.01	0.592	250	**0.610**	**251.05**

$$x_2(k) = x_1(k) + 2e^{-60kT}\sin(2\pi k200T) + 3e^{-100kT} + 2, \tag{7.38b}$$

$$x_3(k) = x_1(k) + 1.5\sin(0.69k\omega T) + \sin(1.8k\omega T), \tag{7.38c}$$

$$x_4(k) = \cos(k\omega T) + 1.5\cos(k(\omega - \Delta\omega)T + \theta) + 1.5\cos(k(\omega + \Delta\omega)T - \theta), \tag{7.38d}$$

$$x_5(k) = 5\sin(2\pi k49.8T) + 2\sin(2\pi k149.4T) + \sin(2\pi k249T), \tag{7.38e}$$

where $A_1 = 3$, $A_3 = 1$, $A_5 = 0.6$, $\omega = 2\pi 50$ rad/s, $\Delta\omega = 2\pi 10$ rad/s, $\theta = 30°$, $f_s = 1$ kHz. These signals are shown in Fig. 7.1 for 60 msec of duration. The Matlab® code that generates the signals can be founded in [35]. All the signals were processed simultaneously and using a window $N = 60$ samples (60 msec or three cycles). This size was used to ensure that DMD can process all the signals simultaneously. The Padé and DMD methods were compared with other methods, such as matrix pencil (MPM), and FFT methods [34]. Table 7.1 shows the results for the synthetic signal (38*a*). It is shown that the amplitudes and frequencies obtained with the Padé method are near to the original parameters, even in noisy conditions; in this case, the FFT gives the worst approximation for the third harmonic. Table 7.2 shows the comparison for (7.38b); again note that both methods properly identify the decaying terms and the offset value. On the other hand, Table 7.3 presents the results for the signal (7.38c). As can be observed, the proposed method identifies properly the interharmonics of the signal.

Finally, the results obtained for signal with intermodulation conditions as well as for off-nominal harmonic conditions are given in Table 7.4 and Table 7.5, respectively. From Table 7.4, all methods give information of phase, except the FFT method. Through these results, it can be concluded that methods of DMD and Padé can be applied with different kinds of harmonic phenomena.

7.4 Application to simulated data

The methods of DMD and Padé were used to exploring the spatiotemporal process of harmonic pollution when harmonics are injected in the IEEE 14-bus test system. The one-line diagram is showed on Fig. 7.2, and the impedance data was obtained from [36]. In the simulations, the generators were represented as source of voltage with constant fundamental frequency plus harmonic distortion and respecting the total

Table 7.2 Comparative results for x_2.

Component	MPM			FFT			DMD			Padé		
	A	$f\,(Hz)$	σ	A	$f\,(Hz)$	σ	A	$f\,(Hz)$	σ	A	$f\,(Hz)$	σ
1st	3	50	0	2.847	49.80	N.A		**50**	**0**	**3**	**50**	**0**
3rd	1	150	0	0.758	150.40	N.A		**150**	**0**	**1**	**150**	**0**
5th	0.6	250	0	0.601	250.00	N.A		**250**	**0**	**0.6**	**250**	**0**
Damped sinusoid	2	200	−60	N.A	N.A.	N.A		**200**	**−60**	**2**	**200**	**−60**
Decaying DC term	3	0	−100	N.A	N.A	N.A		**0**	**−100**	**3**	**0**	**−100**
Constant DC term	2	0	0	N.A	N.A	N.A		**0**	**0**	**2**	**0**	**0**

N.A means "not available".

Table 7.3 Comparative results for x_3.

Component	MPM		FFT		DMD		Padé	
	A	*f (Hz)*	A	*f (Hz)*	A	*f (Hz)*	A	*f (Hz)*
1st	3	50	2.800	49.80	3.158	50	**3**	**50**
3rd	1	150	0.773	150.40	1.043	150	**1**	**150**
5th	0.6	250	0.600	250	0.592	250	**0.6**	**250**
0.69^{th}	1.5	34.5	1.266	34.18	1.5	3.45	**1.5**	**34.5**
1.80^{th}	1	90	0.956	89.94	1	90	**1**	**90**

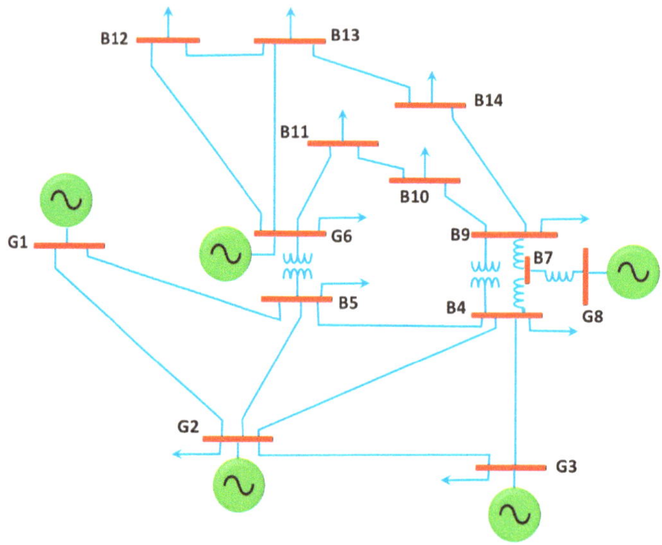

Figure 7.2 One-line diagram of the IEEE 14-bus test system.

harmonic distortion (THD) established in the standard IEEE [37], whereas loads were modeled as constant power. Two cases are studied. Table 7.6 summarizes Case I; here each generator supplies balanced sub, inter- and supra-harmonics at different time instants. The Case II considers injection of triplen harmonics at generators G3, G6, and G8, presented in Table 7.7. Case I can be reproduced by running the Simulink file given in [35]. The time of simulation was of 30 cycles of the fundamental frequency (60 Hz) at $f_s = 5$ kHz for both cases. In this case, the matrix of observation is composed by 84 signals and 2500 samples, i.e., $\mathbf{X} \in \Re^{84x2500}$, where the first 42 rows are the voltages at each bus, and the rest corresponds to current signals of each bus.

Fig. 7.3 and Fig. 7.4 show the voltages and currents signals for the generators at the instants of time where they are injecting the harmonics for cases Case I and Case 2, respectively. In this case, the signals for voltages are stationary, whereas the current signals for G1 and G3 present transient behavior. The signals have low levels of harmonic distortion and are noiseless.

Table 7.4 Comparative results for x_4.

Component	MPM			FFT			DMD			Padé		
	A	$f\,(Hz)$	$\theta°$	A	$f\,(Hz)$	$\theta°$	A	$f\,(Hz)$	$\theta°$	A	$f\,(Hz)$	$\theta°$
0.8^{th}	1.5	40	30	1.5	40.03	N.A	**1.5**	**40**	**30**	**1.5**	**40**	**30**
1st	1	50	0	1	49.8	N.A	**1**	**50**	**0**	**1**	**50**	**0**
1.2^{th}	1.5	60	−30	1.5	59.57	N.A	**1.5**	**60**	**−30**	**1.5**	**60**	**−30**

N.A means "not available".

Table 7.5 Comparative results for x_5.

Component	MPM		FFT		DMD		Padé	
	A	*f (Hz)*	A	*f (Hz)*	A	*f (Hz)*	A	*f (Hz)*
1st	5	49.8	5	49.8	4.5	49.8	**5**	**49.8**
3rd	2	149.4	2	149.4	2.18	149.4	**2**	**149.4**
5th	1	249	1	249	1.004	249	**1**	**249**

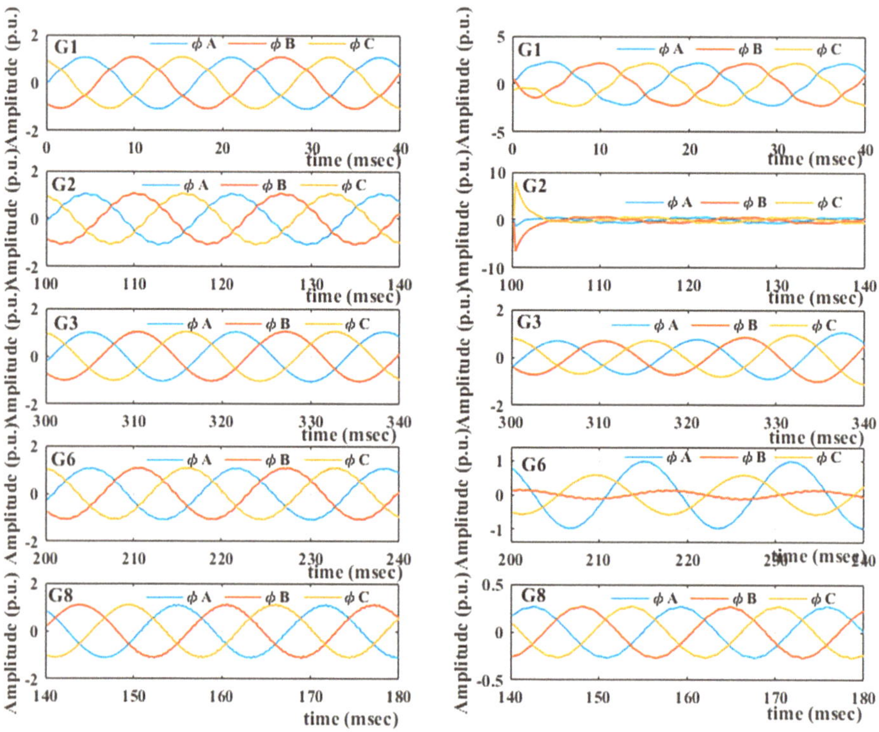

Figure 7.3 Time domain voltages (*left*) and current (*right*) responses at generator buses for Case I.

Due to the temporal evolution of the frequencies injected by the generators, the procedure to identify the harmonics consisted in choosing a small window of 200 samples (40 msec) and sliding this window over the entire matrix of observation (24 windows). The sliding window considers an overlapping of 100 samples (20 msec) between two consecutive windows, allowing thereby capturing the transient events. Next are presented the main findings encountered with both spatiotemporal approaches for these two cases of study.

Table 7.6 Voltage harmonics injected at generators buses in pu for Case I.

Bus	Time		0.85^{th}	5th	7th	11th	13th	17th	19th	25.3^{rd}	29th	35th	%*THD*
	From	to											
1	0	0.20	–	0.03	0.02	–	–	–	–	–	–	–	3.40
2	0.10	0.30	–	–	–	0.012	0.04	–	–	–	–	–	3.99
3	0.25	0.45	0.03	–	–	–	–	–	–	–	0.01	–	3.13
6	0.17	0.38	–	–	–	–	–	0.02	0.01	–	–	–	2.09
8	0.15	0.35	–	–	–	–	–	–	–	0.02	–	0.03	3.30

Table 7.7 Voltage harmonics injected at generators buses in pu for Case II.

Bus	Time		0.85^{th}	3rd	5th	7th	9th	11th	13th	15th	19th	25.3^{rd}	%*THD*
	From	to											
1	0	0.20	–	–	0.03	0.02	–	–	–	–	–	–	3.40
2	0.10	0.30	–	–	–	–	–	0.012	0.04	–	–	–	3.99
3	0.25	0.45	0.03	–	–	–	0.05	–	–	–	–	–	5.77
6	0.17	0.38	–	0.03	–	–	–	–	–	–	0.01	–	3.13
8	0.15	0.35	–	–	–	–	–	–	–	0.03	–	0.02	3.30

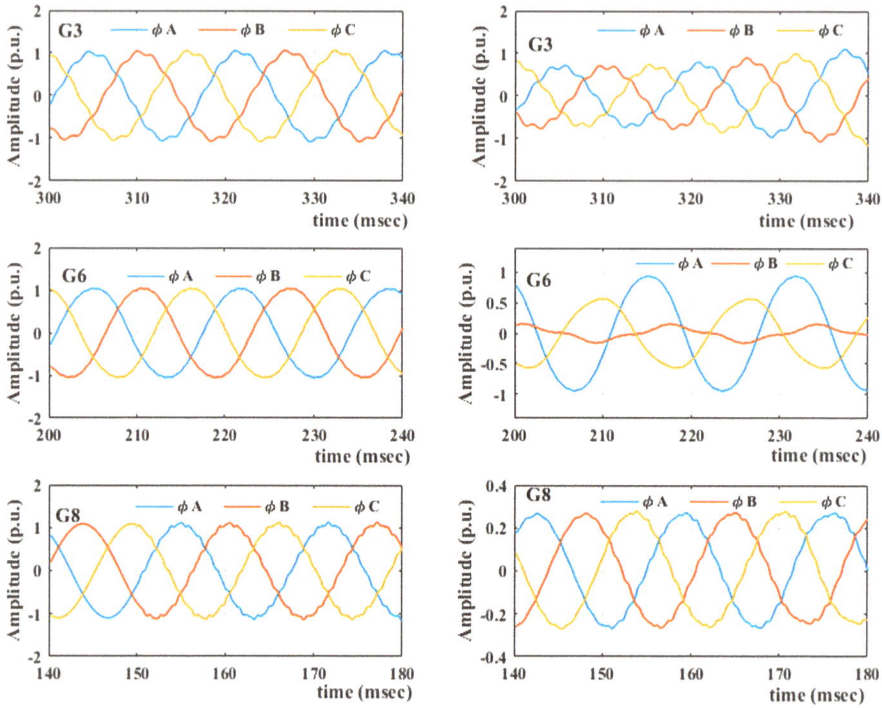

Figure 7.4 Time domain voltages (*left*) and current (*right*) responses at generator buses for Case II.

7.4.1 Application of DMD

The first step of the DMD implementation is to show the effectiveness of the reconstruction. The time series of the voltage affected by the harmonics were processed simultaneously, selecting windows 2, 6, 11, and 18 to carry out the analysis of Case I. In Fig. 7.5 a comparison of the DMD reconstructed signals (dotted line) with the three different phases of the original signal from G1, in each of the different time windows. DMD can optimally reconstruct the signal with its respective modal characteristics of frequency, amplitude, and damping.

Table 7.8 shows the harmonic distortion extracted from the DMD modes of the voltages together with the dominant frequencies in the different time windows, where there is a contribution of the corresponding harmonic. The results show all harmonic components as expected, since DMD globally extracts the frequencies belonging to the harmonics of the analyzed time series. To identify the harmonic source, it is necessary to obtain the participation factors, as shown below. DMD associates the energy with the real magnitude of the harmonics.

Another relevant feature of the DMD algorithm is the ability to calculate participation factors. This information makes it possible to identify the most active sections of the system that are affected by the different harmonic sources in terms of energy input. In addition, the strength of the spatial contributions of each bus to the observed

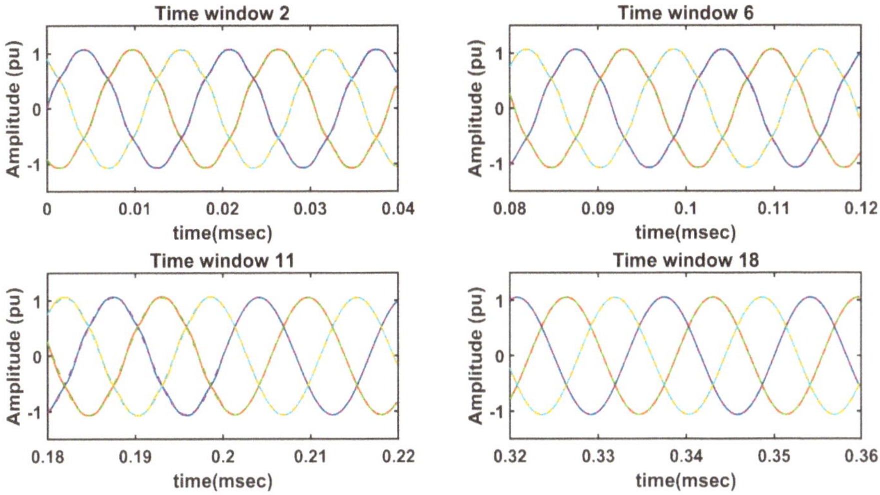

Figure 7.5 Voltage reconstruction G1 for Case I corresponding to the windows 2, 6, 11, and 18; dotted lines represent the signal obtained with DMD.

Table 7.8 Voltage harmonics obtained of DMD for Case I.

0.85^{th}	5th	7th	11th	13th	17th	19th	25.3^{rd}	29rd	35th
0.034	0.029	0.019	0.011	0.048	0.021	0.048	0.019	0.012	0.036

time series can be characterized and visualized. Fig. 7.6 shows a representation of the participation measurements as a function of bus locations and frequency for phases A. For Case I, it can be concluded that the 13th harmonic (780 Hz) and the 11th harmonic (660 Hz) are strongly observable on G2. On the other hand, the 17th harmonic (1020 Hz) is observable on B10–B13 of the system with different intensity. This phenomenon can indicate harmonic propagation in the system.

Fig. 7.7 shows the mode shape obtained from the DMD modes for the 7th, 11th, and 13th order harmonic, respectively. The magnitude and phase are affected by the harmonic source element placed at G1 for 7th harmonic and bus #2 for 11th and 13th order, where the higher contribution of the harmonics represents where G1 and G2 were injected.

In an analogous way, the Case II voltage time series with harmonic sources in windows 2, 6, 11, and 18 were processed simultaneously. Fig. 7.8 shows a comparison of the different phases of the original G1 signal with the reconstructed DMD signals (dotted line) in the selected time windows.

Table 7.9 contains the expected dominant frequencies of the harmonics for Case II extracted from DMD that are shown in Fig. 7.9 with the respective participation factors. The locations of the buses affected by the harmonic frequencies for the phases A in Case II are observed where the 9th harmonic (540 Hz) and the 0.85^{th} harmonic (51 Hz) are clearly observable in G3. Additionally, the 13th harmonic (780 Hz) prop-

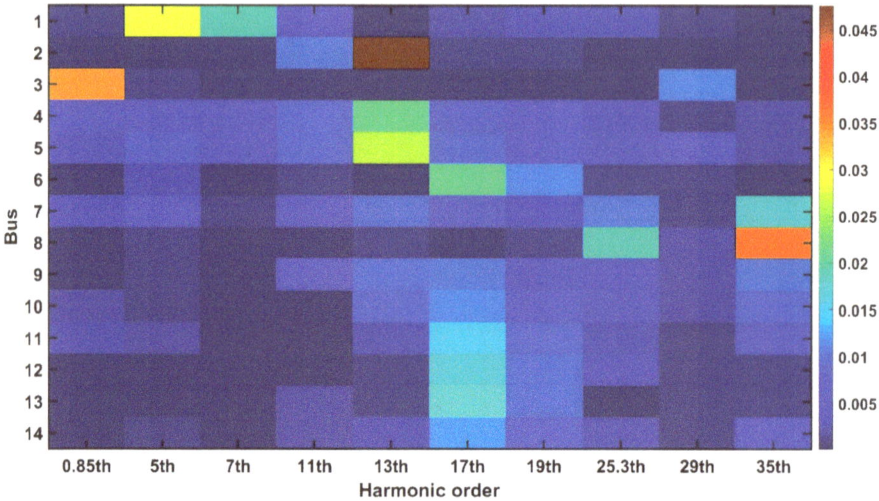

Figure 7.6 DMD participation factors dominant harmonics for the "A's phases" of the signals for Case I.

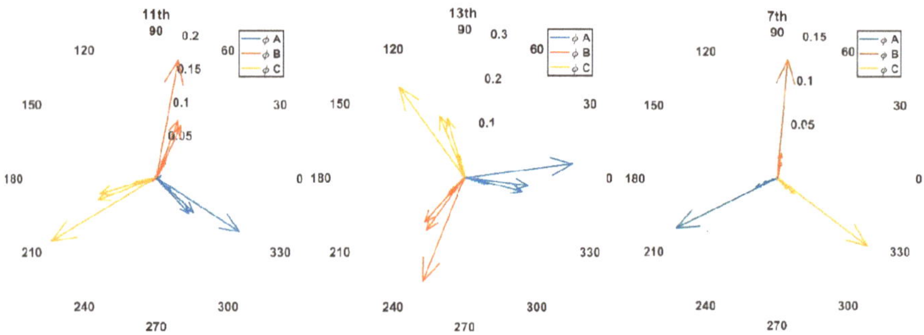

Figure 7.7 DMD mode shape dominant harmonics 7th, 11th, and 13th for the phases of the signals for Case I.

Table 7.9 Voltage harmonics obtained of DMD for Case II

0.85^{th}	3rd	5th	7th	9th	11th	13th	15th	19th	25.3^{rd}
0.030	0.036	0.037	0.023	0.052	0.023	0.043	0.035	0.011	0.023

agates on B4, B5, and B7, and the 13th harmonic (780 Hz) propagates on buses B4, B5, and B9-B14, in different states of intensity on each bus.

On the other hand, Fig. 7.10 indicates the 3^{hrd}, 9th, and 15th triplen harmonics mode-shape. For the 3^{hrd}, the highest contributions come from the G6; additionally, 9th and 15th harmonics are observed in the bus G3 and G8, respectively. Therefore the location of the injected harmonic sources and their contribution to the system is confirmed.

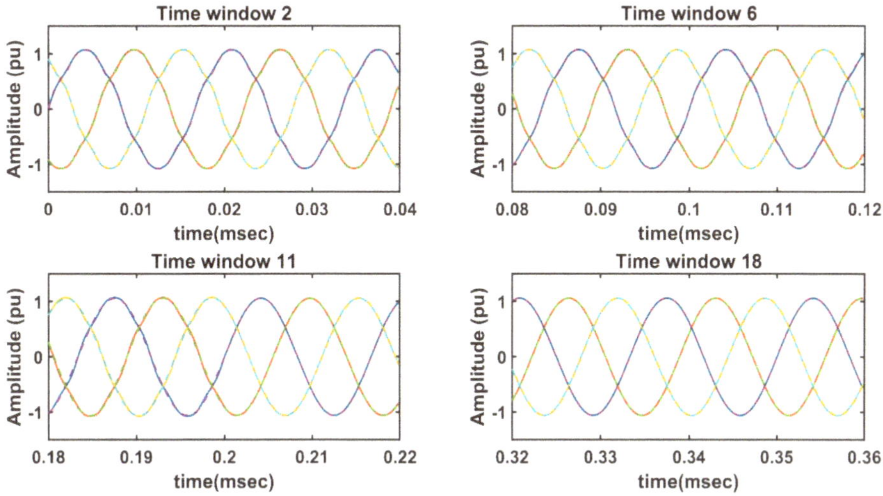

Figure 7.8 Voltage reconstruction for Case II corresponding to the windows 2, 6, 11, and 18 of G1; dotted lines represent the signal obtained with DMD.

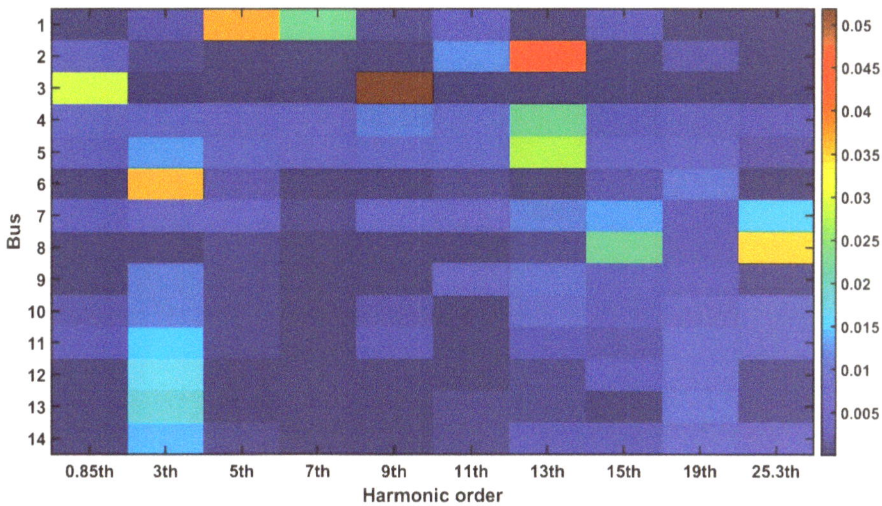

Figure 7.9 DMD participation factors dominant harmonics for the "A's phases" of the signals for Case II.

7.4.2 Application of Padé method

7.4.2.1 Reconstruction of signals

The signals under study were processed in simultaneous form for several windows; of particular interest are those where some generators begin to inject harmonics to the network. As the first step, it was verified that the Padé approach was able to reconstruct the processed signals. Fig. 7.11 shows the reconstruction of selected signals of

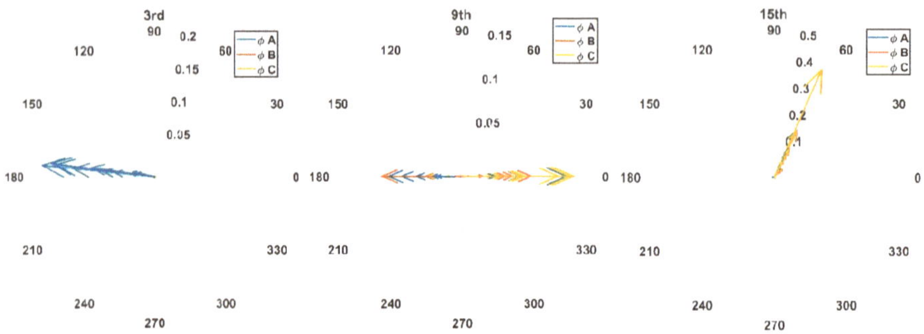

Figure 7.10 Mode shapes DMD dominant harmonics for the phases of the signals for Case II.

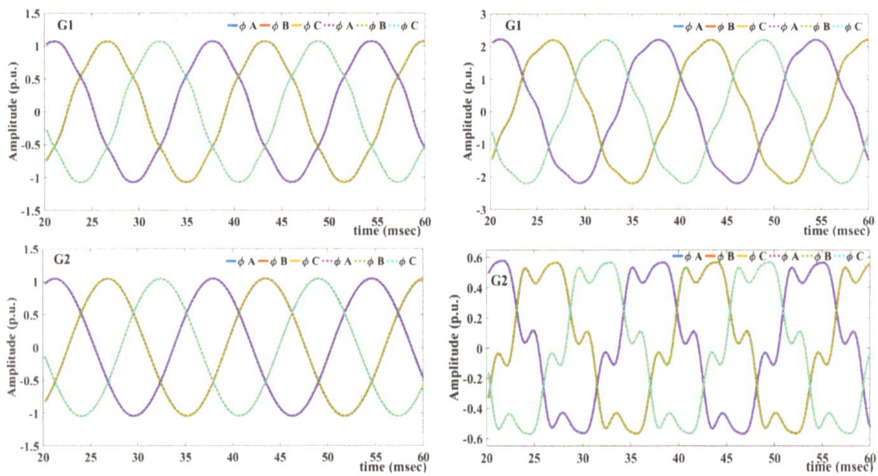

Figure 7.11 Reconstructed signals (*dotted line*) for voltage and current signals for Case I

Case I and the window 2 (200–240 msec) (dotted line). In the same way, Fig. 7.12 depicts the reconstruction of signals of for Case II and window 1. As can be seen, it can be concluded that the Padé method recover the signals in properly, ensuring with this that the information obtained of frequencies, amplitudes, and mode shape is accurate.

7.4.2.2 Spatiotemporal visualization of harmonic pollution

The harmonic pollution process can be understood by projecting the THD onto the one-line diagram. With this purpose, first were computed the participation factors A_m^i, and with this, the THD was obtained by standard expressions. Thus, for example in Fig. 7.13, is displayed how is the THD for Case I, where the THD and the windows 2, 6, 11, and 18, respectively, were observed. As can be seen, as the sliding window advances the THD and the buses that are affected change in value. For the window 1, the voltages are affected in a localized region of G1, and the harmonics are not propagated to the network. In the counterpart, the THD for the currents is present at the region

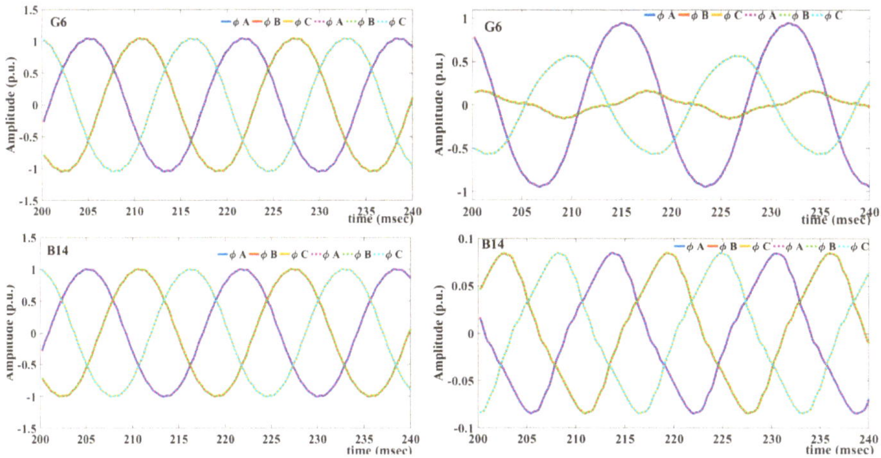

Figure 7.12 Reconstructed signals (*dotted line*) for voltage and current signals for Case II

of G2 and higher value of THD, near to 20%. As the analysis window progresses, the THD changes in value and location; this is the case of window 11, where it is observed that the voltage suffers a notable distortion in a vast area.

In respect to the THD for Case II, the affectation becomes more complex due the presence of unbalanced conditions. Fig. 7.14 shows the contour plot obtained, where similar findings are encountered for window 2 and 6 since the injected harmonics are identical. For windows 11 and 18, the affected region changes due the presence of triplens.

7.5 Conclusions

In this chapter were described and applied the DMD and Padé methods for processing signals with harmonic distortion. Both methods became accurate and efficient with the use of SVD for obtaining the dominant frequencies contained in the signals. Comparison with other techniques indicates that both methods extract accurately the amplitudes, frequencies, and phases of signals with different kinds of harmonics. The findings encountered with DMD in a typical electrical network indicates that the process of harmonic pollution has spatiotemporal characteristics and are properly identified in an efficient way. In the same way, the Padé method represents an alternative methodology in the determination of source of harmonics in interconnected network and can be process voltage and current signals simultaneously. The Padé method obtains a transfer function that can be used later for filtering determined harmonics. Projection of participation factors on the single-line diagram gives an immediate visualization of the major affected areas when a determined harmonic appears in the system. These methods are efficient for the study of other oscillatory phenomena and are very promising to be incorporated in the stage of feature extraction and data compression of the unsupervised ML techniques.

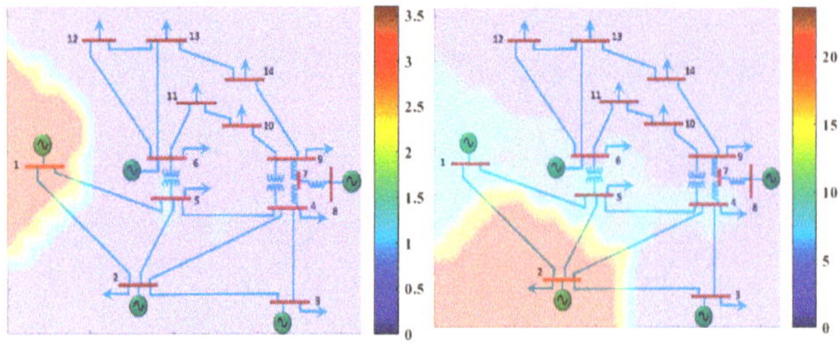

a) THD for voltages and currents at Window 2

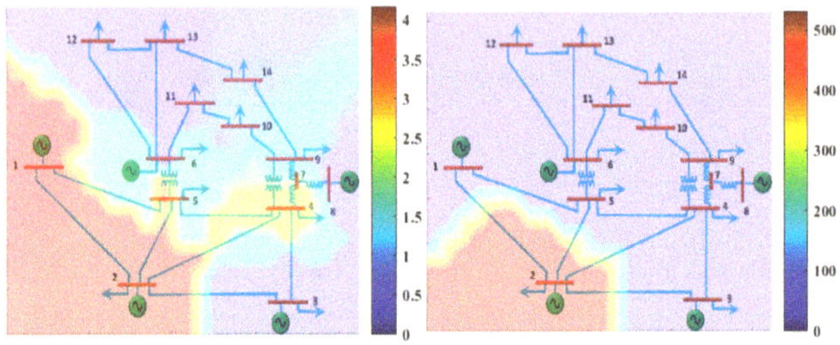

b) THD for voltages and currents at Window 6

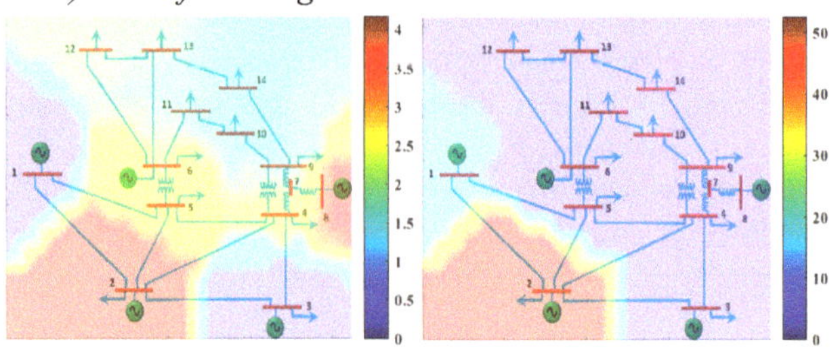

c) THD for voltages and currents at Window 11

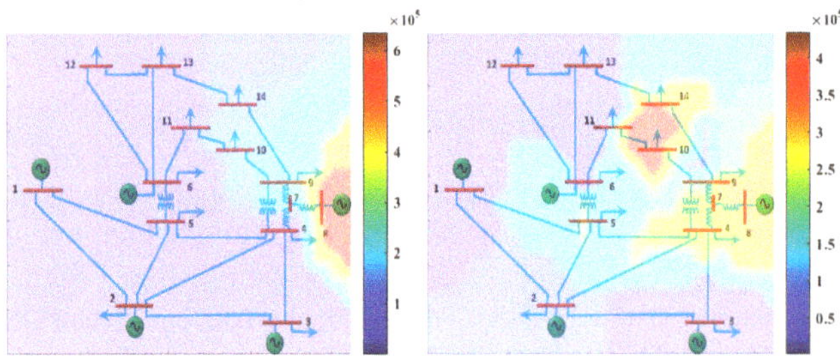

d) THD for voltages and currents at Window 18

Figure 7.13 THD contour plots for Case I

Acknowledgments

Authors with CVU numbers 88193, 824045, 30951, and 230004 express their gratitude to the Mexican National Council for Science and Technology (CONACYT) for the support of this work through the program "Sistema Nacional de Investigadores." The authors are also gratefully with CONACYT by the support provided by the program fordecyt-pronaces under grant 1311344/2020.

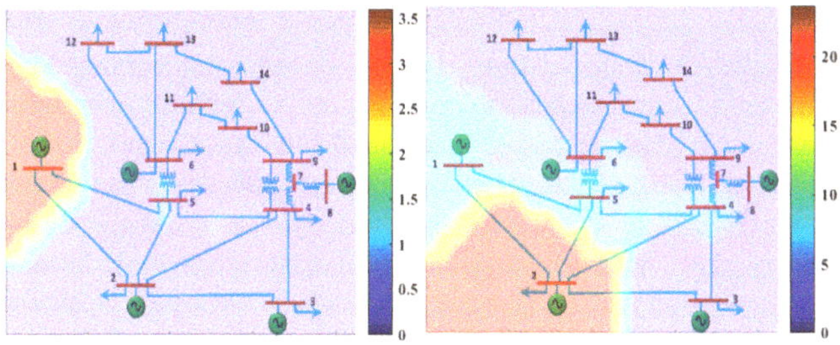

a) THD for voltages and currents at Window 2

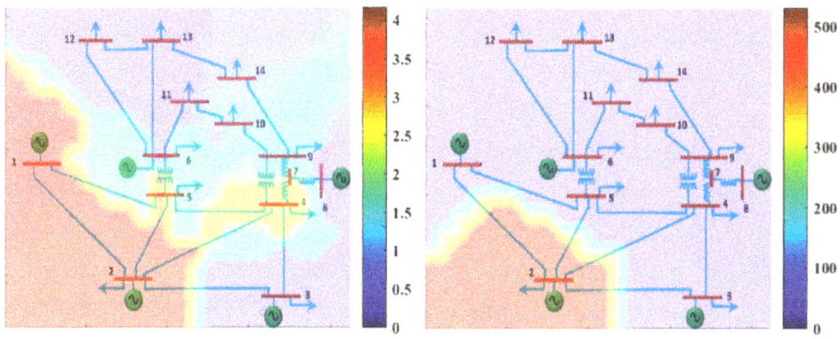

b) THD for voltages and currents at Window 6

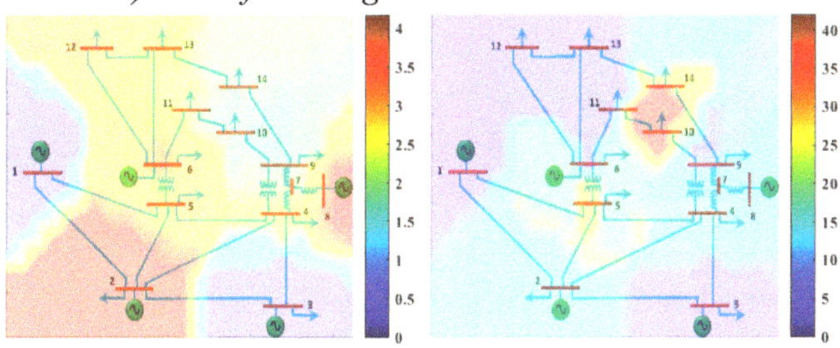

c) THD for voltages and currents at window 11

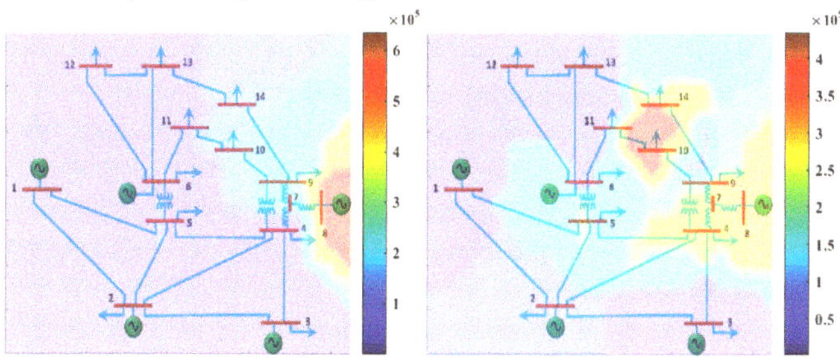

d) THD for voltages and currents at window 18

Figure 7.14 THD contour plots for Case II

References

[1] Arturo Roman Messina, Wide Area Monitoring of Interconnected Power Systems, vol. 77,

IET, 2015.
[2] Muhammad Sohail Ibrahim, Wei Dong, Qiang Yang, Machine learning driven smart electric power systems: current trends and new perspectives, Applied Energy 272 (2020).
[3] Ahmadreza Eslami, Michael Negnevitsky, Evan Franklin, Sarah Lyden, Review of AI applications in harmonic analysis in power systems, Renewable and Sustainable Energy Reviews 154 (2022).
[4] Emilio Barocio, et al., Wide-area monitoring of large power systems based on simultaneous processing of spatio-temporal data, in: Wide Area Power Systems Stability, Protection, and Security, Springer, Cham, 2021, pp. 189–228.
[5] Emilio Barocio, et al., A dynamic mode decomposition framework for global power system oscillation analysis, IEEE Transactions on Power Systems 30 (6) (2014) 2902–2912.
[6] X. Liang, Emerging power quality challenges due to integration of renewable energy sources, IEEE Transactions on Industry Applications 53 (2) (2017) 855–866.
[7] A.Q. Al-Shetwi, M.A. Hannan, K.P. Jern, M. Mansur, T.M.I. Mahlia, Grid-connected renewable energy sources: review of the recent integration requirements and control methods, Journal of Cleaner Production 253 (2020) 1–17.
[8] C. Sankaran, Power Quality, CRC Press, New York, 2017.
[9] Z. Leonowicz, J. Rezmer, T. Sikorski, J. Szymanda, P. Kostyla, Wide-area system of registration and processing of power quality data in power grid with distributed generation: part I. System description, functional tests and synchronous recordings, in: Proceedings: 14th International Conference on Environment and Electrical Engineering, Krakow, Poland, 10–12 May 2014, 2014, pp. 175–181.
[10] Z. Leonowicz, J. Rezmer, T. Sikorski, J. Szymanda, P. Kostyla, Wide-area system of registration and processing of power quality data in power grid with distributed generation: part II. Localization and tracking of the sources of disturbances, in: Proceedings: 14th International Conference on Environment and Electrical Engineering, Krakow, Poland, 10–12 May 2014, 2014, pp. 414–417.
[11] P. Schimd, Dynamic mode decomposition of numerical and experimental data, Journal of Fluid Mechanics 656 (2010) 5–28.
[12] K. Fujii, N. Takeishi, B. Kibushi, et al., Data-driven spectral analysis for coordinative structures in periodic human locomotion, Scientific Reports 9 (2019) 16755.
[13] F.G. Georg, A. Gottwald, Detecting regime transitions in time series using dynamic mode decomposition, Journal of Statistical Physics (2019) 1–23.
[14] J.N. Kutz, Dynamic mode decomposition for financial trading strategies, Quantitative Finance 16 (11) (2016) 1643–1655.
[15] S. Mizuno, F. Oneyama, M. Sugino, H. Semba, Y. Jimbo, K. Kotani, Analysis of echocardiographic video by dynamic mode decomposition, Electron and Communications in Japan 104 (2021) 65–73.
[16] Y. Susuki, A. Chakrabortty, Introduction to Koopman mode decomposition for data-based technology of power system nonlinear dynamics, IFAC Paper OnLine 51 (28) (2018) 327–332.
[17] M.A. Hernandez, A.R. Messina, Nonlinear power system analysis using Koopman mode decomposition and perturbation theory, IEEE Transactions on Power Systems 33 (5) (2018) 5124–5134.
[18] J.J. Ramos, J.N. Kutz, Dynamic mode decomposition and sparse measurements for characterization and monitoring of power system disturbances, Nonlinear Science (2019).
[19] Y. Susuki, I. Mezic, F. Raak, T. Hikihara, Applied Koopman operator theory for power systems technology, IECE Nonlinear Theory and Applications 7 (4) (2016) 430–459.

[20] M. Netto, Y. Susuki, L. Mili, Data-driven participation factors for nonlinear systems based on Koopman mode decomposition, IEEE Systems & Control Letters 3 (1) (2018) 198–203.

[21] A. Saldaña, E. Barocio, A.R. Messina, J.J. Ramos, R.J. Segundo, G.A. Tinajero, Monitoring harmonic distortion in microgrids using dynamic mode decomposition, in: IEEE Power & Energy Society General Meeting, Chicago, IL, USA, 2017.

[22] N. Mohan, K.P. Soman, S.S. Kumar, A data-driven strategy for short-term electric load forecasting using dynamic mode decomposition model, Applied Energy 232 (15) (2018) 229–244.

[23] Ramón J. Betancourt, et al., A spatio-temporal processing Padé approach for visualizing harmonic distortion propagation on electrical networks, Electric Power Systems Research 203 (2022) 107643.

[24] H. Arbabi, I. Mezić, Ergodic theory, dynamic mode decomposition and computation of spectral properties of the Koopman operator, SIAM Journal on Applied Dynamical Systems 16 (4) (2017) 2096–2126.

[25] B. Koopman, Hamiltonian systems and transformation in Hilbert space, Proceedings of the National Academy of Sciences of the United States of America 17 (5) (1931) 315–318.

[26] Wide Area Power Systems Stability, Protection, and Security, Springer, 2020, pp. 189–228.

[27] J.N. Kutz, J.L. Proctor, S.L. Brunton, Applied Koopman theory for partial differential equations and data-driven modeling of spatio-temporal systems, Hindawi Complexity 2018 (2018) 1–17.

[28] J.H. Tu, C.W. Rowley, D.M. Luchtenburg, S.L. Brunton, J.N. Kutz, On dynamic mode decomposition: theory and applications, Journal of Computational Dynamics 1 (2014) 391–421.

[29] George A. Baker, et al., Pade Approximants: Encyclopedia of Mathematics and It's Applications, Cambridge University Press, Cambridge UK, 1996.

[30] R.J. Betancourt, L. Contreras, J.M. Gonzalez, E. Barocio, C.M. Rergis, Parameter identification of low frequency oscillations by Pade method, in: Proceedings of the 2016 IEEE PES Transmission & Distribution Conference and Exposition-Latin America (PES T&D-LA), Morelia, Michoacan Mexico, 20–24 September 2016, 2016, 6 pp.

[31] G.H. Golub, C.F. Van Loan, Matrix Computations (Vol. 3), Johns Hopkins University Press, Baltimore, MD, USA, 2013.

[32] P. Barone, R. March, Some properties of the asymptotic location of poles of Pade approximants to noisy rational functions, relevant for modal analysis, IEEE Transactions on Signal Processing 46 (1998) 2448–2457.

[33] D.G. Jablonski, A pole-zero deletion method for calculating Pade approximates, Proceedings of the IEEE 74 (1) (Jan. 1986) 217–218, https://doi.org/10.1109/PROC.1986.13434.

[34] Keyhan Sheshyekani, et al., A general noise-resilient technique based on the matrix pencil method for the assessment of harmonics and interharmonics in power systems, IEEE Transactions on Power Delivery 32 (5) (2017) 2179–2188.

[35] R. Betancourt, et al., Synthetic signals and IEEE 14 bus test system for harmonic identification, https://github.com/RamonFIE/Chapter-Book-Elsevier-DMD-and-Pade.git, 2022.

[36] IEEE14 bus test system, Power Systems TCA, http://www.ee.washington.edu/research/pstca/pf14, 2022.

[37] IEEE Standard Conformance test procedures for equipment interconnecting distributed energy resources with electric power systems and associated interfaces, in: IEEE Std 1547.1-2020, 2020, pp. 1–282.

Power quality harmonic monitoring by the O-splines-based multiresolution signal decomposition

8

José Antonio de la O Serna[a], Gabriela Castillo-García[b], Mario Roberto Arrieta-Paternina[b], and Alejandro Zamora-Méndez[c]
[a]Universidad Autónoma de Nuevo León, Monterrey, Mexico, [b]National Autonomous University of Mexico, Mexico City, Mexico, [c]Michoacan University of Saint Nicholas of Hidalgo, Morelia, Mexico

8.1 Introduction

In the power quality monitoring literature, one of the major concerns is the ability to monitor the harmful harmonic content in power systems highly interfaced with power electronic converters and to the presence of nonlinear loads, since harmonics may lead electrical equipment and devices to irreparable damage conditions [1–3]. To counteract this concern, a plethora of harmonic estimation methods has been envisioned [4] due to their usability on different power quality applications, such as harmonic state estimation, harmonic source allocation, and harmonic control [5–7].

Among these, power quality engineers have adopted the use of the fast Fourier transform (FFT) to process and analyze voltage and current channels by taking its Fourier-based estimates into measured indices, such as the total harmonic distortion (THD), which is regulated by the IEEE/ANSI and IEC standards using their maximum limit values [8–11]. Despite the massive applications of the FFT in engineering, its use is bounded to sinusoidal signals with constant amplitude and phase, which is not a realistic assumption in electrical networks, since they have been changing, and their electrical variables are not sinusoidal anymore [12–14]. Thus novel resolution techniques have been advocated to reinforce the harmonic monitoring, such as wavelet transform, Kalman and Prony-based methods, artificial neural network-based approaches, or their combination [15–19]. However, these techniques require prior knowledge of the system, high sampling rate capabilities, preprocessing stages under noisy conditions, data-assisted training, or feedback errors. Despite these improvements, they exhibit the presence of spurious components, leading them to poor performance.

To address the abovementioned problems, this chapter implements the power quality harmonic monitoring by means of the O-splines-based multiresolution signal decomposition method, preventing the power quality monitoring from signal or data

Monitoring and Control of Electrical Power Systems Using Machine Learning Techniques
https://doi.org/10.1016/B978-0-32-399904-5.00014-4

dependency. The O-splines are derived from the discrete-time Taylor–Fourier transform (DTTFT), achieving a high rate of reduction of the computational burden, which makes them suitable for their implementation in microcontrollers or digital signal processors with a reduced cost [20]. They are adopted to design a desired set of finite impulse response (FIR) filters, attaining the best Taylor–Fourier (TF) coefficients that correspond to the harmonic estimates at each frequency. The nature of these filters is characterized by maximally flat pass-bands and zero flat stop-bands. Thus the proposed approach has the potential to precisely extract the amplitude at each harmonic band and capture the energy concentrated in its spectrum. These are the main distinctions that make O-splines unique, optimal, useful, and effective for use in practical applications [21].

DTTFT has already been applied to improve results with machine learning (ML) techniques in the analysis of electrocardiogram (EEC) and electroencephalogram (EEG) signals in biomedical applications to detect arrhythmia and epilepsy, respectively, in [22,23]. We expect DTTFT may also support PQ monitoring via features-assisted ML techniques. This fact motivates us to include it in the PQ monitoring and its application to ML. It is noteworthy to remark that the DTTFT also provides derivatives that has not previously been used in these DTTFT-based ML applications. It is well known that this additional information is crucial for detection and prediction of transients, since the first derivative measures the rate of change (speed), and the second one indicates its trend (acceleration).

8.2 Taylor–Fourier transform and O-splines for harmonic estimation

This section is devoted to describe the DTTFT fundamentals for power quality monitoring. Herewith, an O-splines-based fast Taylor–Fourier transform (FTFT) implementation is proposed as a multiresolution technique to precisely achieve odd and even harmonic estimations with a very low computational complexity. At the end of this section, nonlinear device are theoretically modeled to demonstrate the DTTFT performance, even under the presence of dynamic compliance in signal phase.

8.2.1 Taylor–Fourier transform

Taylor–Fourier transform was introduced in matrix form in [21] and its closed form solution in [20]. O-splines help to reduce the computational complexity of the Taylor–Fourier transform, and provide a series of functions whose spectra converge to the ideal lowpass filter. Harmonic bandpass filters designed with O-splines can be obtained by modulating them at a particular harmonic frequency. These filters tend to the ideal bandpass filter (with maximally flat unit gain at their bandpass) as the order of the O-spline goes to infinity.

The key idea for obtaining the Taylor–Fourier signal model of the transform is to relax the complex constant Fourier coefficients ξ_h of the Fourier series:

$$x(t) = \sum_{h=-\infty}^{\infty} \xi_h e^{j2\pi h F_1 t}, \quad -\frac{T_1}{2} \leq t \leq \frac{T_1}{2} \tag{8.1}$$

to complex time functions of the form $\xi_h(t) = a_h(t)e^{j\varphi_h(t)}$, which carry the *complex envelope* of each hth harmonic frequency. Hence, the extended Fourier signal model becomes

$$x(t) = \sum_{h=-\infty}^{\infty} \xi_h(t) e^{j2\pi h F_1 t}, \quad -\frac{CT_1}{2} \leq t \leq \frac{CT_1}{2}, \tag{8.2}$$

where F_1 is the fundamental frequency in Hz, T_1 is its period ($T_1 = \frac{1}{F_1}$) in seconds, and C is a constant number of cycles.

In this way, harmonics become narrowband signals [24] or *dynamic harmonics*. The second step for obtaining the Taylor–Fourier transform is to approach the complex envelope $\xi_h(t)$ of each harmonic by best Taylor polynomial of K-th degree,

$$\xi_h^{(K)}(t) = \xi_h(t_0) + \dot{\xi}_h(t_0)(t - t_0) + \cdots + \xi_h^{(K)}(t_0)\frac{(t-t_0)^K}{K!}, \tag{8.3}$$

obtained by the least mean squares (LMS) algorithm in the observation interval, centered at the time instance t_0.

Notice that the Taylor–Fourier signal model in (8.3) and (8.2) improves the Fourier one of the form $c_h e^{jh\omega_1 t}, h \in \mathcal{Z}$ in (8.1). It corresponds to the response of a linear system with repeated singularities of multiplicity $K+1$ at $s_h = -j2\pi h F_1, h \in \mathcal{Z}$ in the Laplace plane. The Taylor–Fourier signal model is even more flexible than the Laplace (or Prony) model of the form $c_h e^{-\sigma_h t} e^{j\omega_h t}$, since the attenuated exponential $e^{-\sigma_h t}$ is replaced by a more flexible polynomial than the exponential one determined by the constant σ_h. In consequence, the space provided by the Taylor–Fourier model is more complete than those provided by the compared signal models.

8.2.2 Discrete time Taylor-Fourier transform

The matrix formulation of the signal model of the discrete time Taylor–Fourier transoform (DTTFT) is obtained simply by sampling (8.2) and (8.3) with N samples per fundamental period [20] as follows:

$$x = \Phi\xi$$

$$= \left(I \begin{pmatrix} W_N \\ W_N \\ \vdots \\ W_N \end{pmatrix} \quad T \begin{pmatrix} W_N \\ W_N \\ \vdots \\ W_N \end{pmatrix} \quad \cdots \frac{1}{K!}T^K \begin{pmatrix} W_N \\ W_N \\ \vdots \\ W_N \end{pmatrix} \right) \begin{pmatrix} \xi_N \\ \dot{\xi}_N \\ \vdots \\ \overset{(K)}{\xi}_N \end{pmatrix}, \tag{8.4}$$

where the $(K+1)N \times (K+1)N$ matrix Φ contains the basis vectors, and ξ their corresponding coefficients, which can be arranged in $N \times 1$ subvectors $\xi_N^{(k)}$, $k = 0, 1, \ldots K$, each one containing the kth derivative of the harmonic dynamic phasors in $\xi_N(h)$, $h = 0, \ldots, N-1$. The submatrix W_N is a $N \times N$ Fourier matrix of the discrete Fourier transform (DFT) with harmonic columns $w_h = e^{j\frac{2\pi}{N}hn}$; $h, n = 0, \ldots, N-1$. And finally, $(K+1)N \times (K+1)N$ diagonal matrix T contains the samples of the first Taylor term in the centered time interval of $C = K+1$ fundamental cycles. See [21] for more details of this formulation.

The parameters of the best polynomial approximation to a given signal x are given by the analysis equation:

$$\widehat{\xi} = \widetilde{\Phi}^H x, \tag{8.5}$$

where $\widehat{\xi}$ contains the best parameters of the approximation, and $\widetilde{\Phi}$ is the dual matrix of Φ, given by

$$\widetilde{\Phi} = \Phi(\Phi^H \Phi)^{-1} \tag{8.6}$$

and such that $\widetilde{\Phi}^H \Phi = I$. Notice that the Hermitian of the dual matrix $\widetilde{\Phi}^H$ is simply the inverse of Φ. The modulated O-splines, contained in the first matrix, are biorthogonal to the modulated Taylor terms, contained in the second one.

8.2.3 Taylor and Fourier operators

O-splines are obtained through the rearrangement of (8.4), in which $CN \times CN$ diagonal matrix Y is separated into C $N \times N$ diagonal submatrices Y_c, $c = 1, 2, \ldots, C$ arranged in a column, each one of them containing samples of a one-cycle segment of the first Taylor term. In terms of this cyclic partition, we obtained the following factorization of the Φ_K matrix:

$$\Phi_K = \Upsilon_K \Omega_K$$
$$= \begin{pmatrix} I & Y_1 & \ldots & \frac{1}{K!}Y_1^K \\ I & Y_2 & \ldots & \frac{1}{K!}Y_2^K \\ \vdots & \vdots & \ddots & \vdots \\ I & Y_C & \ldots & \frac{1}{K!}Y_C^K \end{pmatrix} \begin{pmatrix} W_N & 0 & \ldots & 0 \\ 0 & W_N & \ldots & 0 \\ \vdots & \vdots & \ddots & \vdots \\ 0 & 0 & \ldots & W_N \end{pmatrix} \tag{8.7}$$

in which the matrix Υ_K contains in each column the diagonal submatrices with successive cyclic pieces of a different Taylor term included in the polynomial, and the matrix Ω_K contains C Fourier matrices arranged diagonally. In this way, all the cyclic pieces of a Taylor term are modulated at the full set of harmonics included in the vectors of the Fourier matrix.

This factorization of Φ_K into two operators, one for Taylor and the other for Fourier is not only crucial for obtaining the O-splines in closed form, but also reduces the computational complexity of the digital solution, since the elements of the matrix Υ_K are real and the modulations can be calculated with the fast Fourier transform.

In addition, the dual matrix can also be simplified through this factorization. The dual matrix of Φ is

$$\widetilde{\Phi} = \Upsilon(\Upsilon^H \Upsilon)^{-1} \frac{\Omega}{N} = \widetilde{\Upsilon} \frac{\Omega}{N}, \tag{8.8}$$

which is again a modulation of vectors in $\widetilde{\Upsilon}$ at set of harmonic frequencies included in W_N. Notice that again $\widetilde{\Upsilon}^T \Upsilon = I$, and in consequence

$$\widetilde{\Upsilon} = \Upsilon^{-T}. \tag{8.9}$$

Finally, we have

$$\widetilde{\Upsilon} = \frac{Adj(\Upsilon)^T}{|\Upsilon|} = \frac{Cof(\Upsilon)}{|\Upsilon|}. \tag{8.10}$$

Matrix $\widetilde{\Upsilon}$ inherits the structure of subdiagonal matrices of Υ in (8.7). This means that diagonal submatrices in the first column of $\widetilde{\Upsilon}$ contain the polynomial pieces of the O-spline, and those of its successive columns contain the pieces of its successive derivatives. One interesting reduction in the computational burden is obtained when the analysis equation is applied with the former Taylor and Fourier operators. We have

$$\widehat{\xi} = \widetilde{\Phi}^H x = \frac{\Omega^H}{N} \widetilde{\Upsilon}^H x. \tag{8.11}$$

The vector of the product $\widetilde{\Upsilon}^H x$ contains $K + 1$ segments of N data. Each one of them contains the addition of Hadamard products of the cyclic pieces of the O-splines with the corresponding pieces in signal x, in the first one with the pieces of the O-spline, in the second one the pieces of its first derivative, and so on. The product of this vector with the Omega matrix $\frac{\Omega^H}{N}$ is nothing else than the discrete Fourier transform (DFT), divided by N, of those cyclic pieces, containing the sum of cyclic Hadamard products of the signal with the pieces of the O-spline and its derivatives.

Each DFT renders the Fourier coefficients of those cyclic sums. The first one provides the Fourier coefficients, or the harmonic dynamic phasors; the second one provides the first derivative of the harmonic phasors, and so on, up to their Kth derivative. Usually, only up to the second derivative are needed. As can be seen, the reduction in computational burden of this factorization is remarkable overall if the number of samples per cycle N is an integer power of 2, in which case the algorithm of the fast Fourier transform (FFT) can be applied. Notice also that this reduction exploits the fundamental periodicity of the Euler complex sequences in the Fourier matrix W_N.

8.2.4 O-splines

Now we are ready to find the O-splines in closed form. The key idea for obtaining the inverse matrix of the Taylor operator is to consider that submatrices in the

Taylor matrix in (8.7) operate each other as a Hadamard multiplication between the corresponding segments of the Taylor terms in the considered pair of multiplied submatrices.

For the 0th-order O-spline, we have

$$\Upsilon_0 = I, \tag{8.12}$$

and then $\widetilde{\Upsilon}_0 = I$. The Taylor–Fourier transform of zeroth order corresponds to the DFT; this means that the zeroth-order O-spline is a rectangular pulse of one cycle centered at zero.

For the 1st-order O-spline, we have

$$\Upsilon_1 = \begin{pmatrix} I & Y_1 \\ I & Y_2 \end{pmatrix}. \tag{8.13}$$

Now the subdiagonal matrices in the operator are changed by scalars to replace their Hadamard products by scalar products and dramatically reduce the dimension of the matrix Υ_1. We define the segments $t_1 = t_{[-T_1,0)}$, and $t_2 = t_{[0,T_1)} = t_1 + T_1$, where T_1 is the fundamental period, then

$$\Upsilon_1 = \begin{pmatrix} 1 & t_1 \\ 1 & t_2 \end{pmatrix}, \tag{8.14}$$

with $|\Upsilon_1| = t_2 - t_1 = T_1$. Now we simply have

$$\widetilde{\Upsilon}_1 = \frac{\begin{pmatrix} t_2 & -1 \\ -t_1 & 1 \end{pmatrix}}{T_1} = \begin{pmatrix} u_1 + 1 & -F_1 \\ -(u_2 - 1) & +F_1 \end{pmatrix}, \tag{8.15}$$

where u_n is the normalized time: $u = t_n / T_1$. Its fist vector is a triangular pulse:

$$\widetilde{\upsilon}_1(u) = \begin{cases} u + 1 & \text{for} \quad -1 \le u < 0, \\ 1 - u & \text{for} \quad 0 \le u < 1, \\ 0 & \text{otherwise}, \end{cases} \tag{8.16}$$

and the second one, the scaled Haar wavelet: $-F_1 \dot{\widetilde{\upsilon}}_1(u)$.

The proof provides interesting properties of the O-splines.

$$\widetilde{\Upsilon}_1^T \Upsilon_1 = \begin{pmatrix} \int_{-1}^{1} \widetilde{\upsilon}_1(u) du = 1 & \int_{-1}^{1} u \widetilde{\upsilon}_1(u) du = 0 \\ \int_{-1}^{1} \dot{\widetilde{\upsilon}}_1(u) du = 0 & -F_1 \int_{-1}^{1} u \dot{\widetilde{\upsilon}}_1(u) du = 1 \end{pmatrix} = I \tag{8.17}$$

For $K = 2$, we have the matrix of scalar functions

$$\widetilde{\Upsilon}_2 = \begin{pmatrix} 1 & t_1 & t_1^2/2 \\ 1 & t_2 & t_2^2/2 \\ 1 & t_3 & t_3^2/2 \end{pmatrix}, \tag{8.18}$$

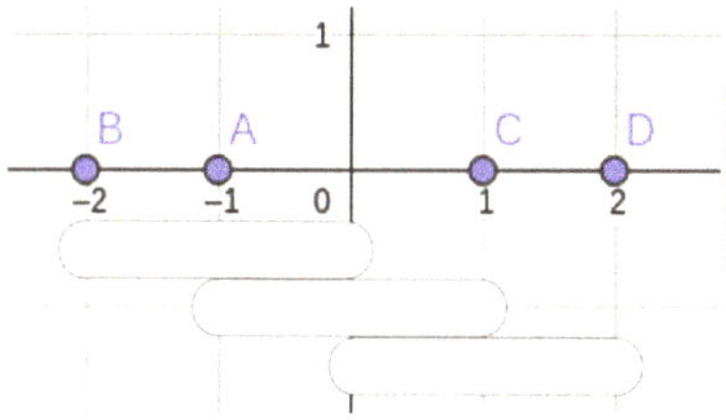

Figure 8.1 Roots of three polynomials in first column of $\widetilde{\Upsilon}_2$.

with $t_1 = t_{[-\frac{3T_1}{2}, -\frac{T_1}{2})}$, $t_2 = t_1 + T_1$, and $t_3 = t_1 + 2T_1$. In this case $|\widetilde{\Upsilon}_2| = T_1^3$, and

$$\widetilde{\Upsilon}_2 = \begin{pmatrix} \frac{1}{2}(u_1+2)(u_1+1) & -F_1(u_1+\frac{3}{2}) & F_1^2 \\ -(u_2+1)(u_2-1) & 2F_1u_2 & -2F_1^2 \\ \frac{1}{2}(u_3-1)(u_3-2) & -F_1(u_3-\frac{3}{2}) & F_1^2 \end{pmatrix}. \tag{8.19}$$

The rule of formation for higher-order O-splines O-spline can already be found in this small order. Fig. 8.1 shows the roots involved in the second-order O-spline. They are the integers k in $[-K, K]_{k\neq 0}$. The first polynomial piece has the roots in a window of size $K+1$ placed at the left ($[-2,-1]$), and it is normalized so that the polynomial is one at the origin. The second piece has the roots in the window shifted to the right by one. Finally, the roots of the last piece are the roots in the window, shifted again to right by one. Then the following vectors in $\widetilde{\Upsilon}_2$ are simply the consecutive derivatives of the first vector (the O-spline) multiplied by the factor $(-F_1)^n$ for the nth derivative.

As a proof of the rule for the O-spline formation, we consider the order $K = 3$. We have $t_1 = t_{[-2T_1, -T_1)}$, and $t_n = t_1 + (n-1)T_1$ $n = 2, 3, 4$. We have $|\widetilde{\Upsilon}_3| = T_1^6$. Its first dual column is

$$\tilde{\upsilon}_3(u) = \begin{cases} \frac{1}{6}(u+3)(u+2)(u+1) & \text{for} -2 \le u < -1, \\ -\frac{1}{2}(u+2)(u+1)(u-1) & \text{for} -1 \le u < 0, \\ \frac{1}{2}(u+1)(u-1)(u-2) & \text{for} \quad 0 \le u < 1, \\ -\frac{1}{6}(u-1)(u-2)(u-3) & \text{for} \quad 1 \le u < 2, \\ 0 & \text{otherwise} \end{cases} \tag{8.20}$$

and the following ones are $-F_1\dot{\tilde{\upsilon}}_3(u)$, $F_1^2\ddot{\tilde{\upsilon}}_3(u)$, and $-F_1^3\dddot{\tilde{\upsilon}}_3(u)$, with the following properties:

$$\begin{array}{llll} \int_{-2}^{2} \tilde{\upsilon}_3(u)du = 1, & \int_{-2}^{2} u\tilde{\upsilon}_3(u)du = 0, & \int_{-2}^{2} \frac{u^2}{2}\tilde{\upsilon}_3(u)du = 0, & \int_{-2}^{2} \frac{u^3}{3!}\tilde{\upsilon}_3(u)du = 0; \\ \int_{-2}^{2} \dot{\tilde{\upsilon}}_3(u)du = 0, & -F_1\int_{-2}^{2} u\dot{\tilde{\upsilon}}_3(u)du = 1, & \int_{-2}^{2} \frac{u^2}{2}\dot{\tilde{\upsilon}}_3(u)du = 0, & \int_{-2}^{2} \frac{u^3}{3!}\dot{\tilde{\upsilon}}_3(u)du = 0; \\ \int_{-2}^{2} \ddot{\tilde{\upsilon}}_3(u)du = 0, & \int_{-2}^{2} u\ddot{\tilde{\upsilon}}_3(u)du = 0, & F_1^2\int_{-2}^{2} \frac{u^2}{2}\ddot{\tilde{\upsilon}}_3(u)du = 1, & \int_{-2}^{2} \frac{u^3}{3!}\ddot{\tilde{\upsilon}}_3(u)du = 0; \\ \int_{-2}^{2} \dddot{\tilde{\upsilon}}_3(u)du = 0, & \int_{-2}^{2} u\dddot{\tilde{\upsilon}}_3(u)du = 0, & \int_{-2}^{2} \frac{u^2}{2}\dddot{\tilde{\upsilon}}_3(u)du = 0, & -F_1^3\int_{-2}^{2} \frac{u^3}{3!}\dddot{\tilde{\upsilon}}_3(u)du = 1. \end{array} \tag{8.21}$$

The former set of properties means that O-splines and its derivatives are biorthogonal to Taylor terms included in the signal model and perform as *maximally flat differentiators*. In this case, this means that if the input to the O-spline filter bank is a 3rd-order polynomial, each filter extracts exactly the coefficient of each Taylor term in the polynomial, and is orthogonal to the rejected ones. That is why filters perform the inverse operation of the composition. The first filter (the O-spline filter) estimates the zeroth coefficient of the polynomial, while rejecting the rest; the second one (the first derivative) estimates the first coefficient, while rejecting the other ones, and so on. Each filter admits the term in the unit dot product in the corresponding row of the matrix, while rejecting the terms in the *vanishing moments* outside the diagonal. Of course, if the input signal is of higher order, then errors will appear in the estimated coefficients, but despite this, those coefficients provide the best polynomial approximation (in the LMS sense) to the input signal.

Finally, as a last example, we present the polynomial pieces of the 5th O-spline to corroborate the rule:

$$\widetilde{\varphi}_1^{(5)}(u) = \begin{cases} \frac{1}{5!}(u_1+5)(u_1+4)(u_1+3)(u_1+2)(u_1+1), \\ -\frac{1}{4!}(u_2+4)(u_2+3)(u_2+2)(u_2+1)(u_2-1), \\ \frac{1}{3!*2!}(u_3+3)(u_3+2)(u_3+1)(u_3-1)(u_3-2), \\ -\frac{1}{3!2!}(u_4+2)(u_4+1)(u_4-1)(u_4-2)(u_4-3) \\ \frac{1}{4!}(u_5+1)(u_5-1)(u_5-2)(u_5-3)(u_5-4) \\ -\frac{1}{5!}(u_6-1)(u_6-2)(u_6-3)(u_6-4)(u_6-5) \\ 0, \text{otherwise.} \end{cases} \tag{8.22}$$

At this degree, O-splines are recognized as the *Lagrange central interpolation kernels* used in interpolation. Notice that in our application, O-splines are used as filters, the dual operation of the interpolation one.

Fig. 8.2 shows in its first row the first O-splines up to 3rd order, and below each one its successive derivatives [21].

The integer roots of the polynomials guarantee zero crossings at the nonzero integers inside the domain of the O-splines. This explains why the O-splines of odd order are continuous, whereas those of even order are discontinuous, since their polynomial intervals are cut at integers (i.e., at the roots) in the first case, and at half integers in the second case.

8.2.5 Derivatives as linear combination of O-splines of lower order

We have the following relationships between O-splines and its derivatives. Derivative of the first-order O-spline:

$$\dot{\widetilde{\upsilon}}_1(u) = F_1(\widetilde{\upsilon}_0(u+\frac{1}{2}) - \widetilde{\upsilon}_0(u-\frac{1}{2})). \tag{8.23}$$

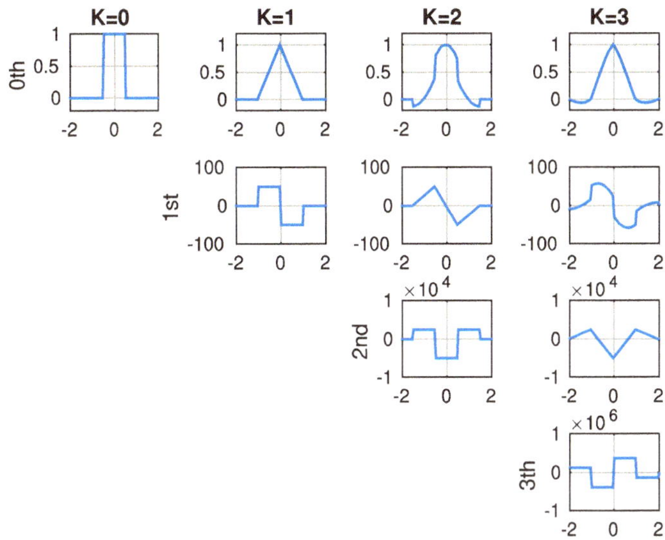

Figure 8.2 O-splines and derivatives up to $K = 3$. At the top, the O-splines of order K, for $K = 0, 1, 2, 3$. Below each one, its successive derivatives.

Derivatives of the second O-spline:

$$\dot{\tilde{\upsilon}}_2(u) = F_1(\tilde{\upsilon}_1(u + \frac{1}{2}) - \tilde{\upsilon}_1(u - \frac{1}{2})), \tag{8.24}$$

and

$$\ddot{\tilde{\upsilon}}_2(u) = F_1^2(\tilde{\upsilon}_0(u + 1) - 2\tilde{\upsilon}_0(u) + \tilde{\upsilon}_0(u - 1)). \tag{8.25}$$

Derivatives of the third O-splines

$$\dot{\tilde{\upsilon}}_3(u) = F_1(\tilde{\upsilon}_2(u + \frac{1}{2}) - \tilde{\upsilon}_2(u - \frac{1}{2})), \tag{8.26}$$

$$\ddot{\tilde{\upsilon}}_3(u) = F_1^2(\tilde{\upsilon}_1(u + 1) - 2\tilde{\upsilon}_1(u) + \tilde{\upsilon}_1(u - 1)), \tag{8.27}$$

and

$$\dddot{\tilde{\upsilon}}_3(u) = F_1^3(\tilde{\upsilon}_0(u + \frac{3}{2}) - 3\tilde{\upsilon}_0(u + \frac{1}{2}) + 3\tilde{\upsilon}_0(u - \frac{1}{2}) - \tilde{\upsilon}_0(u - \frac{3}{2})). \tag{8.28}$$

O-splines are stable functions or multiresolution analysis (MRA) *generators*. If the O-spline $\varphi(u) \in V_0$, the set of integer translates $\{\varphi(u - n)\}_{n \in \mathcal{Z}}$ is an *unconditional basis* (or Riesz basis) of V_0, and forms a multiresolution analysis on $\mathcal{R}$.

The algorithm to obtain O-splines of order K is provided in Appendix 8.A. Fig. 8.3 illustrates the 101st O-spline together with the *Sinc* function, in its top plot, together with its spectrum at the bottom. As can be seen in both plots, it is already a close approximation to the ideal filter.

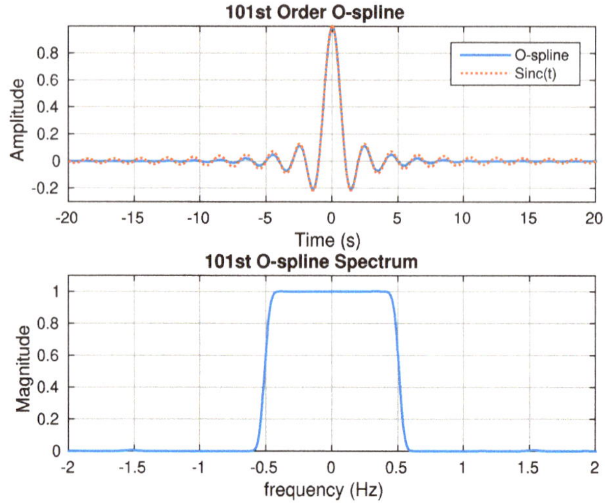

Figure 8.3 O-spline $K = 101$.

8.2.6 Obtention of harmonic estimates

From harmonic phasor estimates, we can obtain the corresponding estimates of amplitude and phase angle and its derivatives. We have $\overset{(k)}{\xi}_h \rightarrow \overset{(k)}{a}_h, \overset{(k)}{\varphi}_h$.

For $h = 0$:

$$\begin{aligned} &a_0(t_0) = \xi_0(t_0), \quad \varphi_0(t_0) = 0, \\ &\dot{a}_0(t_0) = \dot{\xi}_0(t_0), \quad \dot{\varphi}_0(t_0) = 0, \\ &\ddot{a}_0(t_0) = \ddot{\xi}_0(t_0) \quad \ddot{\varphi}_0(t_0) = 0, \end{aligned} \tag{8.29}$$

and for $h \neq 0$ the following set of equations:

$$\begin{aligned} &a_h(t_0) = 2|\xi_h(t_0)|, \\ &\varphi_h(t_0) = \measuredangle \xi_h(t_0), \\ &\dot{a}_h(t_0) = 2\mathrm{Re}\{\dot{\xi}_h(t_0)e^{-j\varphi_h(t_0)}\}, \\ &\dot{\varphi}_h(t_0) = \frac{2}{a_h(t_0)}\mathrm{Im}\{\dot{\xi}_h(t_0)e^{-j\varphi_h(t_0)}\}, \\ &\ddot{a}_h(t_0) = 2\mathrm{Re}\{\ddot{\xi}_h(t_0)e^{-j\varphi_h(t_0)}\} + a_h(t_0)\dot{\varphi}_h(t_0)^2, \\ &\ddot{\varphi}_0(t_0) = \frac{2}{a_h(t_0)}(\mathrm{Im}\{\ddot{\xi}_h(t_0)e^{-j\varphi_h(t_0)}\} - \dot{a}_h(t_0)\dot{\varphi}_h(t_0)). \end{aligned} \tag{8.30}$$

Notice that amplitude derivatives come from the real part, and phase angle and derivatives come from the imaginary part of the harmonic phasor and derivative estimates. Appendix 8.B provides the algorithm to obtain that mapping. Take care of possible divisions by zero in the first two derivatives of the phase angle.

8.2.7 Total phasor error (TPE)

From the synthesis equation, we have the approximated signal

$$\widehat{x} = \Phi\widehat{\xi} = \Phi\widetilde{\Phi}^H x. \tag{8.31}$$

The Pythagorean theorem holds for the approximation error $e = x - \widehat{x}$, this is,

$$||e||^2 = ||x||^2 - \widehat{\xi}^H \Phi^H \Phi\widehat{\xi} \tag{8.32}$$

and normalizing with respect to the signal energy, we have

$$||\epsilon||^2 = 1 - \frac{\widehat{\xi}^H \Phi^H \Phi\widehat{\xi}}{||x||^2}, \tag{8.33}$$

and its normalized norm

$$||\epsilon|| = \sqrt{1 - \frac{\widehat{\xi}^H \Phi^H \Phi\widehat{\xi}}{||x||^2}}. \tag{8.34}$$

8.2.8 O-splines and first two derivatives

Fig. 8.4 shows the O-splines for $K = 1, \ldots, 11$ together with the cardinal sine function ($K = \infty$). As can be seen, even-order O-splines are discontinuous at half integers, where are located the extremes of their time intervals. Notice their polynomial pieces approach the ideal Sinc function pivoting at their zero-crossings placed at integers, whereas the odd-order O-splines are continuous, with their zero-crossings at integers, which are the extremes of their polynomial pieces. Then they approach the lobes of the *Sinc* function with polynomial arcs.

In turn, Fig. 8.5 shows the spectra of the corresponding O-splines in Fig. 8.4. We can see that they approach the rectangular ideal filter corresponding to the *Sinc* function with flat passband gain and flat gains at the stopbands centered at the harmonic frequencies. The spectra of the even O-splines have higher sidelobes than those of the odd ones due to their discontinuities.

Finally, 8.6 shows the spectra of odd-order O-splines in dB. As we can see, they tolerate (with −25 dB) halfharmonic frequencies, while they reject (with −325 dB) the harmonic frequencies with wider stopbands as their order increases. Odd O-spline Spectra

Fig. 8.7 illustrates the impulse response and the frequency response of the first derivatives of the O-splines. Notice that impulse responses converge to the first derivative of the *Sinc* function, whereas their frequency responses converge to that of the ideal passband differentiator, with a maximally flat linear response about the origin.

Finally, Fig. 8.8 illustrates the impulse response and the frequency response of the second derivatives of the O-splines. Impulse responses converge to the second derivative of the *Sinc* function, whereas their frequency responses converge to that of the

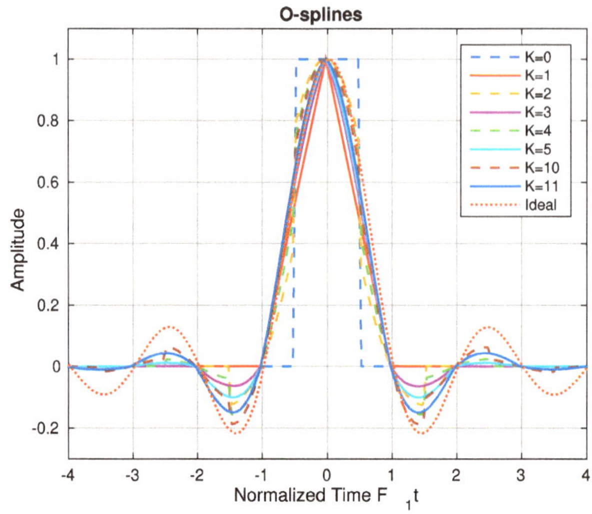

Figure 8.4 O-splines for $K = 0, 1, \ldots, 5, 10$ and 11.

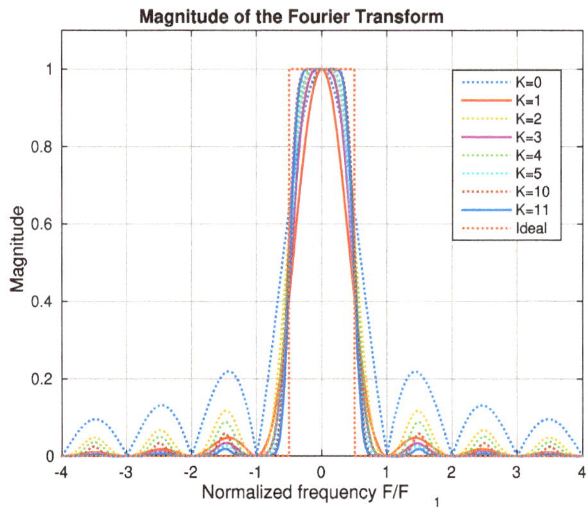

Figure 8.5 Magnitude of the O-spline Fourier transforms, $K = 0, 1, \ldots, 5, 10$ and 11.

ideal passband second differentiator, with a maximally flat parabolic response about the origin.

With the ninth O-spline filter bank one can obtain state spectrograms. Fig. 8.9 shows the phase and frequency spectrogram of the signal

$$s(t) = cos(120\pi t + \varphi(t)) \tag{8.35}$$

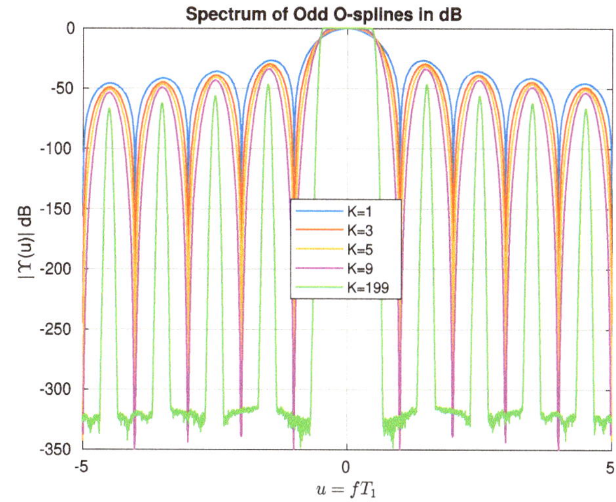

Figure 8.6 Spectra odd O-splines, $K = 0, 1, \ldots, 9$, and 199.

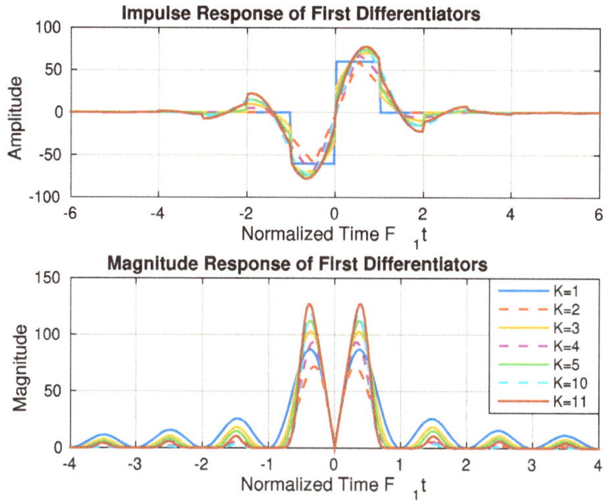

Figure 8.7 Impulse and magnitude responses of the first differentiators, $K = 0, 1, \ldots, 5, 10$, and 11.

obtained with the nonic O-spline and its first two derivatives. Notice that its phase and frequency are the following: with $\varphi(t) = e^{-4t}\cos(10\pi t)$, $\dot{\varphi}(t) = -4e^{-4t}\cos(10\pi t) - 10\pi e^{-4t}\sin(10\pi t)$, and $\ddot{\varphi}(t) = 16e^{-4t}\cos(10\pi t) + 80\pi e^{-4t}\sin(10\pi t) - 100\pi^2 e^{-4t} \times \cos(10\pi t)$. They are ploted in blue, yellow, and green lines, together with the estimates obtained by the filter bank. As can be seen, these plots are very precise time-frequency representations, except for the ROCOF at the beginning, since this filter has a large gain. With the O-splines, it is possible to obtain time-phase and time-ROCOF spectrogams.

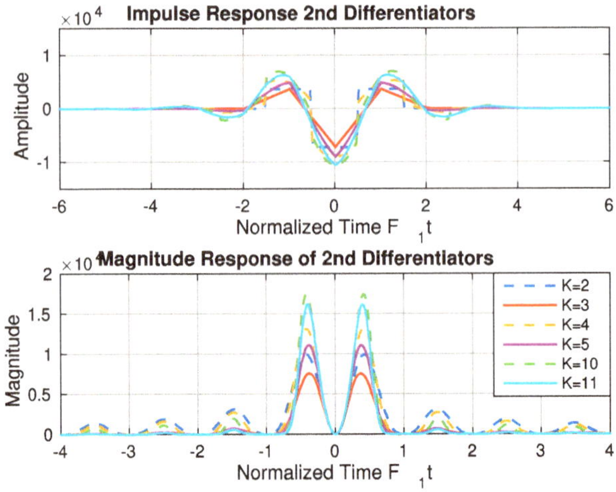

Figure 8.8 Impulse and magnitude responses of the second differentiators, $K = 0, 1, \ldots, 5$, 10 and 11.

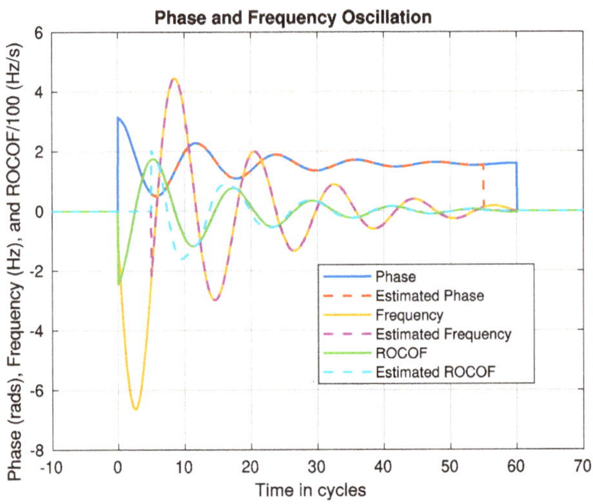

Figure 8.9 Nonic O-spline Spectrogram of $s(t)$.

8.2.9 Harmonic estimation

We consider in this subsection a periodic signal obtained by passing a sinusoidal function at the fundamental at frequency

$$x(t) = \cos(2\pi F_1 t), \tag{8.36}$$

through a nonlinear device governed by

$$s(t) = \tanh 3x(t). \tag{8.37}$$

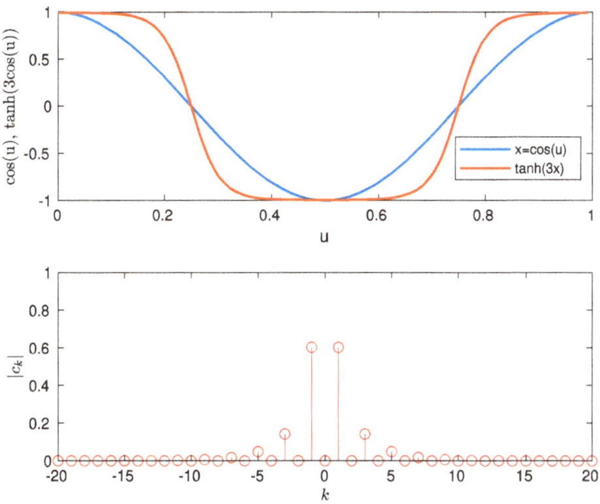

Figure 8.10 Nonic O-spline Spectrogram of $s(t)$.

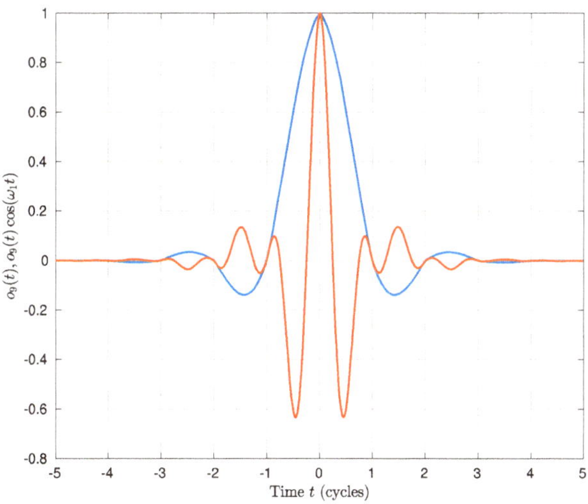

Figure 8.11 Nonic O-spline and Impulse response first harmonic filter.

The signals are sampled at $N = 128$ samples per fundamental cycle. The spectrum of the output is shown in Fig. 8.10. It can be seen that the distorted signal has a 3rd and 5th harmonic. The spectrum of the input signal has only one Fourier coefficient, which is equal to $c_1 = \frac{1}{2}$.

Now, we construct a Taylor–Fourier filter bank with nonic O-splines (of 10 cycles) for the 0th and 1st harmonics, and with one-half contracted nonic O-splines (5 cycles). Fig. 8.11 shows the nonic O-spline and the first harmonic filter. The frequency responses of the filter bank are shown in Fig. 8.12 can be seen filters that have maxi-

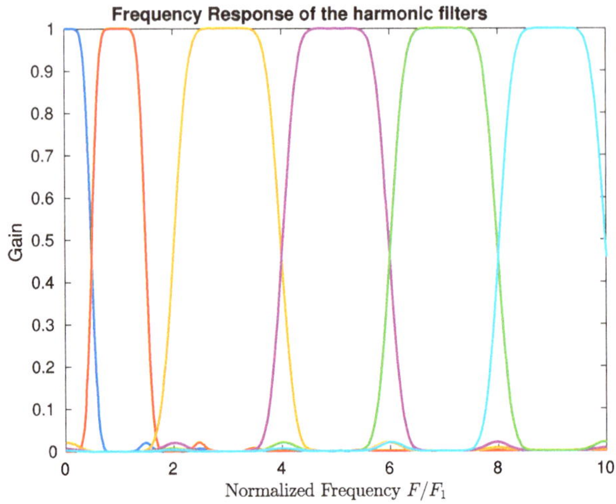

Figure 8.12 Filter bank frequency responses.

mally flat unit gain at the corresponding harmonic and maximally flat zero gain at the rest. Both properties (i.e., no harmonic distortion, nor interference) ensure accurate Fourier coefficient estimates.

The first five plots of Fig. 8.13 show the harmonic decomposition of the periodic signal, and at the bottom the synthetic signal (continuous line) compared with the original one (dotted line). We can see there is a perfect reconstruction after the first five cycles, since for this interval, the signal has a discontinuity at the origin. At the left-hand side of that plot, the transients due to the discontinuity of the input signal is apparent. But once the filters contain a steady-state signal segment, the instantaneous estimates perfectly reconstruct the signal.

Now a dynamic phase is introduced to the sinusoidal signal. We have

$$x(t) = cos(2\pi t + \varphi(t)), \tag{8.38}$$

where

$$\varphi(t) = 1 - e^{-t/10}\cos(\frac{2\pi}{5}t). \tag{8.39}$$

The phase derivatives are as follows:

$$\dot{\varphi}(t) = e^{-t/10}[\frac{1}{10}\cos(\frac{2\pi}{5}t) + \frac{2\pi}{5}\sin(\frac{2\pi}{5}t)], \tag{8.40}$$

and

$$\ddot{\varphi}(t) = \frac{e^{-t/10}}{100}[\cos(\frac{2\pi}{5}t)(16\pi^2 - 1) - 8\pi\sin(\frac{2\pi}{5}t)] \tag{8.41}$$

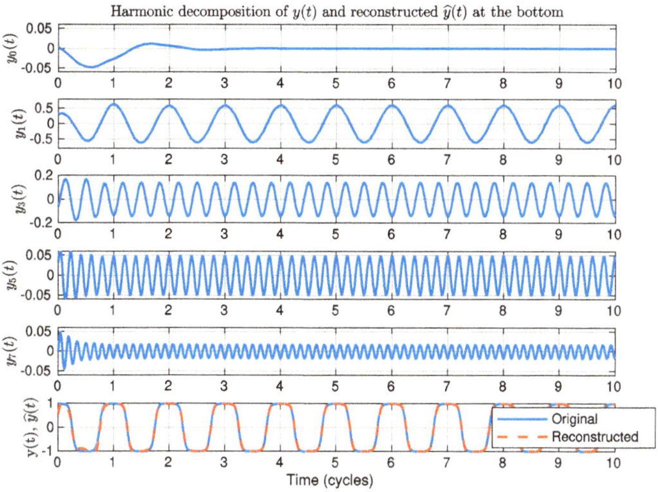

Figure 8.13 Filter bank frequency responses.

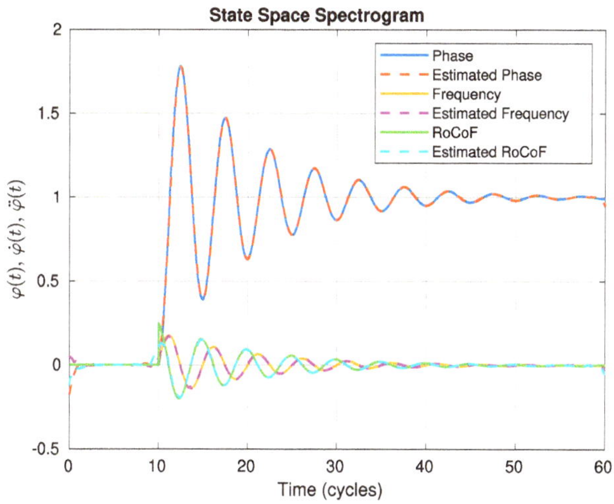

Figure 8.14 Theoretical and estimated values of phase (rads), frequency (Hz), and RoCoF (Hz/s).

in rad/s and rad/s^2, respectively. Fig. 8.14 shows the theoretical and estimated values of phase (rad), frequency (Hz), and Rocof(Hz/s). As can be seen, the estimates are very close to the theoretical values. The dynamic phase starts at the 10th cycle. The apparent error at the beginning is due to the lack of steady-state samples. This error disappears at the 5th cycle, after which the time window is full of significant samples. The accuracy of this state spectrogram is remarkable.

Despite the phase modulated signal is no longer periodic, the DTFT reconstruct the phase modulated signal with good precision. This can be appreciated in the bottom

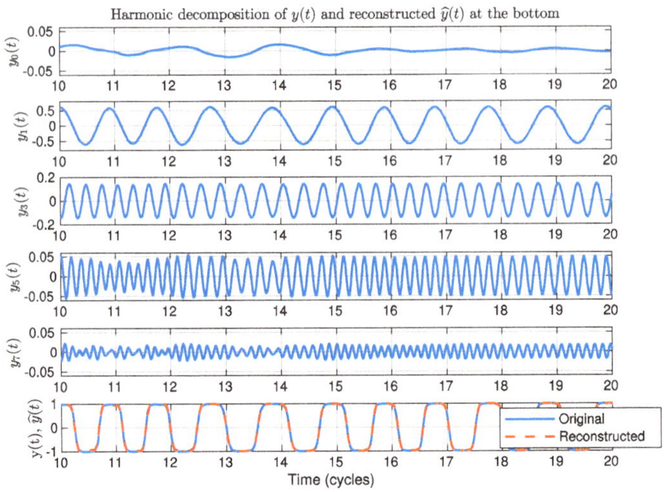

Figure 8.15 Phase modulated signal decomposition and composition.

plot of Fig. 8.15, in which it is apparent that the pulses are no longer centered at integer cycles, while the reconstructed signal is very close to the original one. This means that the projection subspace formed by the harmonic modulated Taylor terms included in the nonic O-spline contains the phase modulated signal.

To improve the demodulation, it is possible to adjust the central frequency of the O-spline bandpass filters to the instantaneous frequency of the signal. In this way, the center frequency of the filters follows the instantaneous harmonic frequencies. This adaptation is all the more important; the higher is the harmonic frequency. Fig. 8.16 shows the synthesis is improved and the apparent changes occur at the 5th and 7th harmonic estimates. The instantaneous absolute error is reduced from 8% to 2%, since the higher harmonics (higher than 7th) are not taken into account in this example.

Finally, it is noteworthy to remark the ability of the FIR filters implemented with the O-splines of the DTTFT for low-energy components and noise sensitivity. This fact has widely been assessed in [25–27]; for instance, for the identification of electromechanical modes [25], the proposal is compared with the Prony analysis, ERA, and matrix pencil method. Thus the proposed method compared with the aforementioned methods demonstrates its effectiveness in presence of noise, where it presents a low-output variance and the best capability for reconstructing signals; i.e., for signals with signal-to-noise ratio (SNR) above 60 dB; the parametric normalized mean squared errors are negligible.

8.3 Power quality assessment based on O-splines

To assess the power quality via O-splines, we choose three typical indicators: individual harmonic distortion (IHD), total harmonic distortion (THD), and voltage

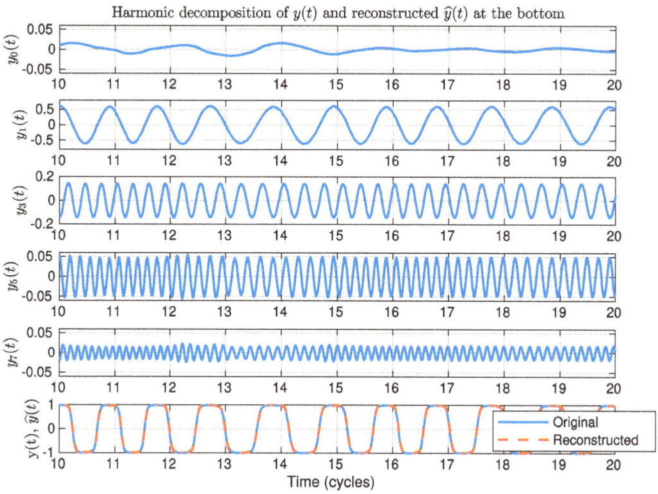

Figure 8.16 Phase modulated signal decomposition and composition with adapted center frequency.

unbalance. According to the IEEE/ANSI and the IEC norms, they are computed as follows [10,11]:

$$\%\text{IHD} = \frac{V_{n,RMS}}{V_{1,RMS}} \times 100 \; \forall n = 2, \cdots, 50, \tag{8.42}$$

$$\%\text{THD} = \frac{\sqrt{\sum_{n=2}^{n=50} V_{n,RMS}^2}}{V_{1,RMS}} \times 100, \tag{8.43}$$

$$\%V_{unb} = \frac{V_{avg}^{max} - V_{avg}}{V_{avg}} \times 100, \tag{8.44}$$

where $V_{1,RMS}$ and $V_{n,RMS}$, respectively, denote the RMS voltage for the fundamental frequency and its n-th harmonic component. Likewise, V_{avg}^{max} and V_{avg} stand for the maximum voltage deviation and average of the three-phase RMS voltages. In turn, all magnitudes are computed for the h-th harmonic, according to Section 8.2.6.

These indicators can be evaluated following the descriptive steps in Algorithm 1 together with Appendices 8.A and 8.B.

8.3.1 Experimental setup for capturing a real transient event

The experimental setup in Fig. 8.17 is carried out to trigger a transient behavior after the connection and disconnection of a capacitor bank in parallel configuration. In this setup, a synchronous generator coupled to a DC motor allows to provide a three-phase system in the *abc* reference frame, having connected in its terminals a three-phase delta load, a three-phase induction motor, and a single-phase capacitor bank, as de-

Algorithm 1 Descriptive methodology for the PQ analysis via O-splines Multiresolution Signal Decomposition.

Require: A set of voltage/current signals S and the sampling frequency.

Ensure: A set of indicators, such as %IHD, %THD and $\% V_{un}$.

1: **Step 1.** Load signals.
2: **Step 2.** Obtain the fundamental frequency F0 and the number of samples per cycle (N0).
3: **Step 3.** Compute samples of the K-th O-splines via *Osplinepp* function in **Appendix 8.A**.
4: **Step 4.** Calculate the impulse responses for the first 50th FIR filters ▷ One FIR filter is required per harmonic frequency.
5: **Step 4.** Obtain the harmonic phasor estimates via *statestimator* function in **Appendix 8.B**.
6: **Step 5.** Compute the indicators in (8.42)–(8.44).

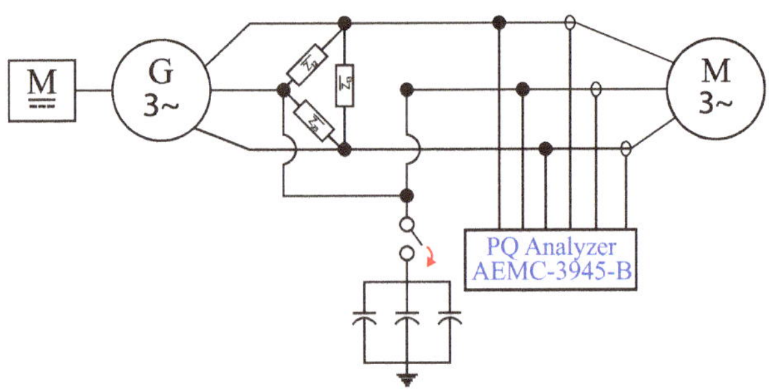

Figure 8.17 Three-phase diagram for the experimental setup.

picted in Fig. 8.17. To record the transient event, a commercial AEMC-3945-B three-phase power quality analyzer is connected to the motor terminals, measuring voltages and currents instantaneously. Fig. 8.18 displays the three-phase voltage waveforms captured by the AEMC-3945-B analyzer. These waveforms are featured by a 4-cycle duration and sampling rate of 15.36 kHz, where (colors) black, red, and blue correspond to the phase named V1, V2, and V3, respectively. These are made available via GitHub in [28].

8.3.2 Analyzing a real transient event

By following the Taylor–Fourier statements and their O-splines, we provide a comprehensible power quality analysis on the real event captured in Section 8.3.1.

First, a set of FIR filter bank with a bandwidth of 60 Hz is designed to estimate the RMS voltage for the fundamental frequency and its harmonic components. This set can be built for either 50 or 60 Hz, and their first 50 harmonics, that is, 2500 or 3000 Hz, as shown in Fig. 8.19. A sampling rate of 256 samples per cycle is adopted for analysis purposes, similar to the one used by the commercial analyzer.

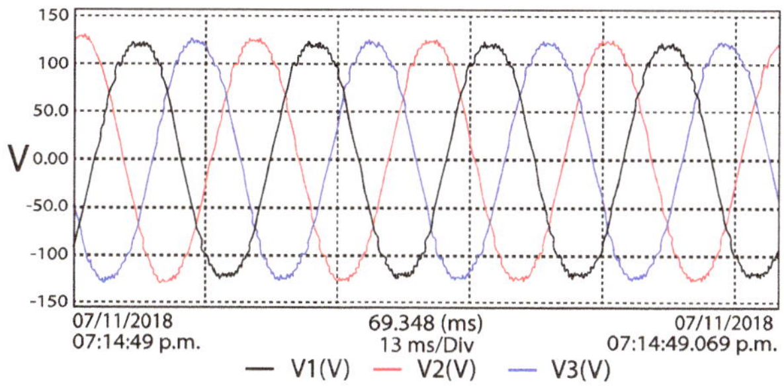

Figure 8.18 Voltage waveforms recorded by the AEMC-3945-B analyzer.

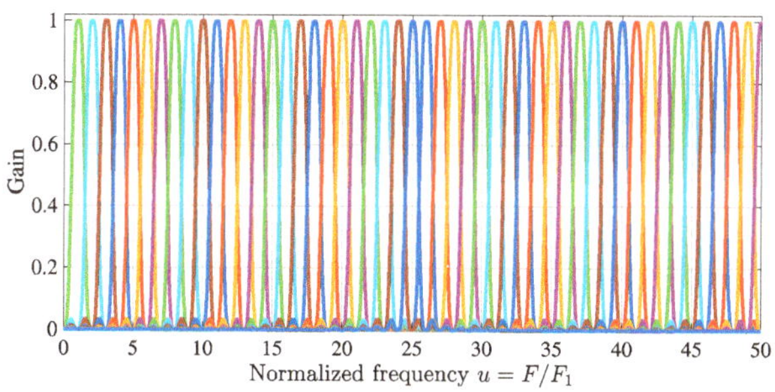

Figure 8.19 Frequency responses for FIR filter designed via a third-order ($K = 3$) O-spline.

Second, amplitudes at the fundamental and harmonic frequencies for voltage waveforms are estimated using a_h in (8.30). This estimation results after passing the four-cycle voltage signals by the analysis equation in (8.11) implemented with O-splines. As expected, the fundamental frequency represented by the first order in the first row of Table 8.1 concentrates the highest energy with RMS voltage values for $V1$, $V2$, and $V3$ of 126.7252, 128.0984, and 127.1772 V, respectively. Meanwhile, from the second up to the 50th harmonic, Table 8.1 evidences the maximum RMS voltages at the 3rd harmonic, also shows the lower voltage values at the even harmonics in comparison with odd harmonics. Since these estimates also facilitate the computation of the voltage unbalance in percentage (8.44), employing the RMS voltages at the fundamental frequency in Table 8.1, we find a V_{unb} of 0.6%.

Once these estimates are available, we can proceed to compute the power quality indicators by using (8.42)-(8.44). Then, their fulfillment is verified according to the harmonic standard. Regarding the individual harmonic distortions, these are illustrated in Fig. 8.20 for $V1$, $V2$, and $V3$ voltages. Here, it is noticeable from the 2nd up to the 50th harmonic that all values report rate below of 4%, where the highest value is at the 2nd harmonic, according to the maximum RMS voltage found in Table 8.1 and other significant odd harmonics, representing almost 1%, are evident at the 5th, 21st, 23rd, and 49th frequencies.

For comparison purposes, we now choose the IHDs for voltage $V1$ to contrast O-splines-based estimation against the commercial PQ analyzer, resulting in the plot in

Table 8.1 Amplitude estimates for voltage waveforms at the fundamental and harmonics frequencies through the analysis provided by O-splines.

Order	*V*1	*V*2	*V*3	Order	*V*1	*V*2	*V*3
1	126.7252	128.0984	127.1772	26	0.0331	0.0376	0.0141
2	0.0527	0.0557	0.0767	27	0.2551	0.2488	0.2779
3	4.9480	2.5621	4.0436	28	0.0088	0.0151	0.0082
4	0.0042	0.0208	0.0134	29	0.0191	0.0507	0.0097
5	1.5796	1.1336	0.7142	30	0.0123	0.0167	0.0073
6	0.0192	0.0439	0.0158	31	0.0130	0.0423	0.0373
7	0.5321	0.3024	0.7461	32	0.0144	0.0131	0.0242
8	0.0230	0.0651	0.0663	33	0.3074	0.3717	0.2456
9	0.6349	0.5636	0.5402	34	0.0279	0.0157	0.0118
10	0.0217	0.0247	0.0101	35	0.2294	0.2276	0.2413
11	0.6683	0.3483	0.6664	36	0.0119	0.0174	0.0172
12	0.0187	0.0143	0.0021	37	0.0245	0.0317	0.0628
13	0.6946	0.4145	0.6913	38	0.0211	0.0305	0.0217
14	0.0199	0.0088	0.0311	39	0.2086	0.1582	0.1919
15	0.4275	0.478	0.4447	40	0.0088	0.0045	0.0235
16	0.0243	0.0665	0.0333	41	0.0039	0.0153	0.0137
17	0.3851	0.2219	0.3458	42	0.0037	0.0323	0.0054
18	0.0008	0.0203	0.0250	43	0.0520	0.0187	0.0175
19	0.0895	0.0687	0.1114	44	0.0231	0.0070	0.0119
20	0.0429	0.0213	0.0194	45	0.2841	0.2332	0.2430
21	1.1307	0.9395	0.9577	46	0.0231	0.0298	0.0108
22	0.0298	0.0366	0.0220	47	0.4213	0.2813	0.3707
23	1.0438	0.6254	0.9644	48	0.015	0.0124	0.0170
24	0.0308	0.0339	0.0077	49	1.1778	0.7911	1.0230
25	0.5576	0.4472	0.5349	50	0.0054	0.0054	0.0241

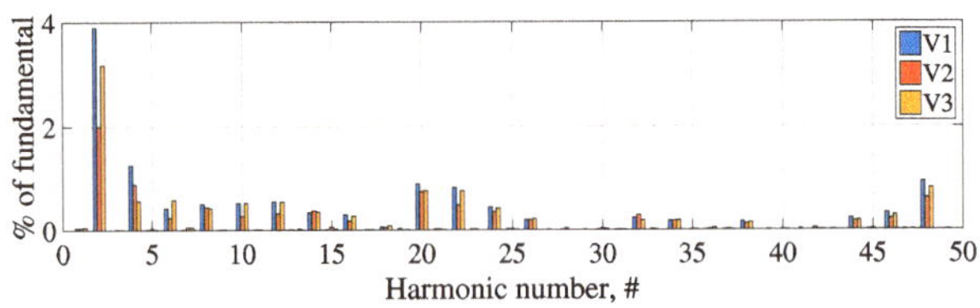

Figure 8.20 Individual harmonic distortions from the 2nd up to the 50th harmonic for $V1$, $V2$ and $V3$ voltage channels provided by O-splines.

Fig. 8.21. Here, the IHDs from the 2nd up to the 50th harmonic for the commercial analyzer, and the estimation provided by O-splines, present high similarity for even and odd harmonics.

Now, the total harmonic distortions in percentage are computed for the voltage channels by using expression (8.43), resulting in Table 8.2. As summarized in this

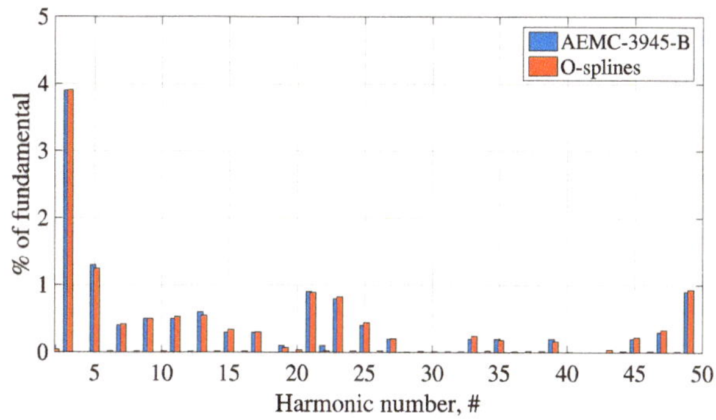

Figure 8.21 Comparison of IHDs for voltage $V1$ between the commercial analyzer and the O-splines.

Table 8.2 Total harmonic distortions for voltage channels through the O-splines analysis.

	% THD_v	
	O-splines	**AEMC-3945-B**
$V1$	4.5694	4.5900
$V2$	2.6328	2.6300
$V3$	3.7375	3.7700

table, the high similarity for the voltage THD, associated with $V1$, $V3$, and $V3$, is noticeable.

8.4 Conclusions

This chapter has demonstrated the effectiveness of the multiresolution analysis provided by the O-splines of the DTTFT for power quality harmonic monitoring. The straightforward implementation of the O-splines is achieved, since it reduces its computational burden by exploiting the fundamental periodicity of the Euler complex sequences in the Fourier matrix. Due to this implementation retaining all Taylor–Fourier filtering properties, the estimates report high accuracy. This makes our proposal suitable for their implementation in microcontrollers or digital signal processors with a reduced cost. Besides that, the attained performance is assessed under synthetic signals and actual signals provided by an AEMC 3945-B three-phase power analyzer, which is equipped with the fast Fourier transform for PQ analysis. Finally, power quality harmonic indicators, such as IHD and THD, exhibited accurate results, making the O-splines a reliable, powerful, and capable processing tool.

Appendix 8.A Matlab function to obtain the samples of the Kth O-spline and its first two derivatives with N0 samples per cycle

```
function [phi,phip,phipp]=Osplinepp(K,N0)
%Order K>3
%N0 Number of samples per cycle
delt=1/N0;
unit=(0:delt:1-delt)';
knots=[flip(-[1:K]) [1:K]];
P=ones(N0,K+1); Pp=ones(N0,K+1);
Ppp=ones(N0,K+1);
ir0=0;
for nint=1:K+1
u=-(K+1)/2-1+nint+unit;
coef= [1, -knots(ir0+1)]/(-knots(ir0+1));
for k=1:K
P(:,nint)=P(:,nint).*(u-knots(ir0+k))/...
(-knots(ir0+k));
if k>1 coef = conv(coef, [1, -knots(ir0+k)]/...
(-knots(ir0+k))); end;
end;
%Derivative Polynomial coefficients
C(:,nint)=coef'; Cp(:,nint)=polyder(coef)';
Cpp(:,nint)=polyder(Cp(:,nint))';
%First derivative by Horner scheme
Pp(:,nint)=Cp(1,nint)*Pp(:,nint);
for k=1:K-1
Pp(:,nint)=Pp(:,nint).*u + Cp(k+1,nint);
end;
%Second derivative by Horner scheme
Ppp(:,nint)=Cpp(1,nint)*Ppp(:,nint);
for k=1:K-2
Ppp(:,nint)=Ppp(:,nint).*u + Cpp(k+1,nint);
end;
ir0=ir0+1;
end;
N=(K+1)*N0;
phi=reshape(P,[N,1]); phip=reshape(Pp,[N,1]);
phipp=reshape(Ppp,[N,1]);
```

Appendix 8.B State estimator

```
hm=phi/N0.*exp(j*2*pi/N0*n);
hpm=F0*phip/N0.*exp(j*2*pi/N0*n);;
hppm=F0^2*phipp/N0.*exp(j*2*pi/N0*n);
%%%%%%%%%%%%%%%%%%%%%%%%%%%%%%%%%%%%%%
function [a,p,ap,pp,app,ppp]=...
statestimator(hm,hpm,hppm,s)
%hm,hpm,hppm impulse response of the
%FIR filter and first two derivatives
%sh samples of the filtered signal
%%%%%%%%%%%%%%%%%%%%%%%%%%%%%%%%%%%%%%%
%a,p        amplitude and phase
%ap,pp      first derivatives of a and p
%app,ppp    second derivvatives of a and p
%%%%%%%%%%%%%%%%%%%%%%%%%%%%%%%%%%%%%%%%
%Amplitude and Phase estimation
k=(0:Nc*N0-1)'; p=conv(s, hm,'same');
ah=abs(p); ph=angle(p);   % rotating phase
phc=angle(p.*exp(-2j*pi/N0*k));% antirotated
a=2*ah; p=phc;
%%%%%%%%%%%%%%%%%%%%%%%%%%%%%%%%%%%%%%
%Frequency estimation
pp=conv(s,hpm,'same');
ppc=pp.*exp(-j*ph);
ahp=real(ppc);
php=imag(ppc)./ah;
ap=ahp; pp=php/(2*pi); %frequency deviation
%%%%%%%%%%%%%%%%%%%%%%%%%%%%%%%%%%%%%
%ROCOF Estimation
ppp=conv(s,hppm,'same');
pppc=ppp.*exp(-j*ph);
ahpp=real(pppc)+ah.*php.^2;
phpp=( imag(pppc)- 2*ahp.*php )./ah;
app=ahpp; ppp=phpp/(2*pi);
%%%%%%%%%%%%%%%%%%%%%%%%%%%%%%%%%%%%%%
```

References

[1] A. Jalilian, S. Samadinasab, Detection of short-term voltage disturbances and harmonics using μpmu-based variational mode extraction method, IEEE Transactions on Instrumentation and Measurement 70 (2021) 1–17.

[2] P. Castello, C. Laurano, C. Muscas, P.A. Pegoraro, S. Toscani, M. Zanoni, Harmonic synchrophasors measurement algorithms with embedded compensation of voltage trans-

former frequency response, IEEE Transactions on Instrumentation and Measurement 70 (2021) 1–10.
[3] S. Toscani, M. Faifer, A. Ferrero, C. Laurano, R. Ottoboni, M. Zanoni, Compensating nonlinearities in voltage transformers for enhanced harmonic measurements: the simplified Volterra approach, IEEE Transactions on Power Delivery 36 (1) (2021) 362–370.
[4] S.K. Jain, S. Singh, Harmonics estimation in emerging power system: key issues and challenges, Electric Power Systems Research 81 (9) (2011) 1754–1766.
[5] G. D'Antona, C. Muscas, P.A. Pegoraro, S. Sulis, Harmonic source estimation in distribution systems, IEEE Transactions on Instrumentation and Measurement 60 (10) (2011) 3351–3359.
[6] A.B. Nassif, W. Xu, W. Freitas, An investigation on the selection of filter topologies for passive filter applications, IEEE Transactions on Power Delivery 24 (3) (2009) 1710–1718.
[7] S. Elphick, V. Gosbell, V. Smith, S. Perera, P. Ciufo, G. Drury, Methods for harmonic analysis and reporting in future grid applications, IEEE Transactions on Power Delivery 32 (2) (2016) 989–995.
[8] Y. Wang, Q. Li, F. Zhou, Y. Zhou, X. Mu, A new method with Hilbert transform and slip-svd-based noise-suppression algorithm for noisy power quality monitoring, IEEE Transactions on Instrumentation and Measurement 68 (4) (2019) 987–1001.
[9] D. Carta, C. Muscas, P.A. Pegoraro, S. Sulis, Identification and estimation of harmonic sources based on compressive sensing, IEEE Transactions on Instrumentation and Measurement 68 (1) (2019) 95–104.
[10] IEEE recommended practice and requirements for harmonic control in electric power systems, in: IEEE Std 519-2014 (Revision of IEEE Std 519-1992) – Redline, 2014, pp. 1–213.
[11] IEEE recommended practice–adoption of IEC 61000-4-15:2010, electromagnetic compatibility (emc)–testing and measurement techniques–flickermeter–functional and design specifications – redline, in: IEEE Std 1453-2011 – Redline, 2011, pp. 1–89.
[12] J. Matas, H. Martín, J. de la Hoz, A. Abusorrah, Y. Al-Turki, H. Alshaeikh, A new thd measurement method with small computational burden using a sogi-fll grid monitoring system, IEEE Transactions on Power Electronics 35 (6) (2020) 5797–5811.
[13] S. Golestan, J.M. Guerrero, J.C. Vasquez, A.M. Abusorrah, Y. Al-Turki, Harmonic linearization and investigation of three-phase parallel-structured signal decomposition algorithms in grid-connected applications, IEEE Transactions on Power Electronics 36 (4) (2021) 4198–4213.
[14] S. Lodetti, J. Bruna, J.J. Melero, V. Khokhlov, J. Meyer, A robust wavelet-based hybrid method for the simultaneous measurement of harmonic and supraharmonic distortion, IEEE Transactions on Instrumentation and Measurement 69 (9) (2020) 6704–6712.
[15] A. Gaouda, M. Salama, M. Sultan, A. Chikhani, Application of multiresolution signal decomposition for monitoring short-duration variations in distribution systems, IEEE Transactions on Power Delivery 15 (2) (2000) 478–485.
[16] C. Chen, G. Chang, R. Hong, H. Li, Extended real model of Kalman filter for time-varying harmonics estimation, IEEE Transactions on Power Delivery 25 (1) (2009) 17–26.
[17] S.K. Jain, S. Singh, Exact model order esprit technique for harmonics and interharmonics estimation, IEEE Transactions on Instrumentation and Measurement 61 (7) (2012) 1915–1923.
[18] G.W. Chang, C.-I. Chen, Y.-F. Teng, Radial-basis-function-based neural network for harmonic detection, IEEE Transactions on Industrial Electronics 57 (6) (2009) 2171–2179.
[19] S.K. Jain, S.N. Singh, Fast harmonic estimation of stationary and time-varying signals using ea-awnn, IEEE Transactions on Instrumentation and Measurement 62 (2) (2013) 335–343.

[20] J.A. de la O Serna, Dynamic harmonic analysis with FIR filters designed with O-splines, IEEE Transactions on Circuits and Systems I: Regular Papers 67 (12) (2020) 5092–5100.

[21] J.A. de la O Serna, Analyzing power oscillating signals with the O-splines of the discrete Taylor-Fourier transform, IEEE Transactions on Power Systems 33 (6) (Nov. 2018) 7087–7095.

[22] J.A. de la O Serna, M.R.A. Paternina, A. Zamora-Mendez, R.K. Tripathy, R.B. Pachori, EEG-rhythm specific Taylor-Fourier filter bank implemented with O-splines for the detection of epilepsy using EEG signals, IEEE Sensors Journal 20 (12) (2020) 6542–6551, [Online]. Available, https://ieeexplore.ieee.org/document/9017999.

[23] R.K. Tripathy, A. Zamora-Mendez, J.A. de la O Serna, M.R.A. Paternina, J.G. Arrieta, G.R. Naik, Detection of life threatening ventricular arrhythmia using digital Taylor-Fourier transform, Frontiers in Physiology 9 (2018), [Online]. Available https://www.frontiersin.org/article/10.3389/fphys.2018.00722.

[24] J.G. Proakis, M. Salehi, Digital Communications, 5th ed., McGraw Hill, Boston, MA, 2008, sec. 2.1.

[25] J.A. de la O Serna, J.M. Ramirez, A. Zamora Mendez, M.R.A. Paternina, Identification of electromechanical modes based on the digital Taylor-Fourier transform, IEEE Transactions on Power Systems 31 (1) (2016) 206–215, https://doi.org/10.1109/TPWRS.2015.2403290.

[26] J.A. De La, O. Serna, M.R. Arrieta Paternina, A. Zamora-Mendez, Assessing synchrophasor estimates of an event captured by a phasor measurement unit, IEEE Transactions on Power Delivery (2020) 3109–3117.

[27] D. Guillen, J.A. de la O Serna, A. Zamora-Méndez, M.R.A. Paternina, F. Salinas, Taylor-Fourier filter-bank implemented with O-splines for the detection and classification of faults, IEEE Transactions on Industrial Informatics 17 (5) (2021) 3079–3089.

[28] J.A. de la O Serna, A. Zamora, M. Paternina, G. Castillo, Power quality harmonic monitoring through o-splines, https://github.com/marioap7/Power-Quality-Harmonic-Monitoring-Through-O-splines, 2022. (Accessed 25 February 2022).

Monitoring system for identifying power quality issues in distribution networks using Petri nets and Prony method

9

Fernando Salinas Salinas[a], *Daniel Guillen*[b], *and Luis Alonso Trujillo Guajardo*[a]
[a]Autonomous University of Nuevo León, San Nicolás de los Garza, Mexico, [b]Tecnológico de Monterrey, Monterrey, Mexico

9.1 Monitoring of distribution grids

9.1.1 Trends and new challenges

Grid monitoring is an essential task to analyze the reliability of the system during a wide range of operating scenarios. To know the dynamic response or the operating point (steady state) of any distribution network, many electrical variables should be measured through the grid's instrumentation and communication systems would be required. Based on the instrumentation and monitoring, it is possible to introduce new digital functionalities that can help to prevent large outages, or help to identify power quality issues according to the electrical regulations defined by grid codes. In this context, the current technology developments offer several devices for monitoring purposes:

a) smart sensors,
b) smart meters,
c) smart transformers,
d) phasor measurement units.

According to the needs of study, one or another device may be more beneficial, for example, phasor measurement units are commonly used to conduct research on grid's state estimation [1]. On the other hand, smart transformers provide more flexibility supported by local or remote control schemes to improve the voltage regulation in distribution networks [2]. In the same context, smart meters offer several functionalities, such as demand measurement, instantaneous measurement, energy measurement, power quality, and harmonics measurement [3].

The use of data (electrical quantities) coming from different measurement devices impose new challenges about what to do with a large amount of data, especially for real-time applications. The need of real-time applications can be divided into on-line and off-line. For example, the regulation according to grid codes for power quality issues does not necessarily require on-line applications; so one of the main challenges can be the identification of the harmonic sources. This entails their detection to avoid

Monitoring and Control of Electrical Power Systems Using Machine Learning Techniques
https://doi.org/10.1016/B978-0-32-399904-5.00015-6

the harmonic propagation into the grid. On the contrary, the utility can apply penalties in the same way as industrial loads with low power factor. The new technological advances allow having web-based platforms supported by the new functionalities of smart sensor, smart meter, smart transformer, and PMUs. This leads to a basic idea of developing and using data-driven applications throughout the websites.

9.1.2 Data-driven and web-based platform

To understand the basis of a web-based monitoring platform, one should be thinking in data-driven applications that help in the decision-making, not only for real-time analysis, but also during other periods, for example, monthly analysis to detect the number of infringements that an industrial load does not comply with the established limits by grid codes. For example, to identify low-power factors in large industrial loads as well issues associated with power quality due to harmonics that may propagate into the electrical grid. Therefore there is a need to identify the root cause of harmonic source that does not need to be an on-line process; this means that the process may not demand fast responses so that algorithms in web-based platforms can be very useful in distribution system aided by the new functionalities of smart sensors, smart meters, and PMUs. Web-based platforms offer flexibility and better understanding to the distribution system operator (DSO), because these platforms are mainly developed by organized and filtered data coming from measurement systems through communication systems. The general basis of a web-based monitoring platform is quite similar to the concept of Internet of things, something like Internet of things for electrical grids, or Internet of energy. Fig. 9.1 shows the general concept of a web-based monitoring platform that is mainly composed of a data collection stage, databases, algorithms based on machine-learning techniques and analysis/visualization algorithms using websites. Data collection requires communication links between the distribution system operator (DSO) and the measurement devices, such as smart sensors, smart meters, or PMUs. All of these devices may come along with phasor data concentrators (PDC), digital fault data recorders (DFR). In general, the aim of using web-based platforms tends to facilitate the diagnosis in real-time during no critical periods. However, the proliferation of high-order harmonics can be studied to learn about new possible regulations or services that all microgrids or distributed generators should comply with harmonic limits to avoid the degradation of the power quality in active distribution grids.

9.1.3 Harmonic propagation

Electrical power grids require robust tools to study the harmonic propagation into the grid due to the use of power electronic converters in distributed generators (DGs) as well as the use of electronic devices in industrial, commercial, and residential loads [4]. The new trends in distribution systems have them engaged in active distribution networks, where the harmonic propagation plays a significant role for grid's analysis. In consequence, a DG not only may be a harmonic source, but also may help to miti-

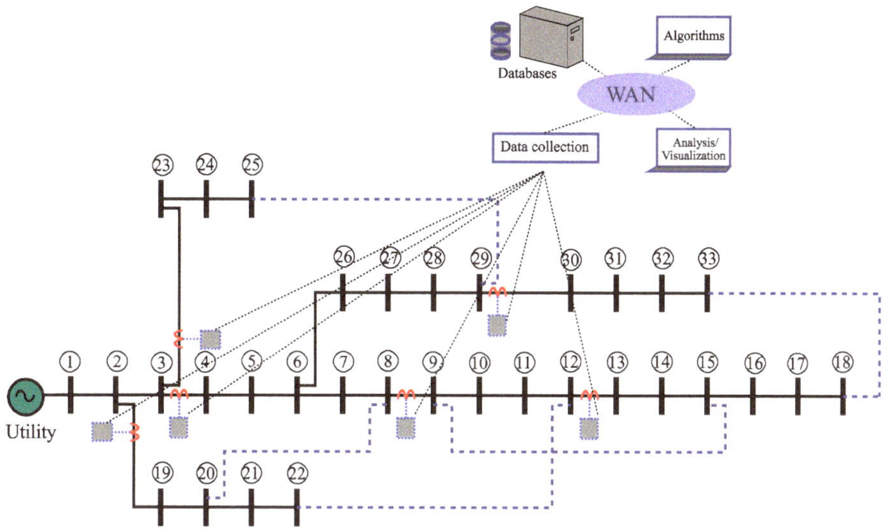

Figure 9.1 Diagram of a smart monitoring system.

gate the harmonic propagation if its control scheme is designed for this purpose. The large proliferation of DGs, battery energy storage systems (BESS), and interconnected microgrids imposes new challenges, because an active distribution grid can have more than one harmonic source, given that most of the DGs use power electronic converters, which can be a source of high-order harmonics [5]. In consequence, the overall power quality may be affected and its propagation should be reduced to avoid the increasing of active power losses associated with these phenomena. Besides, the harmonic propagation should be mitigated, and if it is possible the utility should be able to identify the root cause according to the regulation established by grid codes.

Harmonic propagation uses the principle of harmonic power flow defined by an admittance matrix that includes the harmonic admittance values regarding the network configuration [6]. For example, Fig. 9.2 shows the current injection and its propagation along the distribution lines, because any harmonic current injection generates harmonic voltages. Therefore the harmonic propagation will be dependent on the grid topology and the harmonic impedance (or admittance) between two buses "j" and "k".

In general, the harmonic propagation can be solved by computing the harmonic voltages defined by expression (9.1):

$$\begin{bmatrix} V_h^{(1)} \\ \vdots \\ V_h^{(N)} \end{bmatrix} = \begin{bmatrix} Y_h^{(1,1)} & \cdots & Y_h^{(1,N)} \\ \vdots & \ddots & \vdots \\ Y_h^{(N,1)} & \cdots & Y_h^{(N,N)} \end{bmatrix}^{-1} \begin{bmatrix} I_h^{(1)} \\ \vdots \\ I_h^{(N)} \end{bmatrix}, \tag{9.1}$$

where $V_h^{(k)}$ represents the harmonic voltage at bus "k", $I_h^{(k)}$ stands for the harmonic current injection, and the harmonic admittance between two buses, "j" and "k", is rep-

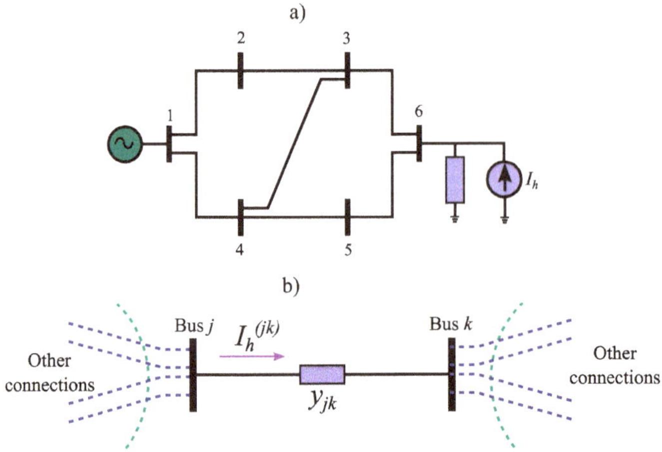

Figure 9.2 a) Harmonic injection and b) harmonic propagation along distribution lines.

resented by $Y_h^{(j,k)}$. In general, the indices "j" and "k" take values from zero up to the total number of buses. Finally, the harmonic propagation can be computed following the fundamentals of power flow following the harmonic domain.

9.1.4 Harmonic distortion indices and limits

The need to quantify the harmonic distortion comes from identifying the harmonic sources to avoid the propagation into the grid, because the system may be exposed to critical scenarios that put its reliability at risk. Additionally, the system is continuously changing due to switching transients that may develop the resonance phenomenon due to harmonic sources. Therefore the quantification of the harmonic distortion becomes crucial to establish limits according to the regulation by each country. For example, Fig. 9.3 displays the results of a typical oscillatory transient so that if one desires to quantify the distortion, the most commonly used index is the total harmonic distortion (THD), which is widely employed by utilities to quantify the harmonic distortion at the point of common coupling (PCC) caused by industrial loads or distributed generators.

The harmonic analysis can be carried out by employing indices in both the frequency and time domain. The frequency domain indices have been widely reported in the literature and some of those expressions are (9.2)–(9.5) [7,8].

Total harmonic distortion (THD) for voltage signals THD_v or current signals THD_i can be expressed as follows:

$$THD_i = \sqrt{\sum_{k=2}^{h}\left(\frac{I_h}{I_1}\right)^2} \times 100; \tag{9.2}$$

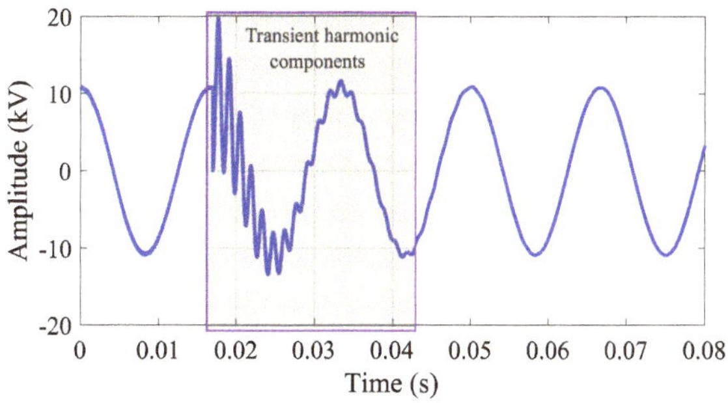

Figure 9.3 Typical response of a capacitor switching.

Table 9.1 Voltage distortion limits at the point of common coupling.

Voltage level	Individual harmonic (%)	THD (%)
$V_{pcc} \leq 1.0$ kV	5.0	8.0
1 kV $< V_{pcc} \leq 69$ kV	3.0	5.0
69 kV $< V_{pcc} \geq 161$ kV	1.5	2.5
$V_{pcc} > 161$ kV	1.0	1.5

total demand distortion (TDD):

$$TDD = \sqrt{\sum_{k=2}^{h} \left(\frac{I_h}{I_L}\right)^2} \times 100; \tag{9.3}$$

total current distortion compared with the transformer rated current:

$$TRD = \sqrt{\sum_{k=2}^{h} \left(\frac{I_h}{I_r}\right)^2} \times 100; \tag{9.4}$$

equivalent harmonic impedance:

$$Zeq = \frac{THD_v}{TRD}, \tag{9.5}$$

where I_1 represents the fundamental component, I_h is the harmonic component, I_L stands for the maximum demand current at the PCC, I_r represents the rated current of the distribution transformer feeding the load, and Zeq corresponds to the equivalent harmonic impedance at the PCC.

In general, the permissible limits depend on the voltage level and may change between one and another utility, because some systems impose limits that are more restrictive. Based on the recommended practices [9], the permissible limits, according to the voltage level at the PCC (V_{pcc}), can be seen in Table 9.1.

Additionally, the distortion limits for current signals also depend on the voltage level. Therefore the permissible current distortion corresponding to distribution net-

Table 9.2 Current distortion limits at the PCC for distribution networks (up to 69 kV).

$R_{PCC} = \frac{I_{SC}}{I_L}$ (%)	**Individual harmonic order (odd harmonics)**					
	3 ≤ h < 11	**11 ≤ h < 17**	**17 ≤ h < 23**	**23 ≤ h < 35**	**35 ≤ h ≤ 50**	**TDD (%)**
R_{PCC}	4.0	2.0	1.5	0.6	0.3	5.0
R_{PCC}	7.0	3.5	2.5	1.0	0.5	8.0
R_{PCC}	10.0	4.5	4.0	1.5	0.7	12.0
R_{PCC}	12.0	5.0	2.0	2.0	1.0	15.0
R_{PCC}	15.0	6.0	6.0	2.5	1.4	20.0

works is shown in Table 9.2. In this case, the current distortion limits are defined by the maximum short-circuit current at the PCC I_{SC} and the maximum demand current (fundamental component) at the PCC I_L. The last relationship should be expressed in percent, which is defined by R_{PCC}. More details about the current distortion limits can be found in IEEE recommended practice and requirements for harmonic control in electric power systems [9].

On the other hand, time-domain features are very useful for classification purposes, especially if we want to identify all possible power quality disturbances. Some of these features are based on statistical indices or by using entropy. For instance, Fig. 9.4 presents the transient information produced by two different signal-processing techniques. Fig. 9.4a shows the high-frequency components produced by the discrete wavelet transform (DWT), whereas Fig. 9.4b displays the results of the intrinsic mode functions (IMFs) resulting from applying the variational mode composition (VMD) technique. Therefore the transient information must be analyzed and quantified to establish the classification rules according to machine learning techniques employed. Based on the coefficients produced by each technique, these will be used to determine the time-domain characteristics of the studied events. For example, Eqs. (9.6)–(9.11) correspond to some of the most typical features used in the time-domain analysis [10]:

$$F_{t1} = \frac{N}{\sum_{j=1}^{N} \frac{1}{d_j}}, \tag{9.6}$$

$$F_{t2} = \sqrt{\frac{\sum_{j=1}^{N} \left(d_j - u\right)}{N-1}}, \tag{9.7}$$

$$F_{t3} = \frac{\frac{1}{N}\sum_{j=1}^{N} \left(d_j - u\right)^4}{\left[\frac{1}{N}\sum_{j=1}^{N} \left(d_j - u\right)^2\right]^2}, \tag{9.8}$$

$$F_{t4} = \sum_{j=1}^{N} log\left(d_j^2\right), \tag{9.9}$$

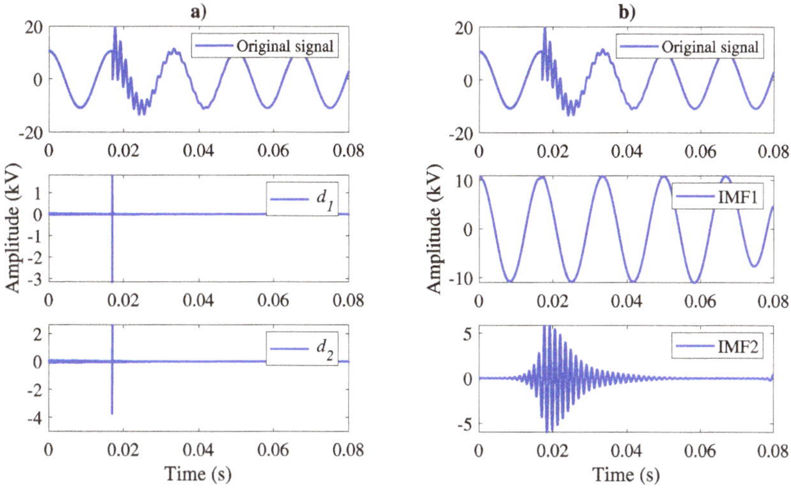

Figure 9.4 Analysis in the time domain by a) discrete wavelet transform and b) variational mode decomposition.

$$F_{t5} = -\sum_{j=1}^{N} d_j^2 * log\left(d_j^2\right), \tag{9.10}$$

$$F_{t6} = \frac{1}{1-\alpha} * log\left(\sum_{j=1}^{N} d_j^{\alpha}\right), \tag{9.11}$$

where F_{t1} is the harmonic mean, F_{t2} represents the standard deviation, F_{t3} is the Kurtosis index, F_{t4} corresponds to the entropy, F_{t5} is the Shannon entropy, and F_{t6} represents the Rényi entropy with a free parameter α. For the last expression, u represents the mean value of the transient components (according to multiresolution approaches), and d_i represents the wavelet coefficients or the produced coefficients by each IMF.

9.2 Harmonic estimation based on signal model

In this section, the parameter estimation technique, Prony method, for its application in a monitoring system for identifying power quality issues, is presented. The purpose of using the Prony method is to obtain the voltage and current signal parameters of the signals being monitored, where in one or more buses of an electrical distribution network, areas that are operating out of the acceptable limits, established by the international power quality standards, could be detected. Prony's method estimation results are considered as an input for the Petri net.

9.2.1 Prony method

Prony method is a signal processing technique based on signal estimation. Prony method extracts desired information from a uniformly sampled signal and builds a series of damped complex exponentials or sinusoids to approximate the sampled signal by solving a set of linear equations for the coefficients of the recurrence equation that the signals satisfy. The method's algorithm and its practical implementation are shown in [11,12], and have been used for power quality analysis [13–15], stability studies applied to power system and nuclear power plants [16], [17], and also has been evaluated for its application in power system protection, mainly in distance relay algorithms [18], [19]. It should be mentioned that the advantage in comparison with traditional estimation methods, such as FFT, is that with less data from a signal, a good parameter estimation of the harmonic signals, such as frequency, amplitude, phase angle, and damping, could be achieved. The general considerations of the Prony method for real-time applications can be found in [18,19], which were developed by one of the authors of this chapter. The Prony method is evaluated for distance protection relay of transmission lines with series compensation. In a similar way, one patent entitled "Digital Distance Relay with Prony Phasor Estimation" describes the considerations and requirements for a real-time implementation of a digital distance protection relay reported in [20]. Prony method estimation results are proposed as an input for the Petri net, where fundamental and harmonic frequencies will be detected. As is well known in Prony method literature, the parameters of a signal model $y\ (t)$ in (9.12) can be obtained by digitizing the signal and obtaining the samples $[y\ (1)\ y\ (2) ...\ y\ (n)]$ using a sampling frequency f_s.

$$y(t) = \sum_{n=1}^{N} A_n e^{\sigma_n t} \cos\left(2\pi f_n t + \theta_n\right) \tag{9.12}$$

The Prony model signal will approximate to the sampled data in (9.12) using the following linear combination of N complex exponentials:

$$y_M = \sum_{n=1}^{N} B_n \lambda_n^M, \tag{9.13}$$

$$B_n = \frac{A_n}{2} e^{j\theta_n}$$

$$\lambda_n = e^{(\sigma_n + j2\pi f_n)T}.$$

The signal $y\ (t)$ in (9.12) has four elements: magnitude A_n, damping factor σ_n, frequency f_n, and the phase angle θ_n. Each exponential term in (9.13) is a unique signal mode of the original signal $y\ (t)$. So, using the Euler theorem and total time t = MT, where M is the length of the signal and T is the time between samples, Eq. (9.12) can be rewritten as (9.13).

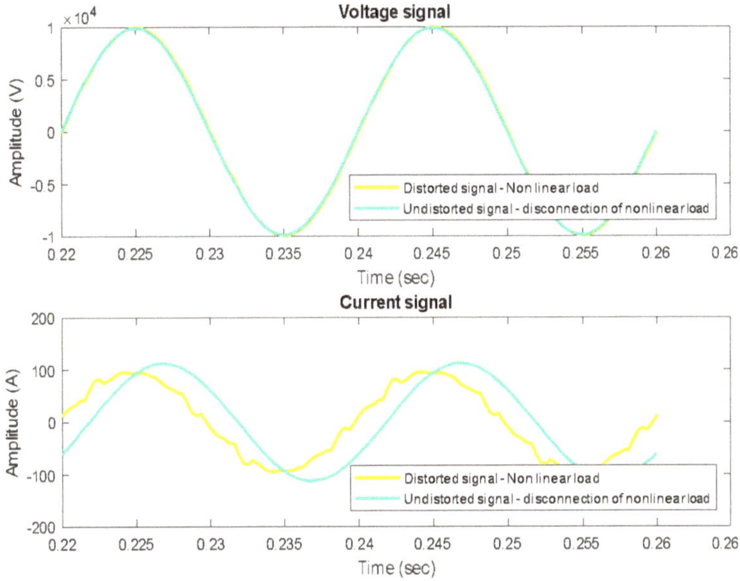

Figure 9.5 Signals under analysis. a) Voltage. b) Current.

9.2.2 Estimated parameters of an analyzed signal

In this section, estimation results of a voltage and current signal are presented, where distorted and undistorted voltage and current signals are used to obtain the estimated parameters of the signals. The sampling of the signals is 64 samples per cycle at 50 Hz, and the observation data window considered for the analysis is 2 cycles.

The estimation is performed using distorted and undistorted voltage and current signals, for example, before and after a connection of a nonlinear load, as presented in Fig. 9.5, so the harmonic frequency content of the signals under analysis can be obtained.

Prony method estimation results: First, the order N of the linear prediction model (LPM) is obtained. The order is obtained evaluating the mean square error (MSE) of the full signal data $N = 1, 2, \ldots, N_s$, where N_s is the total data samples of the full signal. *Second*, the MSE for each value of N in (9.14) needs to be calculated, where MSE is obtained by using the reconstructed signal with the estimated parameters "$\hat{y}$" and the real surge waveform "y," so the MSE of lesser magnitude is selected for the corresponding N value, therefore this value of N is the optimum estimate of the model signal parameters.

$$MSE_N = x_N = \frac{1}{N_s} \sum_{j=1}^{N_s} \left(\hat{y}_j - y_j\right)^2 \tag{9.14}$$

In (9.14), $\hat{y}$ is the reconstructed signal using the estimated parameters of N. It should be mentioned that the accuracy of Prony estimation depends on the length of

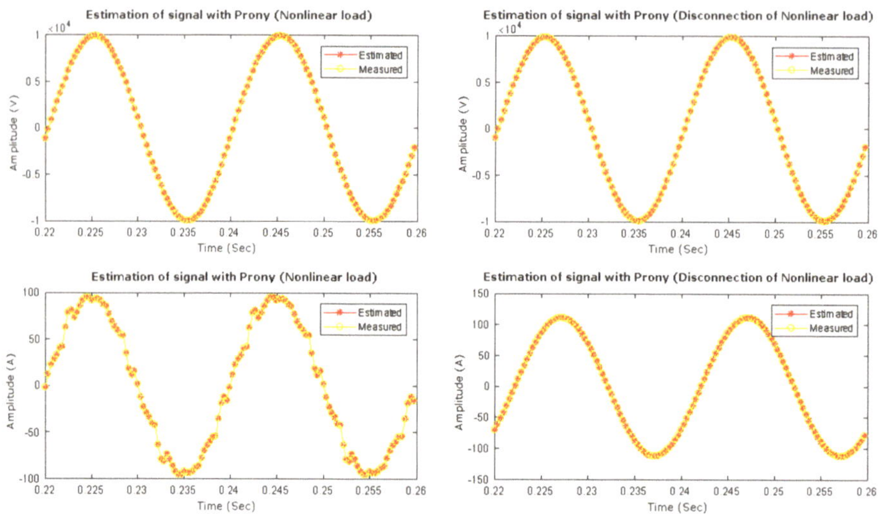

Figure 9.6 Signals estimation validation: a) voltage, b) current.

the observation data window and the number of samples used in the estimation process, as well as the order of the model [21]. The information required for the Petri net is the Prony estimated harmonics of the signal under analysis, voltage or current. The signal parameters of the simulated signals in Fig. 9.5 were estimated using the Prony method, which is described in Section 9.2.1. To validate the results, the simulated and estimated signals are presented in Fig. 9.6.

In Fig. 9.6, it can be observed that the parameters of the signals correspond to the analyzed signals, where the Prony estimated signal model (estimated) is reconstructed using the estimated parameters, which are presented in Table 9.3 and Table 9.4, and then compared with the simulated signal under analysis (measured) in Fig. 9.5. As it is observed, the error between signals has a very small value, and the signals are the same, so this confirms that the estimated parameters correspond to the analyzed signal.

As it could be observed in Table 9.3, the parameters of the signals detected for voltage signals in Fig. 9.5a correspond to the signals being analyzed, where the estimated parameters of undistorted signal and distorted signal correspond to the signals under analysis. The MSE value between simulated and estimated signals in Fig. 9.6a is 2.47×10^{-18} for distorted signal and for undistorted signal is 2.05×10^{-16}, hence, a good estimation is obtained.

In Table 9.4, the parameters of the signals detected for current signals in Fig. 9.5b correspond to the signals being analyzed as in the analyzed voltage signals, where the estimated parameters of undistorted signal and distorted signal correspond to the signals under analysis. The MSE value between simulated and estimated signals in Fig. 9.6b is 2.23×10^{-21} for distorted signal and for undistorted signal is 5.16×10^{-21}, hence, a good estimation is obtained, as in the analyzed voltage signals, and the estimated parameters are validated and could be used as an input to the Petri net.

Table 9.3 Estimated signal parameters of voltage simulated signal.

Estimated signal parameters	**Frequency** (*Hz*)	**Amplitude**	**Phase** (*rad*)	**Damping**	*Estimated signal parameters*	**Frequency** (*Hz*)	**Amplitude**	**Phase** (*rad*)	**Damping**
	50	9957.43	−1.68	−8.92E-10		50	9891.27	−1.66	2.69E-10
	150	3.10	−0.56	−2.36E-06		150	0.02	−1.06	7.46E-06
	250	15.85	−1.03	−2.83E-07		250	0.07	−1.82	7.33E-07
	350	28.01	−1.47	−1.31E-07		350	0.08	−2.65	−6.45E-07
	450	7.57	−2.37	−5.84E-07		450	0.01	2.70	−5.21E-06
	550	43.40	2.03	2.58E-09		550	0.05	−3.06	−1.04E-06
	650	27.45	1.60	−7.24E-07		650	0.10	2.26	−4.25E-07
Distorted signal: Nonlinear load	750	7.71	−2.09	9.61E-07	***Undistorted signal: Disconnection of Nonlinear load***	750	0.03	−1.12	8.18E-07
	850	12.64	−2.45	3.53E-07		850	0.06	−1.60	9.26E-08
	950	10.70	−2.64	2.22E-07		950	0.08	−1.95	−8.76E-09
	1050	1.76	−0.49	−7.26E-07		1050	0.007	−0.63	1.33E-06
	1150	2.12	0.26	−1.24E-08		1150	0.02	0.94	1.89E-07
	1250	1.64	−2.20	7.23E-07		1250	0.01	−2.51	−3.11E-08
	1350	2.94	−0.63	1.50E-06		1350	0.03	−0.27	−1.36E-07
	1450	4.63	−1.30	1.21E-06		1450	0.05	−1.08	−1.05E-07
	1550	4.24	−1.09	−1.77E-06		1550	0.06	−0.99	−4.26E-08

Table 9.4 Estimated signal parameters of current simulated signal.

Estimated signal parameters	**Frequency** (Hz)	**Amplitude**	**Phase** *(rad)*	**Damping**	*Estimated signal parameters*	**Frequency** (Hz)	**Amplitude**	**Phase** *(rad)*	**Damping**
	50	94.72	−1.56	−4.54E-08		50	111.79	−2.24	1.73E-07
	150	0.13	2.12	2.37E-05		150	0.01	−1.12	1.16E-04
	250	0.72	1.16	4.17E-06		250	0.04	−2.02	6.57E-05
	350	1.97	0.39	6.29E-07		350	0.05	−2.99	2.57E-05
	450	0.75	−0.60	−8.91E-07		450	0.01	2.55	1.18E-04
	550	4.70	−2.78	1.90E-07		550	0.02	2.77	6.77E-06
	650	2.92	2.90	4.48E-07		650	0.08	2.01	4.27E-06
Distorted signal – Nonlinear load	750	1.07	−0.94	−9.24E-07	***Undistorted signal – Disconnection of Nonlinear load***	750	0.02	−1.55	2.98E-06
	850	2.06	−1.40	−1.04E-06		850	0.05	−2.01	7.34E-06
	950	1.68	−1.82	−3.06E-07		950	0.07	−2.22	6.33E-06
	1050	0.43	0.87	−1.01E-06		1050	0.01	−0.51	1.47E-05
	1150	0.39	0.86	−1.30E-06		1150	0.02	0.69	−1.33E-05
	1250	0.11	0.08	−1.07E-05		1250	0.01	−2.42	3.71E-05
	1350	0.52	−0.57	−2.00E-06		1350	0.02	−0.53	−5.00E-06
	1450	0.93	−1.05	−1.07E-06		1450	0.05	−1.27	−1.42E-05
	1550	0.81	−1.34	−1.75E-07		1550	0.04	−1.07	1.15E-05

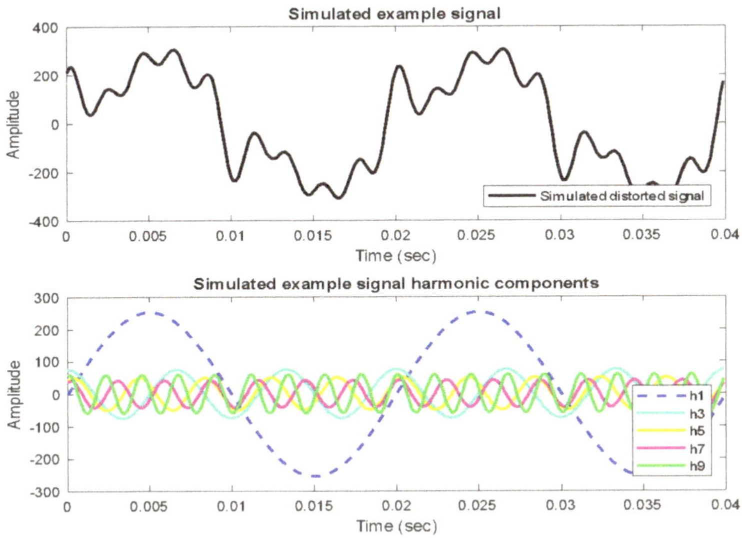

Figure 9.7 Simulated signal example. a) Distorted signal. b) Distorted signal harmonic components.

Table 9.5 Simulated example signal harmonic components.

Harmonic order	Frequency (Hz)	Amplitude	Phase angle (degrees)
1	50	254.04	0
3	150	75	90
5	250	50	55
7	350	42	65
9	450	60	75

9.2.3 Signal parameter estimation example

In this section, an example to obtain the signal parameters using Prony method is presented. Fig. 9.7 shows a simulated signal containing the third, fifth, seventh, and nineth harmonic considering a fundamental frequency of 50 Hz. The amplitudes for each frequency component, including fundamental frequency and its phase angles, are presented in Table 9.5.

The next steps should be followed to obtain the estimated parameters of the simulated distorted signal in Fig. 9.7a (signal harmonic components are presented in Table 9.5) using Prony method; this example should be performed in MATLAB® or GNU octave program:

1. Consider the sampling frequency (F_s) of the signal as 128 samples per cycle at 50 Hz and its sampling period (T_s) is $1.5625 \times 10\text{-}04$.
2. The length of the signal is 2 cycles.
3. Select an initial model order "p," in this example will be $p = 10$.

4. Define a Toeplitz matrix with the data of the signal "x".

$$\begin{bmatrix} x[p] & x[p-1] & \cdots & x[1] \\ x[p+1] & x[p] & \cdots & x[2] \\ \vdots & \vdots & \ddots & \vdots \\ x[2p-1] & x[2p-2] & \cdots & x[p] \end{bmatrix} \cdot \begin{bmatrix} a[1] \\ a[2] \\ \vdots \\ a[p] \end{bmatrix} = - \begin{bmatrix} x[p+1] \\ x[p+2] \\ \vdots \\ x[2p] \end{bmatrix}$$

5. Obtain vector "a" (coefficients of characteristic Eq. (9.4)) from step 4, where "p" is the model order.

$$a = -\boldsymbol{x} \cdot \mathbf{X}^{-1}$$

6. Obtain the roots from vector "a." The resulting roots vector "z" will be used to calculate damping and frequency with T_s defined in step 1, by using

$$\sigma = \frac{ln\,|z|}{T_s},$$

$$f = \frac{1}{2\pi T_s} tan^{-1}\left(\frac{Im(z)}{Re(z)}\right).$$

7. Obtain the Vandermonde matrix of vector "z" in step 6 from

$$\begin{bmatrix} z_1^0 & z_2^0 & \cdots & z_p^0 \\ z_1^1 & z_2^1 & \cdots & z_p^1 \\ \vdots & \vdots & \ddots & \vdots \\ z_1^{p-1} & z_2^{p-1} & \cdots & z_p^{p-1} \end{bmatrix} \cdot \begin{bmatrix} h_1 \\ h_2 \\ \vdots \\ h_p \end{bmatrix} = \begin{bmatrix} x[1] \\ x[2] \\ \vdots \\ x[p] \end{bmatrix}.$$

8. Obtain vector "h" using Vandermonde matrix "Z" and vector "x"

$$h = \boldsymbol{x} \cdot \mathbf{Z}^{-1}$$

9. The resulting vector "h" will be used to calculate amplitude and phase angle, by using

$$A = |h|,$$

$$\theta = tan^{-1}\left(\frac{Im(h)}{Re(h)}\right).$$

10. Form the Prony estimated signal using Eq. (9.12) with the estimated parameters from step 6 and step 9.
11 Compare Prony estimated signal with simulated distorted signal in Fig. 9.7a.

Expected estimation results: In Fig. 9.8, the expected results are shown, where the comparison of estimated and original signal are compared (see Fig. 9.8a), and the estimated spectrum is obtained from the Prony estimation results (see Fig. 9.8b); the estimation results are presented in Table 9.6.

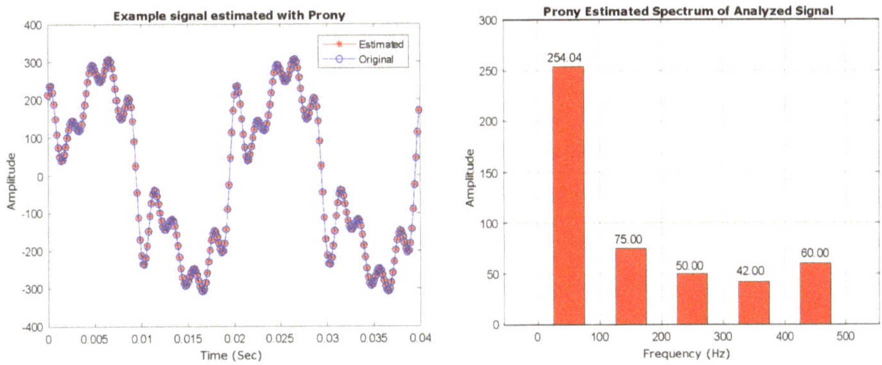

Figure 9.8 Simulated example signal validation. a) Estimated and original signal comparison. b) Estimated Prony spectrum.

Table 9.6 Simulated example signal estimated parameters.

Estimated signal parameters	Frequency (Hz)	Amplitude	Phase (rad)	Phase (degrees)	Damping
Example signal	50	254.04	−0.0002	−0.0169	−4.95E-05
	150	75	1.5705	89.982	0.000166
	250	50	0.9596	54.982	−0.000101
	350	42	1.1341	64.982	3.08E-05
	450	60	1.3087	74.982	−3.39E-06

As it can be shown, the Prony estimation results in Table 9.6 using the simulated example signal in Fig. 9.7a, correspond to the signal amplitude, frequency, and phase angle presented in Table 9.6. It should be mentioned that the damping results, just as the other signal parameters, correspond to the model of the Prony signal presented in Eq. (9.12).

9.3 Petri net application in an energy distribution network

In this section, a monitoring alternative for electrical power grids that allows the identification of specific zones, or buses that operate with a high harmonic content, is presented. This monitoring system is based on Petri nets that is fed with the estimation of the total current and voltage distortion indices obtained using the Prony method. It should be noted that this tool has been used for different applications in which continuous and discrete dynamics are integrated, such as estimation variables in power converters [22–24]. For a better understanding of this approach, in the initial part of this section, the fundamentals of Petri nets are reviewed, and in the second part, a monitoring system for a power grid is proposed. Finally, the operation of the proposal is validated in an electrical system in two test scenarios.

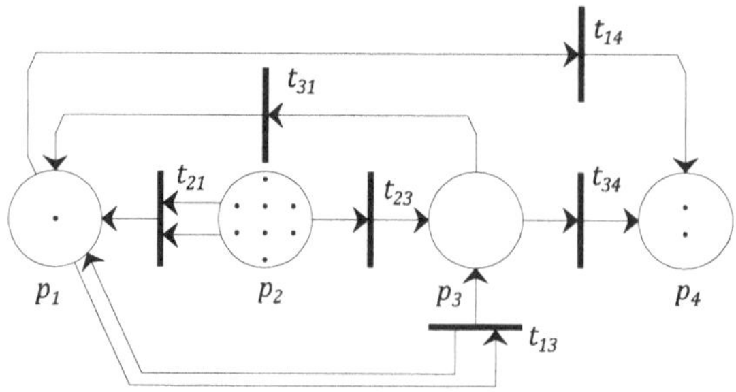

Figure 9.9 Petri net with 4 places and 6 transitions.

9.3.1 *Fundamentals*

For the modeling of a monitoring Petri net, it is essential to know the available information and the configuration of the system to be monitored; in this way, a state transition diagram and the transition conditions will be defined. Subsequently, an initial state is defined, and the dynamics of the scheme is validated. Next, the notation used in the Petri net (PN) approach is defined.

The structure of the PN is composed of the following sets: P, T, A, w, where

P is the set of places that contain the information of the system to be monitored;

T is the set of transitions associated with the system conditions and the defined operating thresholds;

A is the set of arcs from the places to transitions and from transitions to places;

w, is the weight assigned to each arc of set A.

A PN with 4 places and 6 transitions, connected by a set of arcs, is shown in Fig. 9.9.

From Fig. 9.9, the following is obtained:

$$P = \{p_1, p_2, p_3, p_4\}, \tag{9.15}$$

$$T = \{t_{13}, t_{14}, t_{21}, t_{23}, t_{31}, t_{34}\}, \tag{9.16}$$

$$A = \{(p_1, t_{13}), (t_{13}, p_3), (t_{13}, p_1)(p_1, t_{14}), (t_{14}, p_4), (p_2, t_{21}), (t_{21}, p_1)(p_2, t_{23}), (t_{23}, p_3), (p_3, t_{31}), (t_{31}, p_1), (p_3, t_{34}), (t_{34}, p_4)\}, \tag{9.17}$$

$$W = \{w(p_1, t_{13}) = 1, w(t_{13}, p_3) = 1, w(t_{13}, p_1) = 1, w(p_1, t_{14}) = 1, w(t_{14}, p_4) = 1, w(p_2, t_{21}) = 2, w(t_{21}, p_1) = 1, w(p_2, t_{23}) = 1, w(t_{23}, p_3) = 1, w(p_3, t_{31}) = 1, w(t_{31}, p_1) = 1, w(p_3, t_{34}) = 1, w(t_{34}, p_4) = 1\}. \tag{9.18}$$

As can be seen, the weighting w (p_2, t_{12}) has a value of 2, which means that if the condition of the transition t_{12} is fulfilled, two tokens will pass from p_2 to p_1. Each token that appears in each place represents a specific condition of the modeled system, and each transition represents the occurrence of an event. Regarding the specific condition of the system, the number of tokens in the places represents the current condition; this is called the marking of the PN. Thus to define the condition of a system modeled as a PN, a marking vector is added, such as

$$X = \{x(p_1), x(p_2), x(p_3), x(p_4)\} \tag{9.19}$$

where 4, is the number of places in PN.

Then, from Fig. 9.9, the marking described in Fig. 9.9 is

$$X = \{1, 8, 0, 2\}.$$

However, after certain conditions or events occur, there will be a rearrangement of the tokens, which would indicate a change in the condition of the system modeled with PNs. This is known as PN dynamics.

9.3.2 Petri net dynamics

The dynamics of the PN is defined by the occurrence of an event that enables the operation of a transition, which causes the displacement of tokens from one place to another, then the state transition function $\mathrm{N}^n \times \mathrm{T} \longrightarrow \mathrm{N}^n$, of PN structure is

$$x(p_i) \geq w\left(p_i, t_{jk}\right). \tag{9.20}$$

Then, the set $x' = f(x, t_j)$ is defined as

$$x'(p_i) = x(p_i) - w(p_i, t_{jk}) + w(t_{jk}, p_i), \quad \text{for } i = 1, 2, \ldots, n. \tag{9.21}$$

The first condition states that t_{jk} is enabled if the number of tokens in p_i is equal to or greater than the arc weight that connects p_i to t_{jk}. Once the transition is enabled and the event associated to the condition defined in the transition t_{jk} occurs, the transition is fired, leading to the next state or next marking of PN, defined as (9.21). One way to generalize the expression of the next state of the PN $x'(p_i)$ is by using a vector of transitions, u =[0 1 0 1 0 0 0 0 1], where 0 and 1 represents the disabled and enabled t_{jk} transitions. Also, it is possible to define, the incidence matrix A, which represents the "connection" between the states of the system, and whose elements are defined as

$$a_{jki} = w\left(t_{jk}, p_i\right) - w\left(p_i, t_{jk}\right). \tag{9.22}$$

Then, given the construction of matrix A, we can generalize the expression (9.21), which represents the next state of PN as

$$x_r = x_{r-1} + u_r A. \tag{9.23}$$

This approach offers an analytical tool to determine the dynamics of the Petri net when each event occurs.

Table 9.7 Resulting marking for application of u_1 to u_4.

Applied vector (u_r)	**Resulting marking (x_r)**
–	x_0=[1 8 0 2] – Fig. 9.6-a)
u_1=[1 1 1 1 0 0]	x_1=[1 5 2 3] – Fig. 9.6-b)
u_2=[0 0 1 1 1 1]	x_2=[3 2 1 4] – Fig. 9.6-c)
u_3=[1 0 0 1 1 0]	x_3=[4 1 2 4] – Fig. 9.6-d)
u_4=[1 1 0 1 0 0]	x_4=[3 0 4 5] – Fig. 9.6-e)

Example 9.1. Consider the PN of Fig. 9.9, with the initial marking x_0 = [1 8 0 2], and the incidence matrix defined as

$$A = \begin{bmatrix} 0 & 0 & 1 & 0 \\ -1 & 0 & 0 & 1 \\ 1 & -2 & 0 & 0 \\ 0 & -1 & 1 & 0 \\ 1 & 0 & -1 & 0 \\ 0 & 0 & -1 & 1 \end{bmatrix}.$$

If certain conditions of the modeled system are reached according to the vector $u_1 = [1\ 1\ 1\ 1\ 0\ 0]$, the next state x_1 is

$$x_1 = [1\ 8\ 0\ 2] + [1\ 1\ 1\ 1\ 0\ 0] \begin{bmatrix} 0 & 0 & 1 & 0 \\ -1 & 0 & 0 & 1 \\ 1 & -2 & 0 & 0 \\ 0 & -1 & 1 & 0 \\ 1 & 0 & -1 & 0 \\ 0 & 0 & -1 & 1 \end{bmatrix} = [1\ 5\ 2\ 3].$$

If vectors u_2, u_3 and u_4 are applied consecutively, the following marking vectors will be obtained, (see dynamics in Fig. 9.10 and resulting marking in Table 9.7).

Below is the code in Matlab to get the marking x1 when vector u1 is applied:

```
% Example 9.1 – PN dynamics –
xo=[1 8 0 2];
A=[0 0 1 0;-1 0 0 1;1 -2 0 0;0 -1 1 0;1 0 -1 0;0 0 -1 1];
u1=[0 1 1 0 1 0];
x1=xo+u1*A;
```

9.3.3 *Application of PNs for monitoring purposes of a distribution network*

As mentioned, the versatility of the modeling of electrical systems with PNs allows the identification of specific zones that operate with a high distortion content. Thus in this section a monitoring system is proposed.

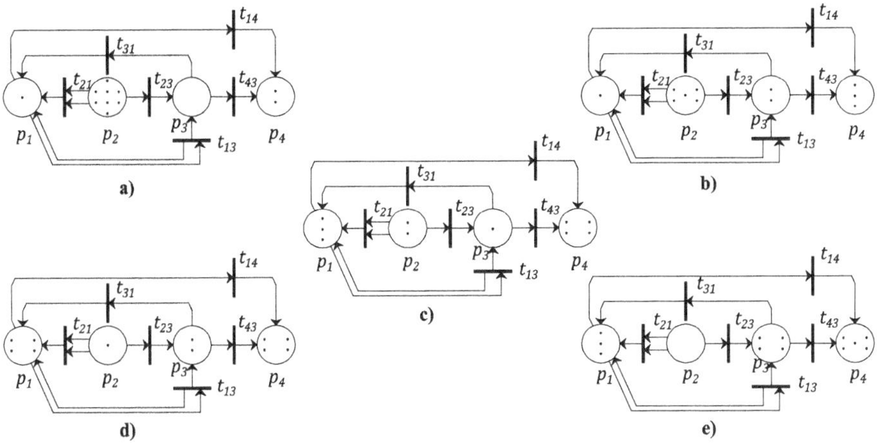

Figure 9.10 Petri Net dynamics. a) Initial condition x_0=[1 8 0 2], b) $u_1 \longrightarrow x_1$=[1 5 2 3], c) $u_2 \longrightarrow x_2$=[3 2 1 4], d) $u_3 \longrightarrow x_3$=[4 1 2 4], e) $u_4 \longrightarrow x_4$=[3 0 4 5].

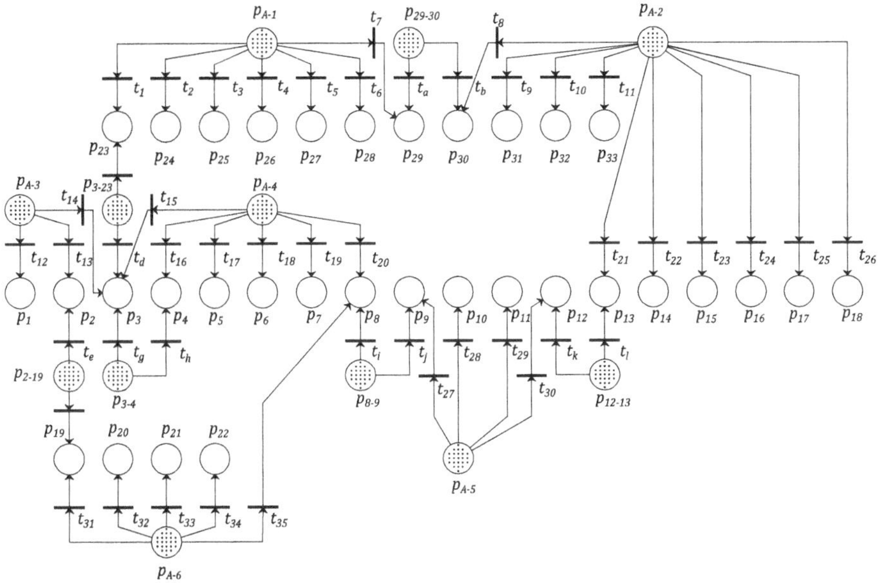

Figure 9.11 Monitoring system modeled with Petri nets for electrical system of Fig. 9.1.

Consider the system in Fig. 9.1 and assume that the 6 current measurement devices between buses 2–19, 3–4, 3–23, 8–9, 12–13, and 29–30 operate normally. Also, consider that voltage meters are enabled on all buses in the electrical system.

The places p_1 to p_{33}, in Fig. 9.11, represent the condition of each bus of the system under analysis, in terms of the presence of distortion in the waveform, that is, if a certain number of tokens appear in any of these places, it represents that there is a high distortion in the current or voltage waveform. The places p_{2-19}, p_{3-4}, p_{3-23},

Table 9.8 Firing conditions for each transition: total harmonic voltage distortion.

Transitions	Firing conditions
t_1	If THD_v bus 23 > THD_v of all buses associated with place $p_{A\text{-}1}$
t_2	If THD_v bus 24 > THD_v of all buses associated with place $p_{A\text{-}1}$
t_3	If THD_v bus 25 > THD_v of all buses associated with place $p_{A\text{-}1}$
t_4	If THD_v bus 26 > THD_v of all buses associated with place $p_{A\text{-}1}$
t_5	If THD_v bus 27 > THD_v of all buses associated with place $p_{A\text{-}1}$
t_6	If THD_v bus 28 > THD_v of all buses associated with place $p_{A\text{-}1}$
t_7	If THD_v bus 29 > THD_v of all buses associated with place $p_{A\text{-}1}$

Table 9.9 Firing conditions for each transition: total harmonic current distortion.

Transitions	Firing conditions
t_a, t_b	If estimated THD_i between buses 29 and 30 > all THD_i estimates
t_c, t_d	If estimated THD_i between buses 3 and 23 > all THD_i estimates
t_e, t_f	If estimated THD_i between buses 2 and 19 > all THD_i estimates
t_g, t_h	If estimated THD_i between buses 3 and 4 > all THD_i estimates
t_i, t_j	If estimated THD_i between buses 8 and 9 > all THD_i estimates
t_k, t_l	If estimated THD_i between buses 12 and 13 > all THD_i estimates

p_{8-9}, p_{12-13}, and p_{29-30}, contains the obtained information from the estimation of the total current distortion index, and the places $p_{A\text{-}1}$ to $p_{A\text{-}6}$, are the auxiliary places that contains information of the total voltage distortion index, both obtained with the Prony method. On the other hand, transitions t_1–t_{35} fire according to Table 9.8, and the transitions t_a to t_l fire according to Table 9.9.

From Table 9.8, since the place $p_{A\text{-}1}$ contains available tokens, any of the transition will fire and give a token to one of the places, p_{23}–p_{29}, when the estimated maximum THD_v corresponds to one of the buses 23–29, whose numbering is associated with place p_{A1}, correspondingly. In the same way, the transitions associated with the auxiliary places, $p_{A\text{-}2}$ to $p_{A\text{-}6}$, will fire if the THD_v of any of the buses associated with these places is maximum. For example, p_{A2} is associated with the group of buses 30, 31, 32, 33, 13, 14, 15, 16, 17, and 18, as can be seen in Fig. 9.11. Therefore, when any of the buses present a maximum THD_v, the corresponding transition will fire, giving a token to the respective place. In the case of t_a–t_l transitions, associated with the places containing the THD_i information, p_{2-19}, p_{3-4}, p_{3-23}, p_{8-9}, p_{12-13}, and p_{29-30}, they will fire when the estimated THD_i is greater with respect to the same measurements, that is, if the maximum THD_i is presented in place p_{2-19}, the transitions t_e and t_f will fire, yielding a token to places p_2 and p_{19}, as shown in Table 9.9.

Then, the structure for the PN of Fig. 9.11 according to Eqs. (9.15)–(9.18), is defined as,

$$P = \{p_1 \text{ to } p_{33}, p_{A-1} \text{ to } p_{A-6}, \text{ and } p_{2-19} \text{ to } p_{29-30}\},$$

$$T = \{t_1 \text{ to } t_{35} \text{ and } t_a \text{ to } t_l\},$$

Table 9.10 Description of events recorded during system operation.

Events	Description
ev1. (0.8 s-1.2 s)	Load disconnection on bus 31
ev2. (1.8 s-2.2 s)	load reconnection on bus 31
ev3. (2.8 s-3.2 s)	Increased load on the bus 25
ev4. (4.8 s-5.2 s)	Capacitor bank connection on bus 24
ev5. (9.8 s-10.2 s)	Capacitor bank disconnection on bus 24
ev6. (11.8 s-12.2 s)	Non-linear load connection on bus 29
ev7. (19.8 s-20.2 s)	Non-linear load disconnection on the bus 29

$$A = \{[(p_{A-1} \text{ to } (t_1 - t_7)) \text{ and, } (t_1, t_7) \text{ to } (p_{23} - p_{29})], [(p_{A-2} \text{ to } (t_8 - t_{11}, t_{21} - t_{26}) \text{ and } (t_8 - t_{11}, t_{21} - t_{26}) \text{ to } (p_{30} - p_{33}, p_{13} - p_{18})], [(p_{A-3} \text{ to } (t_{12} - t_{14}) \text{ and } (t_{12} - t_{14}) \text{ to } (p_1 - p_3)], [(p_{A-4} \text{ to } (t_{15} - t_{20}) \text{ and } (t_{15} - t_{20}) \text{ to } (p_3 - p_8)], [(p_{A-5} \text{ to } (t_{27} - t_{30}) \text{ and } (t_{27} - t_{30}) \text{to } (p_9 - p_{12})], [(p_{A-6} \text{to } (t_{31} - t_{35}) \text{ and } (t_{31} - t_{35}) \text{ to } (p_{19} - p_{22}, p_8)], [(p_{29-30} \text{to } (t_a \text{ and } t_b) \text{ and } (t_a \text{ and } t_b) \text{to } (p_{29} \text{ and } p_{30})], [(p_{3-23} \text{ to } (t_c \text{ and } t_d) \text{ and } (t_c \text{ and } t_d) \text{to } (p_{23} \text{ and } p_3)], [(p_{2-19} \text{to } (t_e \text{ and } t_f) \text{ and } (t_e \text{ and } t_f) \text{to } (p_2 \text{ and } p_{19})], [(p_{3-4} \text{to } (t_g \text{ and } t_h) \text{ and } (t_g \text{ and } t_h) \text{ to } (p_3 \text{ and } p_4)], [(p_{8-9} \text{to } (t_i \text{ and } t_j) \text{ and } (t_i \text{ and } t_j) \text{to } (p_8 \text{ and } p_9)], [(p_{12-13} \text{ to } (t_k \text{ and } t_l) \text{ and } (t_k \text{ and } t_l) \text{to } (p_{12} \text{ and } p_{13})]\},$$

$$W = \{w = 1; \text{ for all arcs of the system}\}.$$

Now, to validate the operation of the monitoring system, it is possible to section the PN in 6 subsystems. These segments are defined by the places $p_{A\text{-}1}$–$p_{A\text{-}6}$ and allow to easily identify the bus with the greatest distortion. To exemplify, consider the following two cases.

9.3.4 Study cases

9.3.4.1 Case 1

From the system of Fig. 9.1, the voltage and current signals were obtained during the operation of the system in 7 scenarios, see Table 9.10. The signals obtained were processed with the described Prony methodology, and the THD_i and THD_v indices were obtained. With this information, the Petri net was fed to identify the bus with the greatest distortion. In the first case, the Petri net was sectioned considering the maximum THD_i values, and once the area was identified, the highest THD_v value of the buses involved was obtained. This process is described in Fig. 9.12.

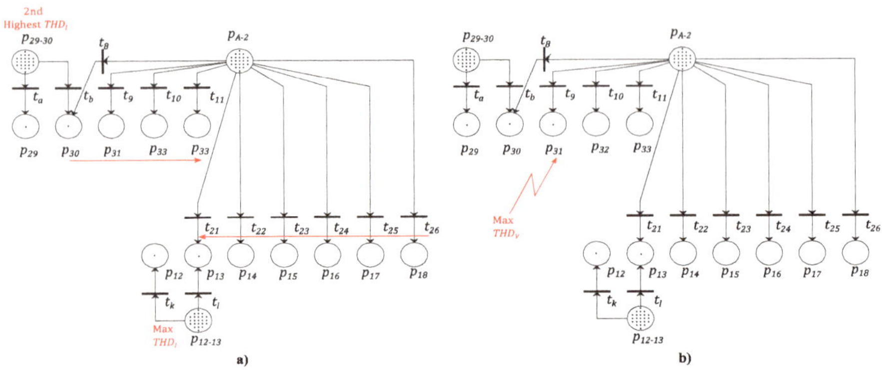

Figure 9.12 Dynamics of the Petri net proposed for the detection of the bus with the highest distortion: a) Since the two maximum values of THD_i were presented in the estimates corresponding to places p_{29-30} and p_{12-13}, the analysis is confined to buses 30–33 and 14–18, see red arrows. Notice the tokens in places $p_{29}-p_{30}$ and $p_{13}-p_{14}$ delimit the analysis area; b) place $p_{A\text{-}2}$, which contains the information on the THDV of the buses under analysis, detects that the maximum value of THD_v is presented on bus 31, which corresponds to event 2 of Table 9.10. See token in place p_{31}.

9.3.4.2 Case 2

Another way to identify the bus with the greatest distortion is to carry out a radial search, starting from the maximum value of THD_i. For example, in another moment of the system operation, a maximum value of THD_i was obtained in one of the places, however, now the search will focus on the maximum value of THD_v among the buses near this place. This process is described in Fig. 9.13.

Finally, it is important to highlight that in this monitoring proposal for the identification of buses with high harmonic content, the THD_v and THD_i indices were used for the identification, however, the estimation of any other index could be used for the detection of buses with high harmonic content or with an operating regime that is harmful to the electrical network.

9.4 Conclusions

In this chapter the integration of two analytical tools to build a monitoring system of an electrical network whose objective is to detect zones or operating points with high harmonic content is presented. The proposed monitoring system is divided into three stages; the first one includes the data acquisition coming from smart sensors. The second stage corresponds to the estimation of the harmonic distortion indices through the Prony method: both voltage and current signals. The last stage includes the detection of zones or points that operate with a high rate of harmonic distortion using a Petri net. The benefits of this approach are listed below:

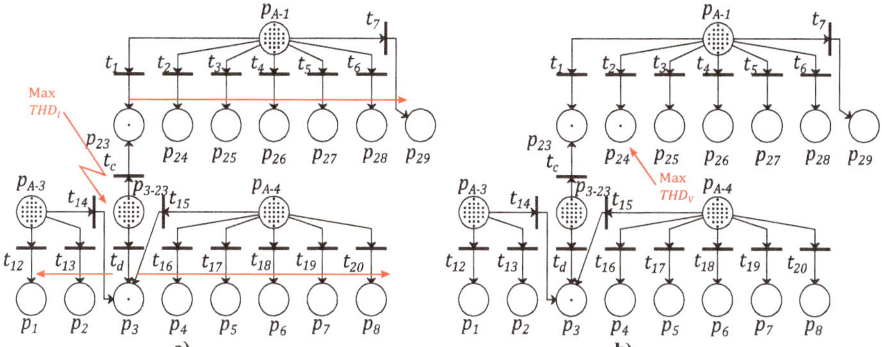

Figure 9.13 Dynamics of the Petri net proposed for the detection of the bus with the highest distortion: a) Given that the maximum value of THD_i was presented in the estimate p_{3-23}, the analysis is confined to buses 23–29, 3–8, and 1–2, see red arrows. Notice the tokens in places p_{23} and p_3, from which the search for the maximum value of THD_v begins; b) place $p_{A\text{-}1}$, which contains the information on the THD_v of the buses under analysis, detects that the maximum value of THD_v is presented on bus 24, which corresponds to event 4 of Table 9.10. See token in place p_{24}.

1. Given the versatility of the Petri net, it is possible to formally build a monitoring system whose inputs are the estimated parameters obtained by the Prony method, which can be adapted to any electrical network, and whose accuracy will be a function of the available measurement points.
2. By using the proposed methodology and aided by statistical analysis, a distribution system operator can obtain critical regions operating out of the permissible harmonic limits and, at the same time, can store the number of infringements to determine the incidence at which specific areas of the electrical network operate outside the harmonic limits allowed by the power quality standards.
3. By using the Prony method, not only THD_v and THD_i can be calculated, it is also possible to obtain different power quality indices that serve as inputs for the Petri net, and thus, sustained electrical power problems could be detected in certain areas of the electrical network.
4. Given the formality of the proposed approach, the use of Petri nets in conjunction with the Prony method can be extended to control systems that allow sectioning the electrical network during failure events to reduce the impact on users, that is, disconnecting the lowest number of users.

References

[1] R. Andreoni, D. Macii, M. Brunelli, D. Petri, Tri-objective optimal PMU placement including accurate state estimation: the case of distribution systems, IEEE Access 9 (2021) 62102–62117, https://doi.org/10.1109/ACCESS.2021.3074579.

[2] A. Zahernia, H. Rahbarimagham, Application of smart transformers in power systems including PV and storage systems under unbalanced and nonlinear load and fault condition, Electric Power Systems Research 201 (2021) 107535, https://doi.org/10.1016/j.epsr.2021.107535.
[3] S. Chakraborty, S. Das, T. Sidhu, A.K. Siva, Smart meters for enhancing protection and monitoring functions in emerging distribution systems, International Journal of Electrical Power & Energy Systems 127 (2021) 106626, https://doi.org/10.1016/j.ijepes.2020.106626.
[4] Y. Baghzouz, et al., Time-varying harmonics. II. Harmonic summation and propagation, IEEE Transactions on Power Delivery 17 (1) (2002) 279–285, https://doi.org/10.1109/61.974219.
[5] W. Zhou, H. Xie, Y. Tong, Propagation mechanisms research of harmonics produced by distributed generations in distribution network, in: 2012 IEEE 15th International Conference on Harmonics and Quality of Power, 2012, pp. 774–777.
[6] Y. Guangzheng, L. Tao, X. Xialing, Comparison of harmonic propagation methods considering uncertainties of harmonic injection, in: International Conference on Renewable Power Generation (RPG 2015), 2015, pp. 1–5.
[7] P. Rodríguez-Pajarón, A. Hernández Bayo, J.V. Milanović, Forecasting voltage harmonic distortion in residential distribution networks using smart meter data, International Journal of Electrical Power & Energy Systems 136 (2022) 107653, https://doi.org/10.1016/j.ijepes.2021.107653.
[8] S. Vlahinic, D. Brnobic, N. Stojkovic, Indices for harmonic distortion monitoring of power distribution systems, IEEE Transactions on Instrumentation and Measurement 58 (5) (2009) 1771–1777, https://doi.org/10.1109/IMTC.2008.4547072.
[9] IEEE recommended practice and requirements for harmonic control in electric power systems, in: IEEE Std 519-2014 (Revision of IEEE Std 519-1992), 11 June 2014, pp. 1–29.
[10] F.A.S. Borges, R.A.S. Fernandes, I.N. Silva, C.B.S. Silva, Feature extraction and power quality disturbances classification using smart meters signals, IEEE Transactions on Industrial Informatics 12 (2) (2016) 824–833, https://doi.org/10.1109/TII.2015.2486379.
[11] T. Lobos, J. Rezmer, J. Schegner, Parameter estimation of distorted signals using Prony method, in: 2003 IEEE Bologna Power Tech Conference Proceedings, vol. 4, 2003, pp. 1–5.
[12] Z. Leonowicz, T. Lobos, J. Rezmer, Advanced spectrum estimation methods for signal analysis in power electronics, IEEE Transactions on Industrial Electronics 50 (3) (2003) 514–519, https://doi.org/10.1109/TIE.2003.812361.
[13] Zbigniew Leonowicz, Parametric Methods for Time–Frequency Analysis of Electric Signals, Wrocław University of Science and Technology, 2006, 15/02/2022. Website https://www.dbc.wroc.pl/dlibra/publication/1877/edition/2021?language=pl.
[14] L. Qi, L. Qian, S. Woodruff, D. Cartes, Prony analysis for power system transients, EURASIP Journal on Advances in Signal Processing 048406 (2007), https://doi.org/10.1155/2007/48406.
[15] M. Meunier, F. Brouaye, Fourier transform, wavelets, Prony analysis: tools for harmonics and quality of power, in: 8th Int. Conf. on Harmonics and Quality of Power ICHQP'98, 1998, pp. 71–76.
[16] M.A. Johnson, I.P. Zarafonitis, M. Calligaris, Prony analysis and power system stability-some recent theoretical and applications research, in: 2000 Power Engineering Society Summer Meeting, 3 , 2000, pp. 1918–1923.
[17] R. Castillo, J.R. Ramírez, G. Alonso, J. Ortiz-Villafuerte, Prony's method application for BWR instabilities characterization, Nuclear Engineering and Design Journal 284 (2015) 67–73, https://doi.org/10.1016/j.nucengdes.2014.11.034.

[18] L.A. Trujillo Guajardo, Prony filter vs conventional filters for distance protection relays: an evaluation, Electric Power Systems Research 137 (2016) 163–174, https://doi.org/10.1016/j.epsr.2016.04.009.
[19] L.A. Trujillo Guajardo, A. Conde Enríquez, G.M. Idarraga Ospina, Prony method implementation in distance relays to mitigate the effect of series-compensated transmission lines, Electrical Engineering 99 (2017) 227–239, https://doi.org/10.1007/s00202-016-0411-4.
[20] L.A. Trujillo Guajardo, Relevador De Protección De Distancia Con Estimador Fasorial De Prony, MX Patent No. 351620 B, México Patent Office, Instituto Mexicano de la Propiedad Industrial, IMPI, 2017, https://vidoc.impi.gob.mx/visor?usr=SIGA&texp=SI&tdoc=E&id=MX/a/2014/012486.
[21] Paulo Fernando Ribeiro, Carlos Augusto Duque, Paulo Márcio Ribeiro, Augusto Santiago Cerqueira, Spectral estimation, in: Power Systems Signal Processing for Smart Grids (254-270), John Wiley & Sons LTD, United Kingdom, 2014.
[22] F. Salinas, M.A. González, M.F. Escalante, Finite control set-model predictive control of a flying capacitor multilevel chopper using Petri nets, IEEE Transactions on Industrial Electronics 69 (9) (2016) 5891–5899, https://doi.org/10.1109/TIE.2016.2578284.
[23] S. Othman, et al., Robust hybrid control based on Petri nets for a multicellular inverter, in: 2020 6th IEEE International Energy Conference (ENERGYCon), 2020, p. 143.
[24] A. Fekik, H. Denoun, A. Azar, M. Hamida, M. Atig, M. Ghanes, N. Benamrouche, J-P. Barbot, Direct power control of a three-phase PWM-rectifier based on Petri nets for the selection of switching states, in: 2018 7th IEEE International Conference on Renewable Energy Research and Applications Conference (ICRERA), 2018, pp. 1121–1125.

Dynamic voltage restorer controlled per independent phases for power quality sags-swells mitigation under unbalanced conditions

10

Vicente Torres-García, Gabriela Castillo-García, Gabriel Mejía-Ruiz, and Mario Roberto Arrieta-Paternina
National Autonomous University of Mexico, Mexico City, Mexico

10.1 Introduction

The most common power quality issues are associated with voltage variations, being known as sags and swells. Voltage variations are triggered by the incorporation of a large amount of power electronic converters that interface distributed energy resources, such as solar and wind. Voltage sags are generated by faults, or unsuitable starting of electromechanical machines, such as motors and transformers, whereas voltage swells are provoked by switching of protecting devices and connection of capacitor banks. Large voltage excursions under unbalanced conditions cause critical events in sensitive loads [1]. These are harmful for industrial, commercial, and residential loads [2,3]. To mitigate these issues, grid codes establish a maximum admissible limits for voltage variations, which must be fulfilled according to the IEC std. 61000-4-30 [4] and the IEEE/ANSI Std. 519 [5].

To solve the voltage variations by using dynamic voltage restorers (DVRs), two approaches have been proposed: artificial intelligence (AI) and traditional linear/nonlinear control structures. The former has recently been explored in combination with nonlinear control techniques for DVRs, using techniques such as machine learning (ML) and deep learning (DL) [6]. Different investigations report the application of AI in nonlinear controllers with ML, controllers based on neural networks [7–9], and controllers with fuzzy logic [10–12]. The latter implement known control structures with topological changes, such as a DVR with automatic balancing of the capacitor voltage [13], a transformerless DVR with predictive controller [14], a DVR with a novel method for detecting sags in less than 2 ms [15], an energy storage system-based DVR that uses a precompensation method for the voltage amplitude in three-phase systems [16], a sliding mode-based control strategy for a 3-phase DVR employing a 12-switch voltage source converter [17], among others [18,19]. Despite their advantages, all of these works present complex control algorithms that are unable to operate in unbal-

Monitoring and Control of Electrical Power Systems Using Machine Learning Techniques
https://doi.org/10.1016/B978-0-32-399904-5.00016-8

anced conditions or require independent hardware for each phase, increasing the cost and complexity of the system.

This work proposes a DVR controlled per independent phases that is able to alleviate power quality sags-swells under unbalanced/balanced grid conditions, employing the widely known two-level, three-phase voltage-source converter (VSC). This proposal implements linear controllers for each phase of the three-phase feeder in the synchronous reference frame *dq0*, regulating the voltage profile through the in-phase compensation method [20]. This approach reduces the amount of energy required from the DC link. The effectiveness of the proposed strategy is demonstrated by carrying out time-domain simulations of benchmark test power systems under steady-state and dynamic conditions, interconnecting DVRs in sensible nodes.

10.2 Dynamic voltage restorer

Its basic operation consists of monitoring the voltage amplitude at the node to be protected. Based on this measurement, the DVR controller calculates the magnitude of the voltage required to compensate the voltage profile. The DVR, connected in series with the line, adds the voltage with the magnitude and phase suitable to restore the voltage level at the point of common coupling (PCC) [21].

10.2.1 Proposed DVR topology

The technical literature reports several topologies for DVR implementation, among these, the following stand out: single-phase two-levels, three-phase two levels, multilevel clamped H-bridge, transformerless multilevel topology, among others [22–27].

These topologies can be classified according to the energy storage element on the DC side. The nonenergy storage topologies are characterized by its smaller volume and simplified hardware. However, this topology requires a complex control structure. The nonenergy storage topology is recommended for robust electrical networks, where an increase in the supply current and the upstream loads may experience a greater voltage variation when a disturbance occurs [28]. Meanwhile the energy storage topologies have an additional hardware cost; however, it has a better dynamic performance and less control complexity than the topologies that do not depend on an energy storage system [28].

Due to its operational advantages in the MV level grids and its control simplicity, the energy storage topology was selected in this work. Its fundamental schematic diagram is depicted in Fig. 10.1(a), and the proposed control scheme is detailed in Fig. 10.1(b).

The proposed DVR is composed by a coupling transformer, a passive filter, a voltage-source converter (VSC), and an energy storage unit. The coupling transformer is used to match the DVR voltage to the voltage levels required in the network. This transformer also galvanically isolate both the DVR and grid systems [29,30]. Its connection can be star-grounded or star-delta; the first allows the injection of positive,

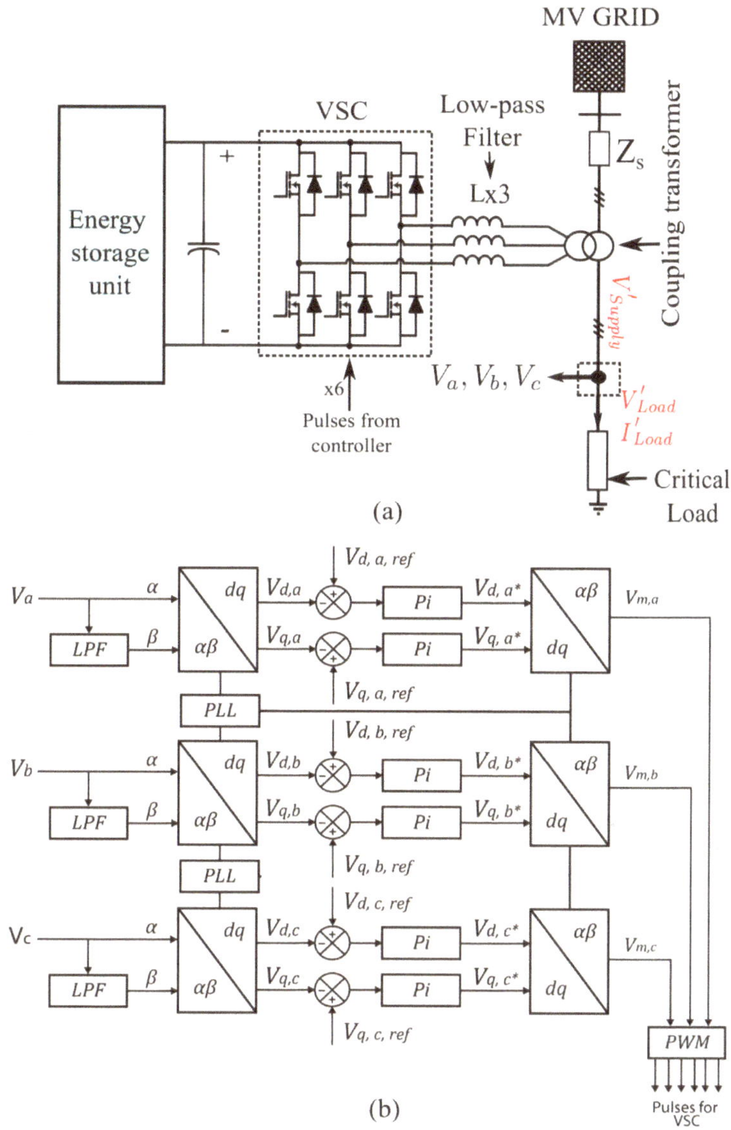

Figure 10.1 Schematic diagram of the proposed DVR.

negative, and zero sequence voltages, whereas the second allows only positive and negative sequence voltages [29–32]. The low-pass passive filters are used to limit high-frequency harmonics from switching power switches, reducing harmonic distortion to provide a near-sinusoidal voltage waveform at the output. These filters can be inductive or inductive-capacitive [30,32,33]. Each voltage-source converter (VSC) can generate a controlled voltage in series with the desired frequency, magnitude, phase and angle [34]. This power electronic device converts the DC voltage of the storage system to an AC voltage useful to compensate the voltage on the PCC, using the suitable switching of six MOSFETs, GTOs, IGBTs, or IGCTs. The energy storage unit transfers power during the voltage compensation [30]. The injection of the voltage into the network will be proportional to the storage unit. Some energy storage devices can be batteries, capacitors, superconducting magnetic energy storage, super capacitors, among others. PCC voltage measurements (V_a, V_b, V_c) and trigger signals from the switches are used to close the control loop.

10.2.2 DVR implementation

In this work, the scheme of the DVR drawn in Fig. 10.1(a) is carried out by using ATP/EMTP [35]. Two main blocks representing the DVR's physical structure and its control are depicted. The former consists of the following blocks: a block representing the grounded wye-connected coupling transformer, a filter to mitigate the harmonics produced by the inverter switching, a voltage inverter, and the energy storage system that supplies the necessary energy to compensate for disturbances. Meanwhile the latter in Fig. 10.1(b) comprises a filtering stage to generate the α and β signal by using two blocks that use the transfer function of the filter, a PLL for control synchronization, transformations blocks from $\alpha\beta$ to *dq0*, a compensation stage to calculate the voltage injection for the disturbance presented in the network with a PI controller, and a *dq0* to $\alpha\beta$ transformation block.

10.2.3 DVR principles

The compensation method determines the magnitude and angle of the voltage to be injected by the DVR. The selection of the compensation method depends on the following criteria: the injection capacity, the energy storage type in the DC link, and the characteristics of the load to be protected [32,36]. Among the traditional compensation methods, the following stand out: pre-sag, in-phase and minimum energy injection compensation [36–38].

10.2.3.1 Pre-sag compensation method

This method compensates magnitude and phase angle; it is based on detecting the amplitude and phase of the voltage before the disturbance occurs. The characteristic phasor diagram is shown in Fig. 10.2, where $V'_{supply-k}$, V'_{load}, V'_{DVR}, and I'_{load} represent the variables after the disturbance, whereas $V_{supply-k}$, V_{load}, and I_{load} are the quantities before the disturbance.

The DVR does not have the same phase as the load current; that is, voltage across the DVR will be demanding both active and reactive powers to compensate the voltage magnitude and phase changes [37]. The equation associated with the voltage injection is given by (10.1), whereas the required angle at the load node is represented in (10.2).

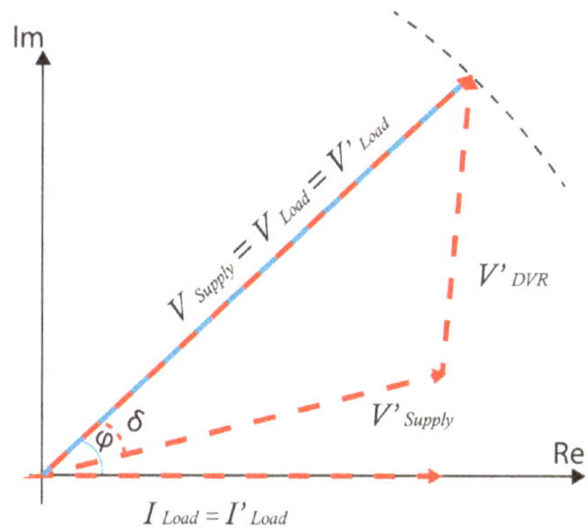

Figure 10.2 Pre-sag compensation method.

$$V'_{DVR} = \sqrt{V_{load}^2 + {V'_{supply-k}}^2 - 2V_{load}V'_{supply-k}\cos(\delta_k)}, \tag{10.1}$$

$$\theta_{DVR} = tan^{-1}\left(\frac{V_{load}\sin(\phi) - V'_{supply-k}\sin(\phi - \delta_k)}{V_{load}\cos(\phi) - V'_{supply-k}\cos(\phi - \delta_k)}\right). \tag{10.2}$$

The apparent power S_{DVR} and the active power P_{DVR} are calculated with (10.3) and (10.4), respectively:

$$S'_{DVR} = I_{load}\sqrt{V_{load}^2 + {V'_{supply-k}}^2 - 2V_{load}V'_{supply-k}\cos(\delta_k)}, \tag{10.3}$$

$$P'_{DVR} = I_{load}(V_{load}\cos(\phi) - V'_{supply-k}\cos(\phi - \delta_k)), \tag{10.4}$$

where the variables are explained in Table 10.1.

This technique require a large source of energy storage to be able to supply the active power injected by the VSC, so its use is recommended for loads that are extremely sensitive to voltage and phase changes.

Table 10.1 Variables in (10.4).

Symbol	Variable
V_{load}	Voltage on the side of the load to be protected
I_{load}	Current on the side of the load to be protected
$V'_{supply-k}$	Power supply voltage
ϕ	Angle between V_{load} y I_{load}
δ_k	Phase shift of $V'_{supply-k}$
k	Phase a, b, and c of power supply voltage

10.2.3.2 In-phase compensation method

In this technique, the supply voltage is in-phase with the injected voltage. The load current and pre-fault voltage are not considered in this method; in consequence, the phase shift is not compensated in this case. Fig. 10.3 shows the diagram of in-phase compensation method [31].

The injected voltage V_{DVR} is calculated with Eq. (10.5), where the difference voltage between the sag/swell voltage at the point of common coupling (PCC) is.

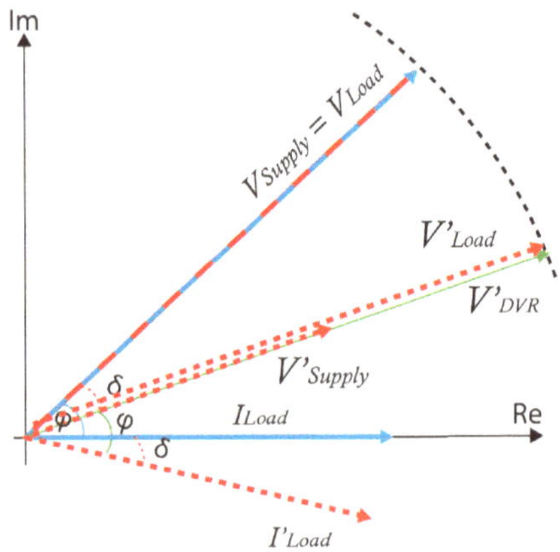

Figure 10.3 In-phase compensation method.

$$|V'_{DVR}| = |V_{load} - V'_{supply-k}| \tag{10.5}$$

$$\theta_{DVR} = \phi \tag{10.6}$$

The apparent power and the injected real power are given by (10.7) and (10.8), respectively.

$$S'_{DVR} = I_{load}(V_{load} - V'_{supply-k}) \tag{10.7}$$

$$P'_{DVR} = I_{load}(V_{load} - V'_{supply-k})\cos\phi \tag{10.8}$$

This method is the most suitable when it comes to loads that are sensitive to voltage variations since it performs the injection of voltage precisely compared with the other methods, since by varying the supply voltage the DVR generates a minimum voltage necessary to mitigate the disturbance with the existing current, which causes a lower demand for active power [29]. As already mentioned, due to the fact that the voltage injection is carried out in phase with the electrical network, the phase changes cannot be compensated, which leads to transients and circulating currents on the side of the protected load.

10.2.3.3 Minimum energy injection compensation method

This strategy seeks to reduce the active power injected; therefore it is based on injecting or absorbing as much reactive power as possible from the supply, and thus reducing the amount of active power of the DC link in the compensation process, in such a way that the voltage injected by the DVR is perpendicular to the load current [30]. Fig. 10.4 shows the phasor diagram for this control strategy, in which the load current and voltage are fixed values, so that only the phase of the mains voltage can be modified.

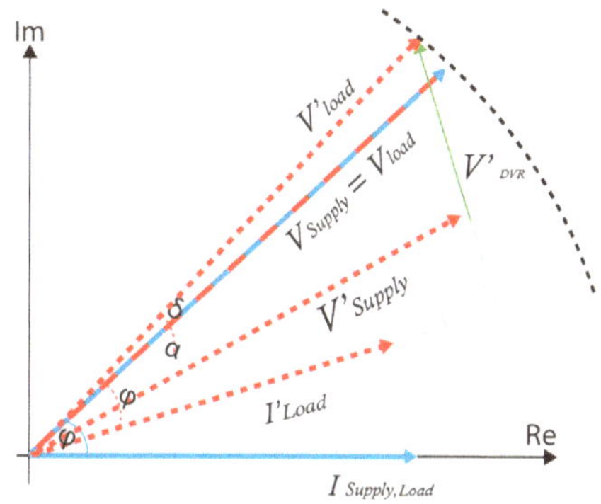

Figure 10.4 Minimum energy injection compensation method.

Eqs. (10.9) and (10.10) determine the voltage injection, however, in unbalanced systems the equations are applied per phase, and it is not considered that the active power between the DVR and the electrical network in each phase is zero, but the total active power exchanged is zero, so it is physically required that the link be common for all three phases.

$$V'_{DVR} = \sqrt{V_{load}^2 + {V'_{supply-k}}^2 - 2V_{load}V'_{supply-k}\cos(\delta_k + \alpha)}, \tag{10.9}$$

$$\theta_{DVR} = tan^{-1}\left(\frac{V_{load}\sin(\phi+\alpha) - V'_{supply-k}\sin(\phi-\delta_k)}{V_{load}\cos(\phi+\alpha) - V'_{supply-k}\cos(\phi-\delta_k)}\right), \tag{10.10}$$

where α is the phase variation of the V_{load} after compensation; this is calculated by (10.11) for balanced systems and (10.12) for unbalanced systems [38].

$$\alpha = \phi - \delta_k - cos^{-1}\left(\frac{V_{load}cos(\phi)}{V'_{supply}}\right), \tag{10.11}$$

$$\alpha = \phi - \delta_k - cos^{-1}\left(\frac{3V_{load}cos(\phi)}{\sqrt{X^2+Y^2}}\right), \tag{10.12}$$

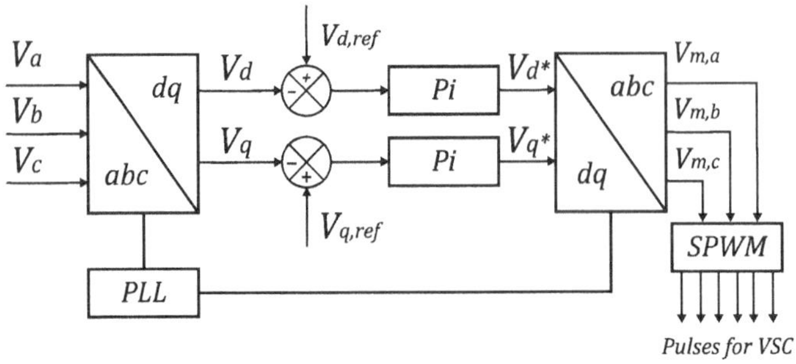

Figure 10.5 Conventional control scheme.

and

$$X = \sum_{k=a,b,c} (V'_{supply} \cos(\delta_k)), \quad (10.13)$$

$$Y = \sum_{k=a,b,c} (V'_{supply} \sin(\delta_k)). \quad (10.14)$$

The active power injected by the DVR is considered (10.15)

$$P'_{DVR} = P_{load} - P'_{supply-k} = 0 \quad (10.15)$$

In (10.11), the following conditions must be met: $V_{load} \cos\phi > V'_{supply}$, for balanced systems, and in (10.12) $3V_{load} \cos\phi > \sqrt{X^2 + Y^2}$ for unbalanced systems to guarantee that the active power injected by the DVR is zero [38]. The advantage of this method is that it does not require active power, but this entails producing a shift phase and that the required voltage is much higher than the previous two methods, the energy storage system being a limitation, in addition to being useful to mitigate shallow disturbances [30,39].

10.3 Proposed controller

Conventionally, battery-based dynamic voltage restorers employ PI controllers to regulate both direct and quadrature voltages that finally generate the modulation signals for three-phase voltage-source converters [34], as shown in Fig. 10.5. Contrary to this fact, our proposed control structure in Fig. 10.1(b) relies on the use of voltage source converters controlled by independent phases, which it makes up as a three-layer control structure that relaxes the dynamics per phase and spans the range of DVR capabilities.

Our proposal harnesses the system transformation from the *abc* reference frame onto the α-β transformation and the *dq0* synchronous reference frame per phase of the three-phase feeder [3,40]. Thus each layer consists of a single-phase control loop that assumes the values of α component in the Clark's transformation, given directly by the control inputs, that is, the monitored voltage at the load. Meanwhile β component arises from a 90 degree shift of α by implementing a low-pass filter (LPF). The transfer function of the LPF is expressed by

$$\frac{V_{out}}{V_{in}} = \frac{1}{s\omega_0 + 1}, \tag{10.16}$$

where V_{out}, V_{in}, and ω_0 stand for the output voltage, input voltage, gain and, angle in radians, respectively.

Subsequently, α-β components are mapped onto d-q components of the Park's transformation to generate the PIs' control errors by means of their comparisons with their respective references. After obtaining the d-q values per phase, these are used through the *In-phase* compensation method that compares them with a reference signal and generates an error signal that indicates the injection voltage necessary to compensate for the present disturbance.

Finally, the inverse of Clark's transformation is applied to the PI controllers output to generate a sinusoidal pulse width modulation for the three-phase voltage-source converter by using the α signal. This is achieved, transforming from α–β onto the direct axis d and the quadrature axis q, which is given by

$$\begin{pmatrix} V_\alpha \\ V_\beta \\ V_0 \end{pmatrix} = \frac{2}{3} \begin{pmatrix} \cos(\omega t) & -\sin(\omega t) & 0 \\ \sin(\omega t) & \cos(\omega t) & 0 \\ 0 & 0 & 1 \end{pmatrix}, \tag{10.17}$$

where ωt is the angular position.

To facilitate this control process, all measurements are taken in per unit, and there is a PLL to synchronize the DVR with the system.

10.4 Independent phases-controlled DVR performance

To assess the performance of the independent phased-driven DVR, the proposed control structure is embedded into the DVR to evaluate in a test system the IEEE 13-node distribution feeder in Fig. 10.6. The system is small, but with a relatively high load, single voltage regulator in substation, overhead and underground lines, shunt capacitors, and unbalanced charge [41]. In this system, the DVR is utilized to protect a critical load; its parameters are found in Table 10.2.

In this work, four scenarios are carried out to verify the versatility of the DVR and to prove its effectiveness to different disturbances associated with single/two/three-phase faults and overvoltage, representing sag and swell conditions, respectively. Along with each scenario, the voltage magnitude is exhibited, which the DVR can compensate. All simulated disturbances are considered that take place at the distribution line. The DVR is setup to protect the load at node 675. According to the IEEE

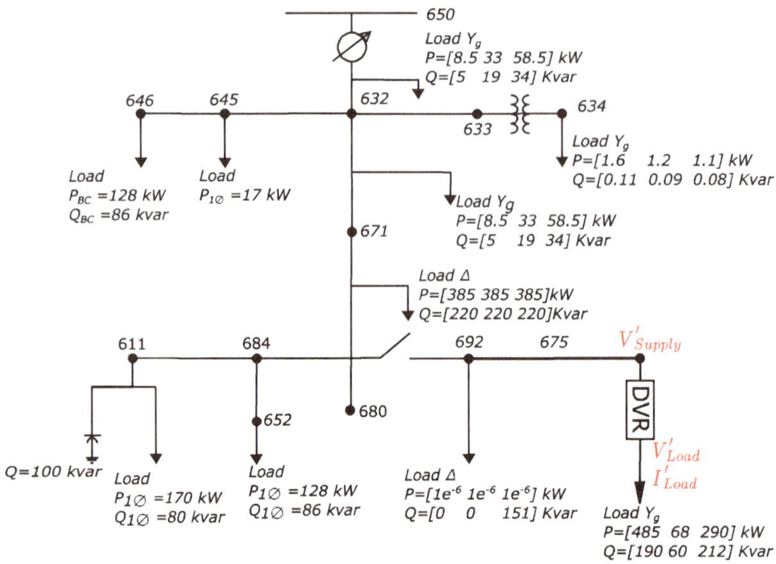

Figure 10.6 IEEE 13-node system.

Table 10.2 DVR parameters.

Parameter	Value
Energy storage unit	3500 V
Filter inductance (L)	1.33 mH
Filter capacitance (C)	5500 μF
Filter resistance (R)	0.3 Ω
Transformer ratio	1:2
Transformer power	0.5 MVA
Switching frequency	1260
Simulation sample time	0.4 ms

test system's base values, the phase B is taken as a reference with Vb=1.0 pu, and the voltage magnitudes in other phases are Va=0.96 pu and Vc=0.95 pu. The ATP/EMTP [35] tool is employed to model and simulate all scenarios and simulations:

Scenario 1. This scenario is carried out by a single-phase fault at the PCC of the load connected at node 675. The fault lasts from $t = 0.1$ s up to $t = 0.3$ s, which yields a 0.7 pu voltage drop in the faulted phase, as shown in Fig. 10.7(a); and in turn, this is compensated by injecting a voltage of 0.23 pu, as exhibited in Fig. 10.7(b). Finally, the voltage at the protected load is displayed in Fig. 10.7(c), which is retained within the permissible voltage variations. It is noteworthy to remark that this fault is the most common in electrical systems, being also considered as asymmetric due to the voltage behavior, as depicted in Fig. 10.7(a). In this graph, the voltage injection in Fig. 10.7(b) specifies that voltages in phases B and C are in phase; this behavior is

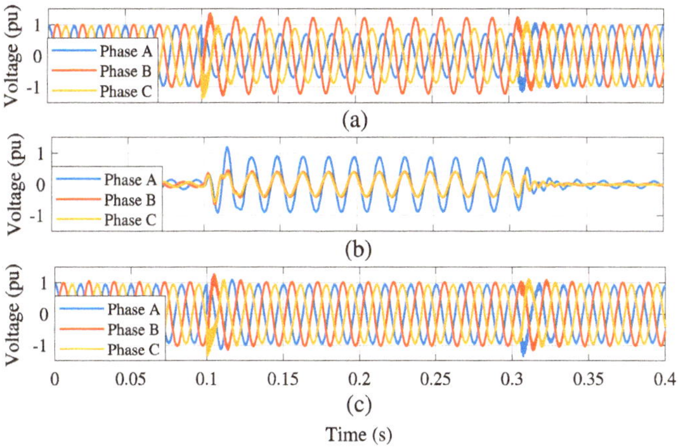

Figure 10.7 DVR response to single-phase fault. (a) Supply voltage. (b) DVR voltage. (c) Load voltage.

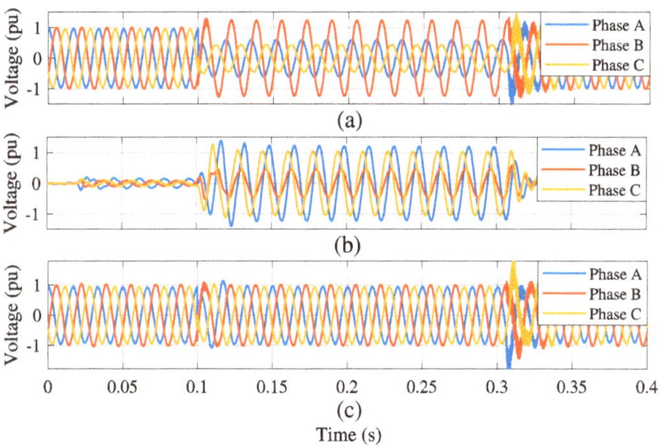

Figure 10.8 DVR behavior to a two-phase fault. (a) Supply voltage. (b) DVR voltage. (c) Load voltage.

due to the phase B increases in voltage, according to the DVR action. Then, a voltage is added to the main voltage, providing an adequate voltage profile for the protected load and resulting in the load voltage indicated in Fig. 10.7(c).

Scenario 2. A two-phase fault taking place in phases *a* and *c* that lasts from 0.1 s up to 0.3 s is simulated. This disturbance produces voltage drops in both phases of 0.4 pu and 0.56 pu, respectively. Also an increase in the healthy phase is evidenced, as shown Fig. 10.8(a). The voltage injection to compensate this drop is depicted in Fig. 10.8(b), producing the voltage in the protected load, as illustrated in Fig. 10.8(c), which it is within the permissible values.

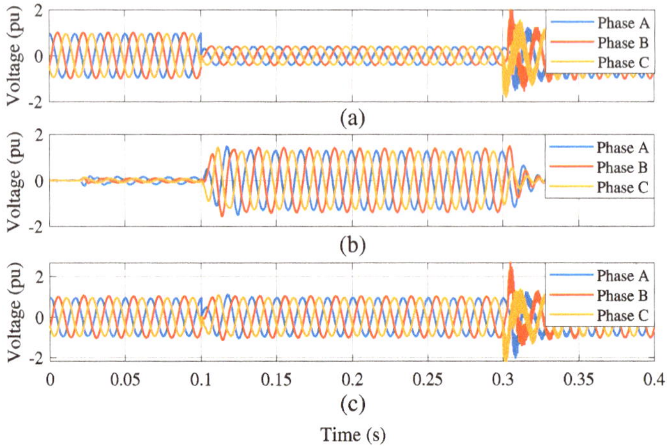

Figure 10.9 DVR response to three-phase fault. (a) Supply voltage. (b) DVR voltage. (c) Load voltage.

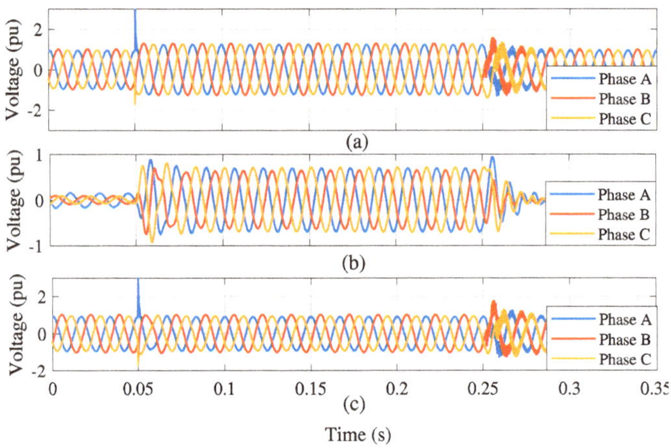

Figure 10.10 DVR response to swell. (a) Supply voltage. (b) DVR voltage. (c) Load voltage.

Scenario 3. This case comprises a 12-cycle three-phase fault at the PCC, affecting the protected load at node 675. Then, a voltage drop of 0.6 pu and a voltage injection can be observed in Figs. 10.9(a)–(b); meanwhile the voltage of the protected load is depicted in Fig. 10.9(c), which retains within the permissible range.

Scenario 4. A voltage swell of 1.5 pu is simulated from t =0.1 s to t =0.3 s, producing an increase of 0.28 pu in the phases, as shown Fig. 10.10(a). The voltage injection compensates this behavior in Fig. 10.10(b), protecting load, as exhibited Fig. 10.10(c).

As evidenced by these numerical results, the DVR controlled per independent phases achieves controlling the short-time voltage variations (sags/swells), even under unbalanced conditions.

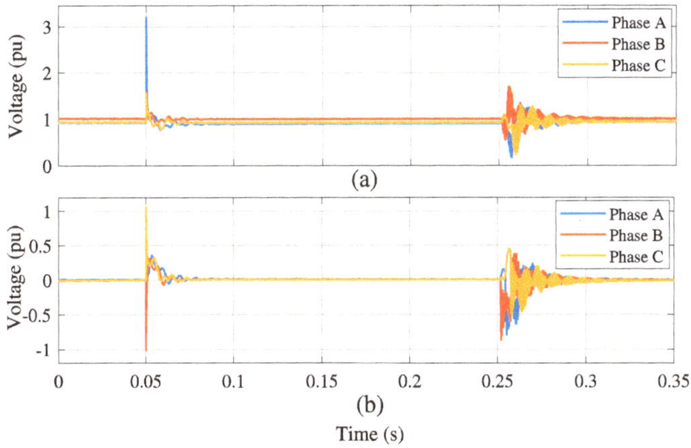

Figure 10.11 Voltages per phase. (a) Direct voltage (V_d). (b) Quadrature Voltage (V_q).

Table 10.3 Voltage magnitudes for all simulated scenarios.

Disturbance	V_{s_a}	V_{s_b}	V_{s_c}	V_{l_a}	V_{l_b}	V_{l_c}
Single-phase fault	0.7	1.21	0.87	0.95	1.02	0.93
Two-phase fault	0.6	1.23	0.44	0.96	1.02	0.94
Three-phase fault	0.4	0.41	0.4	0.96	1.03	0.95
Swell	1.23	1.28	1.25	0.91	1	0.93

To illustrate the internal dynamics in the control structure proposed in Fig. 10.1 (b), the time responses of the direct (V_d) and quadrature (V_q) voltages per phase are depicted in Fig. 10.11 to a swell of 1.5 pu. At the instant $t = 0.05$ s, when the swell disturbance occurs, V_d and V_q experiment transient voltages higher than 3 pu. Then, the swell event is removed at the instant 0.25 s, returning the voltages at their rated values.

10.5 Discussion

The summary of different scenarios is shown in Table 10.3, where the voltage magnitude at the PCC and the voltage of the protected load under different disturbances observed. It can also be seen that the proposed DVR is an effective method to compensate sensitive loads.

A better understanding of the proposed DVR performance can be analyzed in Table 10.4, where the voltage variation in percentage presents the load experiment under the disturbance. Table 10.4 depicts that the DVR has a better performance in the presence of three-phase fault with symmetric sags in all phases and a worse performance when a swell appears; this phenomenon can be caused due to the fact that a greater sizing of a power system is required to counteract these disturbances. In addition, it

Table 10.4 Variation of load voltage.

Disturbance	$V_{l_a}(\%)$	$V_{l_b}(\%)$	$V_{l_c}(\%)$
Single-phase fault	1.4	2	2.1
Two-phase fault	0	2	1.05
Three-phase fault	0	3	0
Swell	5	0	2.1

can be observed from three-phase fault cases that the DVR has a higher performance in the three-phase fault and a worse performance in the single-phase fault. The latter may be due to the fact that both in the two-phase fault and in the three-phase fault there is a phase that rises above its nominal value to be able to counteract the sag voltage.

10.6 Conclusions

Voltage or frequency variations can cause extensive damages to distribution networks, with consequent economic losses. In this context, power electronic converters applied with novel control strategies effectively contribute to mitigate the problem. The results of the simulations show that the proposed controller per phase allows to regulate the voltage amplitude in unbalanced systems with a short response time (50 ms).

The attained results under single/two/three-phase faults evidence the robustness and reliability of the proposed control approach for the DVR over a wide operating range, improving the power quality in the distribution networks.

It is noteworthy to remark that the focus of this proposal chapter is to show a novel alternative solution to the control of electrical power systems, solving the most typical quality power issues, such as sags and swells, raising the opportunity to improve in a future work the control framework through the application of machine learning techniques.

References

[1] P.T. Ogunboyo, R. Tiako, I.E. Davidson, Effectiveness of dynamic voltage restorer for unbalance voltage mitigation and voltage profile improvement in secondary distribution system, Canadian Journal of Electrical and Computer Engineering 41 (2) (2018) 105–115.

[2] A.P.n. Torres, P. Roncero-Sanchez, V.F. Batlle, A two degrees of freedom resonant control scheme for voltage-sag compensation in dynamic voltage restorers, IEEE Transactions on Power Electronics 33 (6) (2018) 4852–4867.

[3] G.E. Mejia-Ruiz, R. Cárdenas-Javier, M.R. Arrieta Paternina, J.R. Rodríguez-Rodríguez, J.M. Ramirez, A. Zamora-Mendez, Coordinated optimal volt/var control for distribution networks via d-pmus and ev chargers by exploiting the eigensystem realization, IEEE Transactions on Smart Grid 12 (3) (2021) 2425–2438.

[4] IEC, International Standard IEC 61000-4-30:2003: Electromagnetic Compatibility (EMC): Part 4-30: Testing and Measurement Techniques – Power Quality Measurement Methods, IEC, 2003.
[5] IEEE/ANSI, IEEE recommended practice and requirements for harmonic control in electric power systems, in: IEEE Std 519-2014 (Revision of IEEE Std 519-1992), 2014, pp. 1–29.
[6] S. Zhao, F. Blaabjerg, H. Wang, An overview of artificial intelligence applications for power electronics, IEEE Transactions on Power Electronics 36 (4) (2021) 4633–4658.
[7] A. M J, S. N, M.S. Shashikala, Power Quality Enhancement using Dynamic Voltage Restorer (DVR) by Artificial Neural Network and Hysteresis Voltage Control Techniques, in: 2019 Global Conference for Advancement in Technology (GCAT), 2019, pp. 1–6.
[8] U. Muthuraman, A. Swetha, J.G. J, P. Annapandi, N.B. Rajesh, J.A. Nesa Priya, Power Quality Improvement Based On Artificial Neural Network Controller and Dynamic Voltage Restorer, in: 5th International Conference on Trends in Electronics and Informatics (ICOEI), 2021, pp. 1646–1652.
[9] F. Jurado, Neural network control for dynamic voltage restorer, IEEE Transactions on Industrial Electronics 51 (3) (2004) 727–729.
[10] H.S. Thaha, T.D. Prakash, Use of fuzzy controller based dvr for the reduction of power quality issues in composite micro-grid, in: 2020 International Conference on Renewable Energy Integration into Smart Grids: A Multidisciplinary Approach to Technology Modelling and Simulation (ICREISG), 2020, pp. 131–136.
[11] K.K. Senapati, M.B. Panda, R.S. Syed, A novel algorithm for power quality improvement using dynamic voltage restorer with fuzzy logic, in: 2016 International Conference on Signal Processing, Communication, Power and Embedded System (SCOPES), 2016, pp. 376–381.
[12] M.H. Sabzalian, K.A. Alattas, M. Aredes, A.K. Alanazi, H.M. Abo-Dief, A. Mohammadzadeh, S. Mobayen, B.W. França, A. Fekih, A new immersion and invariance control and stable deep learning fuzzy approach for power/voltage control problem, IEEE Access 10 (2022) 68–81.
[13] A.V. Khergade, R.J. Satputaley, V.B. Borghate, B. Raghava, Harmonics reduction of adjustable speed drive using transistor clamped h-bridge inverter based dvr with enhanced capacitor voltage balancing, IEEE Transactions on Industry Applications 56 (6) (2020) 6744–6755.
[14] C. Kumar, M.K. Mishra, Predictive voltage control of transformerless dynamic voltage restorer, IEEE Transactions on Industrial Electronics 62 (5) (2015) 2693–2697.
[15] B. Bae, J. Jeong, J. Lee, B. Han, Novel sag detection method for line-interactive dynamic voltage restorer, IEEE Transactions on Power Delivery 25 (2) (2010) 1210–1211.
[16] E.M. Molla, C.-C. Kuo, Voltage sag enhancement of grid connected hybrid pv-wind power system using battery and smes based dynamic voltage restorer, IEEE Access 8 (2020) 003.
[17] S. Biricik, H. Komurcugil, N.D. Tuyen, M. Basu, Protection of sensitive loads using sliding mode controlled three-phase dvr with adaptive notch filter, IEEE Transactions on Industrial Electronics 66 (7) (2019) 5465–5475.
[18] M. Pradhan, M.K. Mishra, Dual *p-q* theory based energy-optimized dynamic voltage restorer for power quality improvement in a distribution system, IEEE Transactions on Industrial Electronics 66 (4) (2019) 2946–2955.
[19] T.A. Naidu, S.R. Arya, R. Maurya, Multiobjective dynamic voltage restorer with modified epll control and optimized pi-controller gains, IEEE Transactions on Power Electronics 34 (3) (2019) 2181–2192.

[20] J. Ye, H.B. Gooi, Phase angle control based three-phase dvr with power factor correction at point of common coupling, Journal of Modern Power Systems and Clean Energy 8 (1) (2020) 179–186.

[21] A.M. Rauf, V. Khadkikar, An enhanced voltage sag compensation scheme for dynamic voltage restorer, IEEE Transactions on Industrial Electronics 62 (5) (2015) 2683–2692.

[22] A.Y. Goharrizi, S.H. Hosseini, M. Sabahi, G.B. Gharehpetian, Three-phase hfl-dvr with independently controlled phases, IEEE Transactions on Power Electronics 27 (4) (2012) 1706–1718.

[23] F. Jiang, S. Cheng, C. Tu, Q. Guo, Q. Li, C. Chen, Optimum control scheme of output voltage based on cascaded h-bridge dvr, CSEE Journal of Power and Energy Systems 6 (2) (2020) 249–258.

[24] F. Jiang, S. Cheng, C. Tu, Q. Guo, R. Zhu, X. Li, M. Liserre, Design consideration of a dual-functional bridge-type fault current limiter, IEEE Journal of Emerging and Selected Topics in Power Electronics 8 (4) (2020) 3825–3834.

[25] T. Koroglu, A. Tan, M.M. Savrun, M.U. Cuma, K.C. Bayindir, M. Tumay, Implementation of a novel hybrid upqc topology endowed with an isolated bidirectional dc–dc converter at dc link, IEEE Journal of Emerging and Selected Topics in Power Electronics 8 (3) (2020) 2733–2746.

[26] J. Ye, H.B. Gooi, X. Zhang, B. Wang, J. Pou, Simplified four-level inverter-based dynamic voltage restorer with single dc power source, IEEE Access 7 (2019) 461–137 471.

[27] G.A. de Almeida Carlos, C.B. Jacobina, J.P.R.A. Méllo, E.C.d. Santos, Cascaded open-end winding transformer based dvr, IEEE Transactions on Industry Applications 54 (2) (2018) 1490–1501.

[28] G.S. Chawda, A.G. Shaik, O.P. Mahela, S. Padmanaban, J.B. Holm-Nielsen, Comprehensive review of distributed facts control algorithms for power quality enhancement in utility grid with renewable energy penetration, IEEE Access 8 (2020) 614–107 634.

[29] A comprehensive review of dynamic voltage restorers, International Journal of Electrical Power & Energy Systems 92 (2017) 136–155.

[30] A.V. Ital, S.A. Borakhade, Compensation of voltage sags and swells by using dynamic voltage restorer (dvr), in: 2016 International Conference on Electrical, Electronics, and Optimization Techniques (ICEEOT), 2016, pp. 1515–1519.

[31] G.S. Pal R, Topologies and control strategies implicated in dynamic voltage restorer (dvr) for power quality improvement, Iranian Journal of Science and Technology, Transactions of Electrical Engineering 44 (2020) 581–603.

[32] S. Ezhilarasan, G. Balasubramanian, Dynamic voltage restorer for voltage sag mitigation using pi with fuzzy logic controller, International Journal of Engineering Research and Applications 3 (2013) 1090–1095.

[33] G.E. Mejía Ruiz, N. Muñoz, J.B. Cano, Modeling, analysis and design procedure of lcl filter for grid connected converters, in: 2015 IEEE Workshop on Power Electronics and Power Quality Applications (PEPQA), 2015, pp. 1–6.

[34] G.E. Mejía-Ruiz, J. Rodríguez, M. Paternina, N. Muñoz-Galeano, A. Zamora, Grid-connected three-phase inverter system with lcl filter: model, control and experimental results, in: 2019 IEEE PES Innovative Smart Grid Technologies Conference – Latin America (ISGT Latin America), 2019, pp. 1–6.

[35] Hans Kristian Høidalen, László Prikler, Francisco Peñaloza, ATP/EMTP, User Manual, https://www.atpdraw.net/, 2021. (Accessed 24 February 2022).

[36] S. Ms, B. Basavaraja, M. Sasikala, An overview of dynamic voltage restorer for voltage profile improvement, vol. 11, 2017.

[37] G.E. Mejia-Ruiz, N. Muñoz, J.M. Ramirez, Transient voltage compensation by a power electronics-based device, in: 2014 IEEE PES General Meeting, Conference Exposition, 2014, pp. 1–5.

[38] A.K. Sadigh, K.M. Smedley, Review of voltage compensation methods in dynamic voltage restorer (dvr), in: 2012 IEEE Power and Energy Society General Meeting, 2012, pp. 1–8.

[39] S.P. Mishra, B. Biswal, J.P. Roselyn, D. Devaraj, Simulation and analysis of dvr for mitigating voltage sags and swells, Procedia Engineering 64 (2013) 341–350, International Conference on Design and Manufacturing (IConDM2013). [Online]. Available https://www.sciencedirect.com/science/article/pii/S1877705813016202.

[40] C. Zhang, M. Molinas, A. Rygg, J. Lyu, X. Cai, Harmonic transfer-function-based impedance modeling of a three-phase vsc for asymmetric ac grid stability analysis, IEEE Transactions on Power Electronics 34 (12) (2019) 552–12 566.

[41] The MathWorks, Inc. Ieee 13 node test feeder. Matlab. Retrieved October 26, 2020. [Online]. Available https://la.mathworks.com/help/physmod/sps/ug/ieee-13-node-test-feeder.html.

AI application for load forecasting: a comparison of classical and deep learning methodologies

11

Ana Karen Apolo Peñaloza, Alexandre Balbinot, and Roberto Leborgne
Federal University of Rio Grande do Sul, Porto Alegre, Brazil

Acronyms

ANN	Artificial Neural Networks
ADAM	Adaptive Moment Estimation
Elu	Exponential Linear Unit
GRU	Gated Recurrent Unit
LSTM	Long Short-Term Memory
MAPE	Mean Absolute Percentage Error
MLP	Multi-layer Perceptron
MSE	Mean Square Error
Nadam	Nesterov Adaptive Moment Estimation
SGD	Stochastic Gradient descent
RNN	Recurrent Neural Network
ReLU	Rectified Linear Unit
RMSProp	Root Mean Squared Propagation
Tanh	Hyperbolic Tangent (tanh)

11.1 Introduction

Electric power demand forecasting has become a major topic in energy management research [35]. Predicting electric power demand is necessary to avoid economic losses, especially in the operational management of microgrids [28]. According to Hong et al., [25] and [53], power demand is influenced by several factors. Namely, among others, climate variables (temperature, relative humidity, wind speed, precipitation, and the duration of sunshine), calendar variables (hour, day, and month) and socioeconomic variables (age, employment, occupation, and income). According to [45], when the climate factor is eliminated, the energy consumption is influenced only by socioeconomic variables. Moreover, residential electricity demand is volatile in nature, because it involves different routines, and different types of energy use [41]. This volatility is more remarkable in microgrids (up to 250 residential consumers) [52]. Therefore residential power demand forecasting plays an important role in microgrids management.

Monitoring and Control of Electrical Power Systems Using Machine Learning Techniques
https://doi.org/10.1016/B978-0-32-399904-5.00017-X

In the last decade, many techniques have been focused on solving the problem of power demand prediction. The most popularly used technique is artificial neural networks (ANN). According to [54], power demand forecasting is a suitable application for ANNs, because it is a nonlinear and weather-dependent function. ANNs are divided into classical neural networks and deep neural networks. The multilayer perceptron (MLP) is the most important classical neural network. MLP is based on the functioning of the human brain [53] and is built with an interconnection of all fully connected neurons.

Recurrent neural networks (RNNs) are types of deep neural networks that are applied to sequential problems, such as time series analysis. Considering load forecasting as a time series analysis. RNNs are used for time series because of their ability to learn long-term dependencies [1]. The main difference between RNNs and the MLP is that MLP needs to learn each feature of the sequence separately, whereas RNNs learn the features of the sequence in each small batch, called mini batch, because it shares the same weights in multiple time steps [21].

RNNs are difficult to train since the gradient can be too large or too small. These problems are called explosive and vanishing gradients, respectively. The solution to explosive and vanishing gradients is to use a controlled gradient [1]. The techniques that use gradient control are long short-term memory (LSTM) and gated recurrent unit (GRU). LSTM is a cell with memory parameters. There are two types of gated: the first memorizes new information, and the second forgets unimportant information [46]. Even though GRU is similar to LSTM, GRU combines two input and output gates of LSTM into a single gate called reset gate [1].

In addition to choosing between classical neural networks and deep neural networks, the accuracy to ANN depends on the selection of its hyperparameters, such as activation function, optimizer function, cell or neuron number, epoch number, and batch number [42]. This hyperparameter selection process is called by literature "tuning hyper-parameters." The tuning can improve the forecast by 10%, according to Kong et al. [33]. Hyperparameters are usually selected by trial and error, rule of thumb, previous experience, and grid search [48].

Tuning hyperparameters can be viewed as a search process. Trial and error, rule of thumb, and previous experience are part of the manual search, which uses the user's experience, in contrast to grid search, which looks at the combination of the hyperparameters as an automated process. However, during grid search, the search space increases exponentially. If three hyperparameters are selected as cells numbers, optimizer function, and batch number, and each of them has 5 sublevels, considering the value 10 for cross-validation, the total number of simulations is 1250 $((5 \times 5 \times 5) \times 10)$ [36]. This number of simulations has a high computation cost, and for this reason, the grid search may not be feasible [4]. It should be pointed out that the hyperparameters are dataset-dependent, so it is necessary to analyze in each dataset the influence of the hyperparameters on the forecasting methodologies.

The purpose of this chapter is to analyze different time series load forecasting methodologies. The methodologies analyzed are MLP, LSTM, and GRU. In this chapter, an Irish public dataset with 250 residential consumers is used; furthermore, these measurements are for three months with a sample frequency of one hour. Also, the hyperparameters such as cells number or neurons number, batches number, optimization function, and activation function are evaluated in MLP, LSTM, and GRU. The epochs number is evaluated with early stopping. Hyperparameter tuning is evaluated by manual search, but using a large space search. This chapter does not use grid search, because the number of simulations is too large (768.000).

11.2 Computational tools

The algorithms used in this chapter are divided into two groups. The first is the algorithm for the preprocessing of the dataset, which consists of cleaning the data using normalization, outliers detection, and missing values correction. The second are the algorithms for forecasting, such as LSTM, MLP, and GRU.

11.2.1 Preprocessing

Real datasets have different problems in the data acquisition, such as missing values, errors in measurements, and outliers [49]. The preprocessing is called dataset cleaning. Cleaning is very important, because it identifies the missing values, and it duplicates measurements and outlier values [30]. Different dataset cleaning algorithms, such as linear interpolation for missing values, "*two sigma rule of thumb*" for outlier detention and normalization by min-max algorithm, are used.

11.2.1.1 Missing values

Missing values are data or measurements that are not stored due to the loss of the connection with the meter [29]. This problem is common in the dataset, so the dataset may contain values such as "NaN", "??", zeros, blanks, and other placeholders [40]. The basic strategy for missing values is to leave them out, but the accuracy of the forecasting algorithm will be affected. Another strategy for missing values is to replace these values with statistical methodologies, such as linear interpolation and polynomial interpolation [40]. Linear interpolation is a statistical method, which is used on one-dimension databases. Linear interpolation estimation for missing values is performed based on two adjacent points, such as two previous data or one previous data and one next data, as shown in Eq. (11.1) [49].

$$f(x) = \begin{cases} \frac{x_{i-1}+x_{i+1}}{2} & x_i \in NaN; x_{i-1}, x_{i+1} \notin NaN \\ \frac{x_{i-1}+x_{i-2}}{2} & x_i \in NaN; x_{i-1}, x_{i-2} \notin NaN; x_{i+1} \in NaN \\ x_i & x_i \notin NaN \end{cases}, \quad (11.1)$$

where x_i represents the measurement data, i is timestamp and *NaN* is not a number [49].

11.2.1.2 Outlier

Outlier is an observation or measurement that appears to be different or inconsistent with other data or observations [23]. The outlier can be thought of as noise, anomalous data, or measurement drift [23]. The outlier detection can be seen as a purification of the dataset, which is important to improve the accuracy of the prediction [23]. Different statistical methodologies are used for outlier detection, such as the two-sigma rule of thumb, and three-sigma rule of thumb. The two-sigma rule of thumb is used on datasets that follow a normal distribution or a unimodal distribution. This rule says that almost all data are within two standard derivations (*std*) of the mean. In the normal distribution, about 95% of the dataset is closer than two standard deviations from the mean [55]. Eq. (11.2) shows this rule of thumb. This rule considers 5% of the values as outliers.

$$f(x_i) = \begin{cases} avg(x) + 2std(x) & if x_i > avg(x) + 2std(x) \\ x_i & otherwise \end{cases}, \quad (11.2)$$

where x is a vector, $avg(x)$ is the average value of x, $std(x)$ is the standard derivation of x [49].

11.2.1.3 Normalization

The normalization of the data is a transformation of each sample to a certain range, such as [0, 1] [40]. The normalization process can be implemented with different algorithms, such as min-max scaler and z-score standardization [50]. The most used algorithm is min-max normalization. This process, which should be done after the outlier detection, is linear scaling and transforms the data to the range [0, 1]. The min-max normalization is shown in Eq. (11.3). It is used when the dataset does not follow the normal distribution, and z-score standardization is more used in the dataset that follows the normal distribution [50].

$$f(x_i) = \frac{x_i - min(x)}{max(x) - min(x)}, \quad (11.3)$$

where $\min(x)$ is the minimum value of x and $\max(x)$ is the maximum value of x.

11.2.2 Forecast

Different techniques to perform load demand forecasting are presented. These techniques are divided into classical machine learning (such as multilayer perceptron (MLP)) and deep machine learning (such as recurrent neural networks (RNN)). The most common RNNs are the long short-term Memory (LSTM) and the gated recurrent unit (GRU).

11.2.2.1 Multilayer perceptron

Multilayer perceptron (MLP) has been widely used to determine energy consumption forecasts since the 1990s [36]. MLP is an intelligent machine learning method

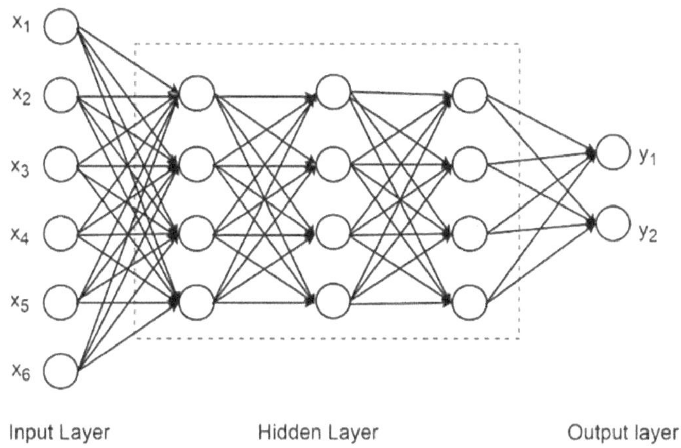

Figure 11.1 Architecture of MLP. (Adaptation of Aggarwal [1].)

based on mathematical models that describe the artificial neuron as an attempt to reproduce the biological neuron by building connections similarly to the human brain [1,42]). MLP is based on a mathematical model that has the advantage of mapping the relationship between input and output through a weighted sum (between the input attributes and the weights of the networks) [42]. This sum is the argument to the activation function [42]. Furthermore, the MLP extracts the nonlinear relationship of the input variables using a training process [24]. MLP is constituted by the interconnection of neurons organized in more than one layer [1]. The back propagation (BP) algorithm is used by MLP. BP works with the chain rule for differential calculations and computes the gradient error in terms of the sum of local gradients, generating multiple paths from an output node [1]. The typical architecture of MLP is shown in Fig. 11.1.

Fig. 11.1 shows that each neuron produces an output y. Each output is generated by a linear combination of the weights matrix $\mathbf{W} = [w_{11}, w_{12},\ldots,w_{1\mathrm{d}}]$ and input vector $\mathrm{X} = [x_1, x_1,\ldots, x_\mathrm{d}]$, d is the length of features. This linear combination $\mathbf{W}\odot\mathrm{X}$ is transformed with a typically sigmoid activation function that produces an output with values in the range of (0, 1) and dominion of $(-\infty, \infty)$ [1]. Eq. (11.4) is the linear combination $\mathbf{W}\odot\mathrm{X}$ that produces an output y for an input x [2].

$$y_j = f\left(\sum_{i=1}^{n} w_{ji}x_i - b\right) \tag{11.4}$$

where y is an output neuron in j layer, w is a weight between i input neuron and j layer, b is a bias, n is total samples, and f is the activation function. Different activation and optimizer functions need to be evaluated for MLP.

The advantage of working with MLP, according to Kuster et al. [34], is the possibility of using different independent input variables, such as income, occupancy, electricity price, weather and other variables found in the literature; but the most used are the weather variables, such as temperature and relative humidity [2,5,10,11,51].

Activation function

Different types of activation functions, such as sigmoid, hyperbolic tangent, rectified linear unit, and exponential linear unit, have been evaluated for MLP. Each activation function, as well as its advantages and disadvantages, is presented in the text sequence.

Sigmoid

In the 90's the most widely used activation function f was the sigmoid, whose equation is shown in (11.5). The sigmoid has a continuous differentiable and nondecreasing behavior [18]. Its main advantage is that it produces good generalization of the inputs [39]. This function is shaped like an inverted s and its derivative flattened the bell shape, as shown in Fig. 11.2. It has a derivative close to zero when the values are in the range of $[-5, 5]$, as shown in Fig. 11.2. Thus MLP can have vanishing gradients. It is not centered on zero. This causes the gradient to propagate in different directions, causing the saturation of the gradient. This saturation is a very small value, close to zero and, as shown in Fig. 11.2(a), is far from the center [1].

$$f(x) = \frac{1}{1 + e^{-x}} \tag{11.5}$$

Hyperbolic tangent (Tanh)

The hyperbolic tangent (Tanh) is a zero-centered function with range of $(-1, 1)$ and dominion of $(-\infty, \infty)$. Tanh is a nonlinear function, whose equation is shown in (11.6) [39].

$$f(x) = \frac{e^x - e^{-x}}{e^x + e^{-x}} \tag{11.6}$$

The main advantage of Tanh is that it is zero-centered, which helps in the backpropagation process. On the other hand, its disadvantage is that it only produces its maximum value when the input is zero, as shown in Fig. 11.2(b) (see derivative Tanh) [39]. The tanh has the shape of an inverted s with range $(-1, 1)$, and the derived tanh has the shape of a bell, as shown in Fig. 11.2(b).

Rectified linear unit (ReLu)

The *rectified linear unit* (ReLU) proposed by [38] is a fast-learning activation function and has a good generalization. This activation function is compared with sigmoid and Tanh activation function [39]. It is a quasilinear function and preserves the properties of linear functions. This makes ReLu easy to use with gradient descent methods [39]. Its equation is shown in (11.7).

$$f(x) = \begin{cases} x_i & x_i \geq 0 \\ 0 & x_i < 0 \end{cases} \tag{11.7}$$

Eq. (11.7) shows that ReLu forces the negative values to zero, which helps the vanishing gradient problem [39]. The derivative of ReLu is zero in the negative domain

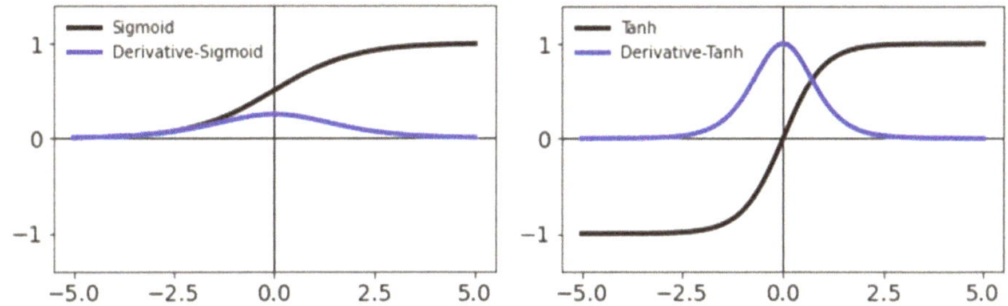

(a) (b)

Figure 11.2 (a) Activation function sigmoid and derivative of sigmoid. (b) Activation function Tanh and derivative of Tanh.

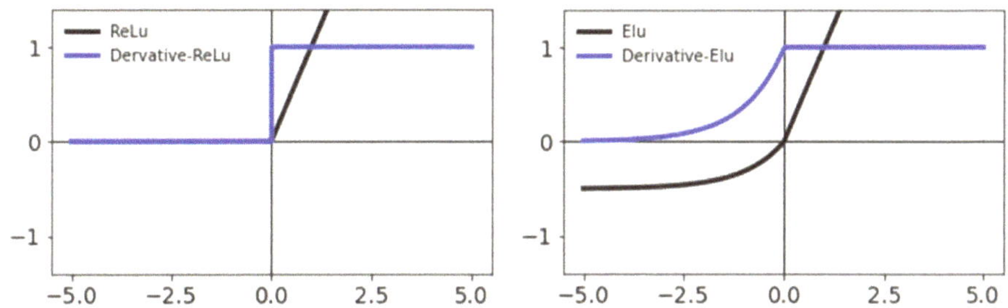

(a) (b)

Figure 11.3 (a) Activation function ReLu and derivative of ReLu. (b) Activation function Elu and derivative of Elu.

and one in the positive domain [39]. This condition is shown in Fig. 11.3(a). ReLu is zero in negative domain, and it is shown in Fig. 11.3(a).

Exponential linear unit (*Elu*)

The exponential linear unit (Elu) is proposed by [9]. Elu is similar to ReLu, because the former is a linear function in the positive dominion. On the other hand, Elu is different from ReLu in the negative dominion, where the first is an exponential function, and the latter is a linear function [39]. This difference can be seen in (11.8), which presents the Elu function.

$$f(x) = \begin{cases} x_i & x_i \geq 0 \\ \alpha e^{x_i} - 1 & x_i < 0 \end{cases} \tag{11.8}$$

Elu was designed to solve the problem of ReLu neuron death, caused by the null derivative of ReLu in the negative domain. Fig. 11.3(b) shows that the derivative of Elu is similar to the derivative of tanh in the negative domain, and the derivative of Elu is similar to the derivative of ReLu in the positive domain. Fig. 11.3(b) shows that Elu is linear in the positive domain and exponential in the negative domain.

Optimizer function

MLP has also been evaluated with different types of optimizer functions, such as stochastic gradient descent, RMSProp, adaptive moment estimation (Adam), and Nesterov adaptive moment estimation (Nadam). Each optimizer function, as well as its advantages and disadvantages, are presented in what follow.

Stochastic gradient descent

Stochastic gradient descent (SGD) results from a simplification of gradient descent. It is an optimization method that minimizes the mean squared error (MSE) loss function. In each interaction, it is necessary to update the weights w, as shown in Eq. (11.9).

$$W_{t+1} = W_t - \alpha \nabla L\left(W_t\right), \tag{11.9}$$

where α is learning rate; $\mathbf{W}$ is weights matrix, ∇L is gradient of loss function (MSE).

SGD is prone to oscillations. For this reason, SGD uses the β moment rate to speed up the method [58]. Learning rate α and moment rate β are hyper-parameters of SGD. These hyperparameters have been the aim of research to define the best α and β values for each database. The learning rate of the step decay was considered to improve the SGD performance. This function updates the learning rate after a certain number of epochs [19]. Step decay needs two parameters: the initial learning rate and the number of epochs in which the learning rate will be reduced [19]. The advantage of using step decay learning is that it starts with a broad search that, as the epochs increase, turns into a more detailed search [19].

RMSProp and AdaGrad

RMSprop was proposed by [12], it has the same motivation as AdaGrad, which is an adaptive subgradient method. The main idea of AdaGrad is adaptive learning rate with the dynamic incorporation of knowledge of the geometry of the observer data [14]. The problem with AdaGrad is that the learning rate drops to values close to zero, because AdaGrad uses the gradient sum, and thus the algorithm stops learning [44]. To solve this problem RMSProp uses a decay factor to accumulate the gradient. In addition, RMSProp considers the adaptive moment rate [1]. The update of the weights with the RMSProp formulation is shown in (11.10).

$$W_{t+1} = W_t - \frac{\alpha \nabla L\left(W_t\right)}{\sqrt{g_{t+1}} + \varepsilon} g_{t+1} = \beta_t g_t + \left(1 - \beta_t\right) \nabla L\left(W_t\right)^2, \tag{11.10}$$

where α is learning rate; $\boldsymbol{W}$ is weights matrix, ∇L is gradient of loss function (MSE), g is the exponentially weighted moving average of the weights, β is moment rate, and ε is hyperparameter [12].

Adaptive moment estimation (Adam), and Nesterov adaptive moment estimation (Nadam)

Adaptive moment estimation (Adam) is proposed by [31], and Nesterov adaptive moment estimation (Nadam) is proposed by [13]. Adam and Nadam work similarly, they differ in the used moment. Nadam uses the Nesterov moment, which consists of the opposite to the gradient direction.

Adam is an optimization algorithm, which is based on a first-order objective function. Typically, this objective function tries to minimize the mean square error (MSE) in the case of regression algorithms [31]. Furthermore, it minimizes the objective function based on adaptive estimation of lower-order moments. Adam is used to calculate the adaptive learning rate [44].

The adaptive learning rate is calculated with the first-order gradient of the objective function (β_1) and the exponential moving average of the quadratic gradient (β_2). It is recommended that β_1 and β_2 be close to 1 by [31].

Adam and Nadam are also similar to AdaGrad; all of them have adaptive updating of the learning rate α. Also, Adam and Nadam are similar to RMSProp; all of them have adaptive moment [31]. The difference between Adam and Nadam and RMSProp is that Adam and Nadam used the first- and second-order optimizer moment.

The update of the weights of Adam and Nadam formulation is shown in (11.11). Adam is the most used optimizer, because it converges fast. Also, Nadam is more unstable than Adam, because the Nesterov moment can lead to an exploding gradient [1].

$$\begin{aligned} \mathbf{W}_{t+1} &= \mathbf{W}_t - \frac{\alpha m_{t+1}}{\sqrt{g_{t+1}}+\varepsilon} \\ m_{t+1} &= \frac{\beta_t+1g_t+\left(1-\beta_{t+1}\right)\nabla L(\mathbf{W}_t)}{1-\beta_{t+1}} \\ g_{t+1} &= \frac{\beta_t^*+1g_t+\left(1-\beta_{t+1}^*\right)\nabla L(\mathbf{W}_t)^2}{1-\beta_{t+1}^*} \end{aligned} \tag{11.11}$$

where α is learning rate; $\mathbf{W}$ is weights matrix ∇L is gradient of loss function (MSE), m is the exponentially average squared of gradient along the weights, g is the exponentially average of gradient along the weights, β is moment rate, β^* is second moment rate, and ε is hyperparameter [31].

Fig. 11.4 shows a practical example for MLP networks with {1, 2, 3, 4, 5, 6, 7, 8, 9} time series as input. The first step in this procedure is to normalize the time series using the minimum and maximum.

With the normalized data, a window of the time series is made in matrix form. In the example, the window interval is 4, so with 4 time series elements, the output is created.

In the following step, the time series is divided into training and testing sets. Finally, the forecast model is built in Keras [7]. To build the model, the number of neurons, the activation function, the number of batches, the number of epochs, and the dropout percentage must be chosen. These parameters can be selected by trial and error.

11.2.2.2 Long short-term memory (LSTM)

Long short-term memory (LSTM) is a type of recurrent neural networks (RNNs) based on data sequencing. Its specialty is time series or sequence data, and these can help to process large amounts of data [1,22,47]. Consequently, the RNNs are widely used in text translation and time series forecasts [21,47]. RNNs have two problems: exploding and vanishing gradients. The former problem generates a gradient so large that it can be close to infinity (exploding gradient), the latter generates a gradient so small that

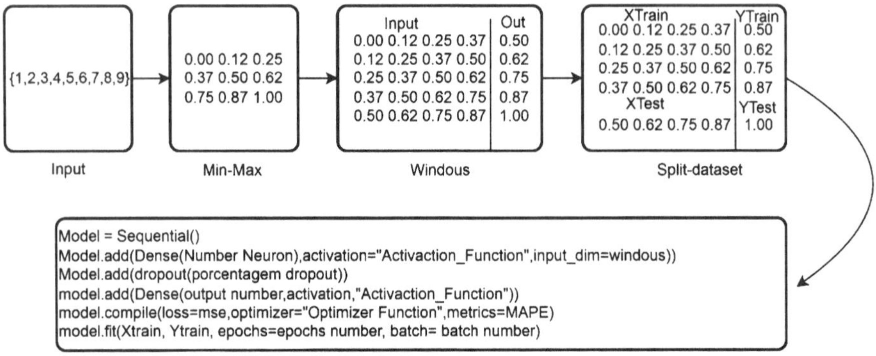

Figure 11.4 MLP example.

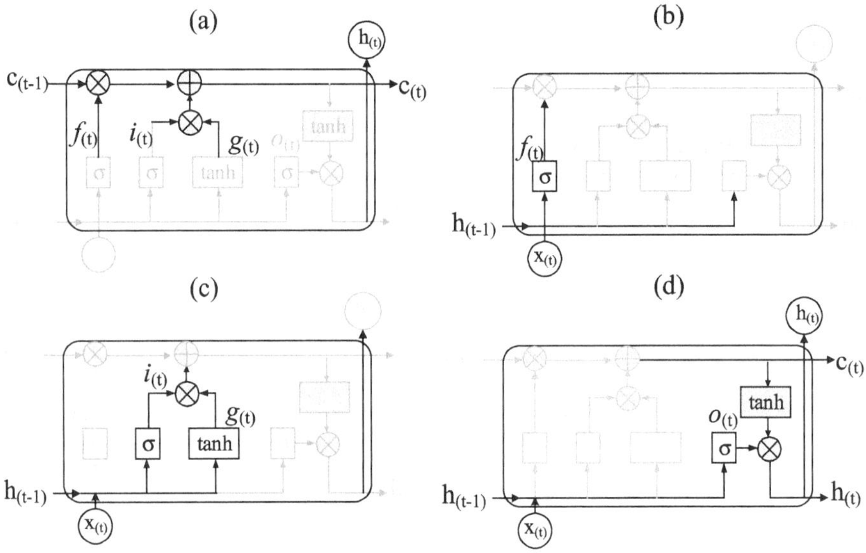

Figure 11.5 Architecture of LSTM cell (Adaptation of [47]). (a) Cell state. (b) Forget gate. (c) Input gate and input activation. (d) Output gate.

it can get close to zero (vanishing gradient) [1,21]. These gradient problems are associated with successive multiplication of the weight matrix **W** [1,21]. The solution to the mentioned gradient problems is through finite control of the gradient by different gates [1,21]. This control is observed in the LSTM architecture [1,21]. Furthermore, the LSTM architecture is formed by a recurrent subset of cells [47]. Fig. 11.5 shows the LSTM cell.

As shown in Fig. 11.5, LSTM cell can be broken down into four steps to be better explained. The cell state (c_t), which is the key to the LSTM cell, can also be seen in Fig. 11.5. This cell state (c_t) is the long-term memory calculated by (11.12). The cell state (c_t) passes the relevant information from the previous time step to the next time step. The equation c_t is divided into two parts. The first part is called the reset of the LSTM, which determines the value that will be saved or removed, because of this, it performs the multiplication of the forget gate (f_t) and the previous cell state, shown in Fig. 11.5(a). The f_t, which determines if each input value x_t will be saved or removed, is shown in Fig. 11.5(b) and Eq. (11.13). The f_t is the sum of the multiplication of the weight matrix **W** by the input matrix and weight matrix **W** by hidden state h [1,21,32,47]. The second part of equation c_t determines the value to be kept by the LSTM cell. This part consists of the multiplication of the input gate (i_t) by the input activation gate (g_t) and is shown in Fig. 11.5(a). The input gate i_t is shown in Fig. 11.5(c) and in Eq. (11.14), which defines the input value x_t to be updated in the following time steps. g_t selects which value of the cell state should be updated and is shown in Fig. 11.5(c) and in Eq. (11.15). Finally, Fig. 11.5(d) shows the output gate (o_t). o_t is found in Eq. (11.16), which is the current input for the next time step. The hidden state update h_t is shown in Eq. (11.17) [1,21,32].

$$c_t = g_t \odot i_t + (c_{t-1} \odot f_t), \tag{11.12}$$

$$f_t=\sigma\left(\mathbf{W}_{f\cdot x}x_t+\mathbf{W}_{f\cdot h}h_{t-1}+b_f\right) \tag{11.13}$$

$$i_t=\sigma\left(\mathbf{W}_{i\cdot x}x_t+\mathbf{W}_{i\cdot h}h_{t-1}+b_i\right) \tag{11.14}$$

$$g_t=\tanh\left(\mathbf{W}_{g\cdot x}x_t+\mathbf{W}_{g\cdot h}h_{t-1}+b_g\right) \tag{11.15}$$

$$o_t=\sigma\left(\mathbf{W}_{o\cdot x}x_t+\mathbf{W}_{o\cdot h}h_{t-1}+b_o\right) \tag{11.16}$$

$$h_t = tanh\,(c_t) \odot o_t \tag{11.17}$$

where c_t is cell state; h is hidden state; f_t is forget gate; i_t is input gate; g_t is input activation gate; o_t is output gate. x is input, **W** is weight matrix, b is bias in gate, σ is the sigmoid function, *and Tanh* is hyperbolic tangent.

Fig. 11.6 shows a practical example with the following time series as input $\{1, 2, 3, 4, 5, 6, 7, 8, 9\}$. The first step in this procedure is to normalize the time series by means of the minimum and maximum. With the normalized data, a windowing of the time series in matrix form is performed. This windowing is called timesteps and it defines the number of elements that will produce an output. In the example, the timesteps is 4, so with 4 elements of the time series the output will be the next value of the time series.

The following step is to divide the timesteps into training and testing sets. Finally, the forecast model is built in Keras [8]. To build the model, the number of LSTM cells, the activation function, the number of batches, the number of epochs, and the dropout percentage must be chosen. These parameters can be selected by trial and error.

11.2.2.3 Gated recurrent unit (GRU)

The gated recurrent unit (GRU) is a variation of LTMS. GRU was proposed by [8]. Its difference from LSTM lies in the fact that GRU simplifies LSTM, since GRU combines the input gate (i_t) and the output gate (o_t) into only one gate, which is called the reset gate (r_t). The GRU architecture, shown in Fig. 11.7, does not use cell state (c_t) [1]. Fig. 11.7(a) can be compared with Fig. 11.5(a). Fig. 11.7 shows the reduction of

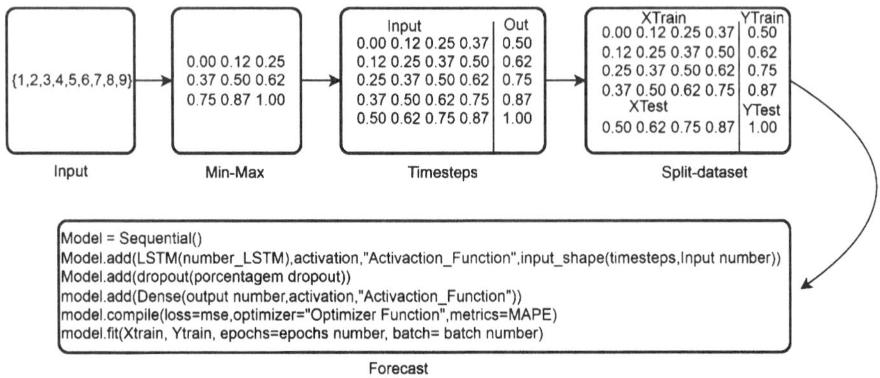

Figure 11.6 LSTM example adaptation [15].

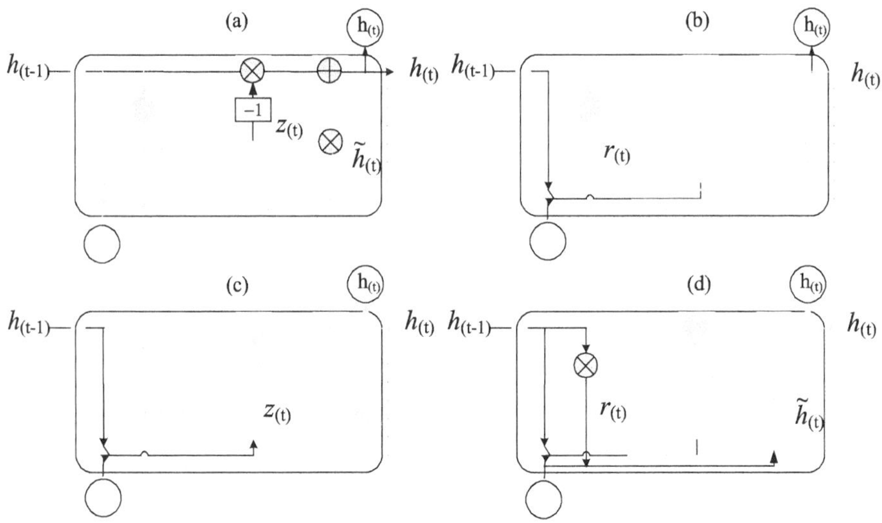

Figure 11.7 Architecture of GRU cell (adaptation of [8]). (a) Update hidden state. (b) Reset gate. (c) Update input gate. (d) Current memory content.

elements such as the cell state c_t and the gates: input i_t and output o_t. On the other hand, the hidden state is maintained in the GRU. Moreover, the reset gate is similar to the LSTM [1]. Aggarwal [1] and Hossen et al. [27] reported that LSTM and GRU showed similar results.

The reset defines how much of the previous hidden state is forgotten and how much should enter from input x. Similar to the reset of the LSTM cell, which is performed by the multiplication of the forget gate by the previous cell state ($c_{t\text{-}1}$), the GRU is reset with *reset gate* r_t, as shown in (11.18). The reset weights matrix **W** is multiplied with the input x_t and the previous hidden state $h_{t\text{-}1}$ [8]. This process is shown in

Fig. 11.7(b). This gate uses the sigmoid activation function (σ) [1].

$$r_t = \sigma\left(\mathbf{W}_r x_t + \mathbf{W}_r h_{t-1}\right) \tag{11.18}$$

LSTM is updated by multiplication of input gate i_t and input activation gate g_t, whereas GRU is updated by update gate (z_t). Update gate is similar to reset gate equation and is shown in (11.19). Fig. 11.7(c) shows the update gate that uses input x and the previous hidden state $h(_{t\text{-}1})$ multiplied by their respective weights matrices, and it uses the sigmoid activation function (σ). This gate decides which contribution from the previous hidden state will be updated in the current hidden state. This type of control prevents all the information from the previous state from passing to the next state, thus helping to decrease vanishing gradient problems [1,47]. The current memory content ($\tilde{h}_t$) of the GRU can be seen in Fig. 11.7(d). This figure represents the input x_t, the reset gate r_t, and the previous hidden state $h(_{t\text{-}1})$, with the hyperbolic tangent activation function having values in the range $[-1, 1]$. This is observed in (11.20). The reset gate is used to store the key information of the hidden state [1,47].

$$z_t = \sigma\left(\mathbf{W}_z x_t + \mathbf{W}_z h_{t-1}\right) \tag{11.19}$$

$$\tilde{h}_t = \tanh\left(\mathbf{W}\left(r_t \circ h_{t-1}\right) + \mathbf{W} x_t\right) \tag{11.20}$$

As shown in Fig. 11.7, the last step of the GRU is to update the hidden state. This figure is composed of the update gate (z_t), the current memory content ($\tilde{h}_t$), and the previous hidden state $h(_{t\text{-}1})$. The update gate is shown in (11.21) [1]. z_t will define what information is important to the predictor, and therefore should move to the next hidden state. On the other hand, z_t also allows for forgetting unimportant information, and therefore that information should not move to the next hidden state [47]. The example case for GRU is very similar to the LSTM example in Fig. 11.6. The difference would be in the use of GRU cells instead of LSTM cells.

$$h_t = (1 - z_t)\, h_{t-1} + z_t \tilde{h}_{t-1} \tag{11.21}$$

11.3 Methodology and case study

The methodology to compare classical machine learning and deep machine learning algorithms for load forecasting is performed by using an Irish residential load database. First the database is preprocessed to clean outliers and missing measurements, and then classical (MLP) and deep (LSTM and GRU) machine learning algorithms using optimized hyperparameters are implemented. All the results of load forecasting are compared with actual values. The metrics to assess accuracy of the algorithms are MSE and MAPE. The preprocessing is called dataset cleaning and includes outlier detection (two sigma rule of thumb), missing value (linear interpolation), and data normalization (min max). The forecast stage is based on time series prediction. The entire process is summarized in Fig. 11.8.

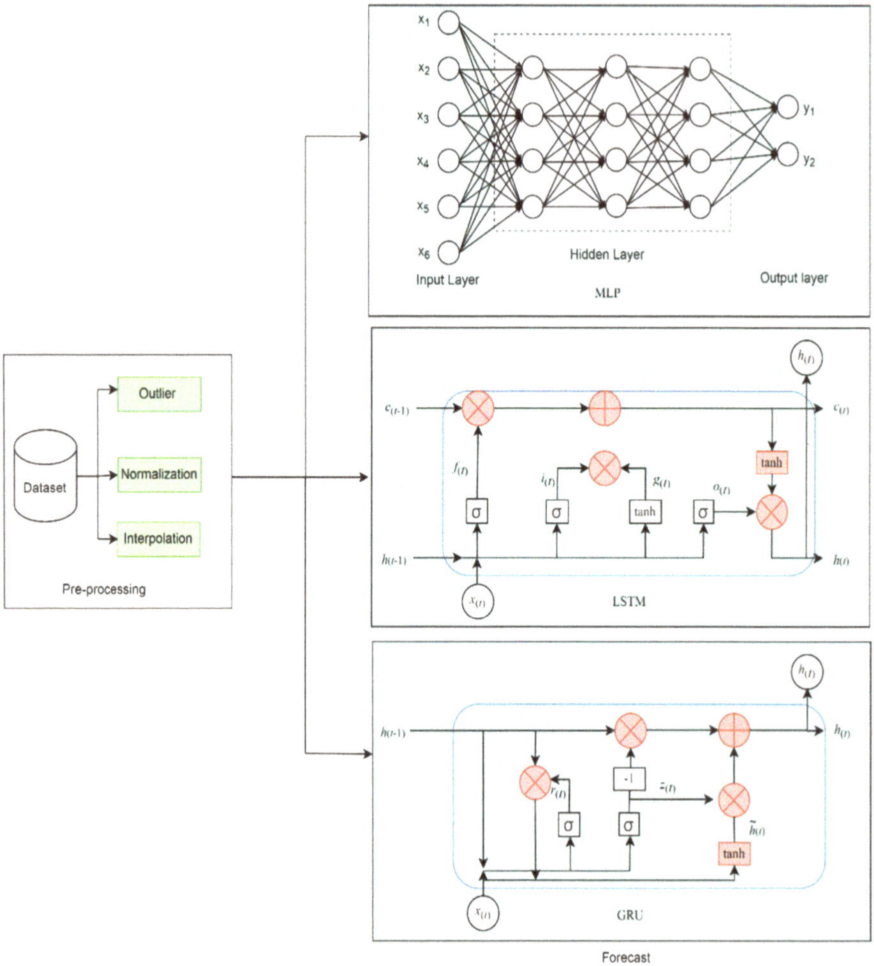

Figure 11.8 Dataset preprocessing and load forecast methodologies.

The dataset used can be found in the Irish Social Science Data Archive (ISSDA)[1] repository. The Irish Commission for Energy Regulation (CER) has set up the project "Smart Metering Electricity Consumer Behavior Trials (CBTs)" [6,16,17]. The project aimed to analyze the behavior of different types of electrical consumers. In this way, the end users were divided into six groups, according to different time-of-use tariffs. The database used in this chapter is called Control and its tariff is constant. The number of residential consumers is 260. The measurements started on July 14, 2009, and ended on December 31st, 2010. The sample period is 30 minutes, but the database was adapted for a sampled period of one hour. The database is divided into 70% for train-

[1] https://www.ucd.ie/issda/data/commissionforenergyregulationcer/.

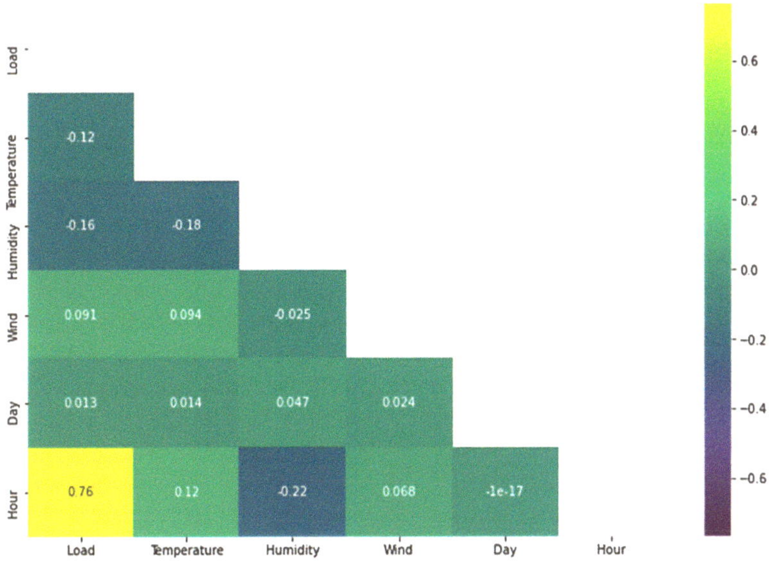

Figure 11.9 Correlation between load and temperature, wind speed, relative humidity, day, and hour.

ing the algorithms and 30% for testing the algorithms. Fig. 11.9 shows the correlation of the aggregated load with different variables, such as weather and calendar.

The weather variables found in the literature are temperature, relative humidity, and wind speed [3,26,36,43,56,57]. On the other hand, calendar variables, such as day, year, and hour are the most common variables used in the literature. Lusis et al. [37], present the analysis of calendar variables in load forecasting. According to the authors, calendar variables have little effect on single household consumption. On the contrary, for aggregated loads, these variables may exert greater effect. Fig. 11.9 shows little correlation of load with temperature, humidity, wind speed, and day; but shows large correlation between hour and load. Therefore climate variables are not included in the forecast algorithms developed in this chapter. And they only consider variables with large correlation.

11.4 Results

11.4.1 Selection of hyper parameters

Since the selection of hyperparameter depends on the dataset [4], MLP, LSTM, and GRU are evaluated with optimized hyperparameters. In the cases of LSTM and GRU, the hyperparameter is the number of cells. The MLP hyperparameters are the number of neurons in each layer, the activation function in each layer, the optimizer function, and the number of batches. The choice of hyperparameter is based on manual search to find which set produces the smallest loss function. All algorithms are evaluated by the mean squared error (MSE) (11.22) as loss function.

$$MSE = \frac{1}{N}\sum_{i=1}^{N}\left(y_i' - y_i\right)^2 \tag{11.22}$$

where y_i' is the target; y_i is the predicted value; N is the number of samples.

The number of neurons or cells is available in the set {8, 16, 32, 64, 128, 256, 512, 756, 1024}. This set is wide, because it is important that a large number of neurons or cells be available. A large number of neurons or cells do not affect generalization, but they have a high computational cost [4]. The activation function is available from the set {Sigmoid, Tanh, Relu, Elu}. Two are nonlinear functions (Sigmoid and Tanh) and two are linear functions (Relu and Elu). The nonlinear functions have good generalization, but have a vanishing gradient problem, because the derivative has a smaller range [−5,5] [20]. On the other hand, linear functions, such as Relu have problems with neuron death [39]. The optimizer functions are chosen from the set {Adam, SGD, Nadam, RMSProp}, where Adam, and RMSProp are the most used in forecasting. Finally, the number of batches are in the set {8, 16, 24, 32, 64, 128}. This set allows the idea of different processing time, because the number of batches influences the computational cost. Fig. 11.10 shows the loss function of the MLP, according to the hyperparameters.

The configuration includes two hidden layers. In these layers the number of neurons (see Fig. 11.10(a–b)) and the activation function are analyzed (see Fig. 11.10(c–e)). Fig. 11.10(f) also shows the optimizer function and Fig. 11.10(g) shows the batch numbers. This figure allows for analysis of the best configuration of the hyperparameter for the MLP. The nonlinear activation function in hidden layers (such as Sigmoid and Tanh) does not allow for learning forecast patterns, because the vanishing gradient does not allow the algorithm to continue learning.

The best hidden layer activation function is Relu, because this function does not present problems with vanishing gradients (see Fig. 11.10(c–d)). The best dense layer activation function is sigmoid (see Fig. 11.10(e)). The larger the number of neurons, the lower the MSE, but also the larger the computational cost. The best neurons number has a balance between computation cost and MSE. For this reason, the select neurons number is 256. In summarizing, Table 11.1 presents the selected hyperparameter for the hidden layers of the MLP algorithm tuned for load demand forecasting.

A number of batches greater than 24 does not allow the algorithm to learn the pattern, and a number below 24 batches shows very similar MSE results. Therefore 8 batches is an optimal choice (see Fig. 11.10(g)). The Nadam optimizer does not allow this algorithm to converge. Adam and RMSProp have similar results, but Adam is faster than RMSProp (see Fig. 11.10(f)).

Fig. 11.11 shows the loss function (MSE), according to the selection of LSTM hyperparameters. The configuration includes number of cells (see Fig. 11.11(a)), activation function (see Fig. 11.11(c)) in LSTM layer, and activation function in dense layer (see Fig. 11.11(d)). Number of cells greater than 128 has similar MSE; the main difference is related to the number of epochs. 1024 cells have fewer epochs than 756

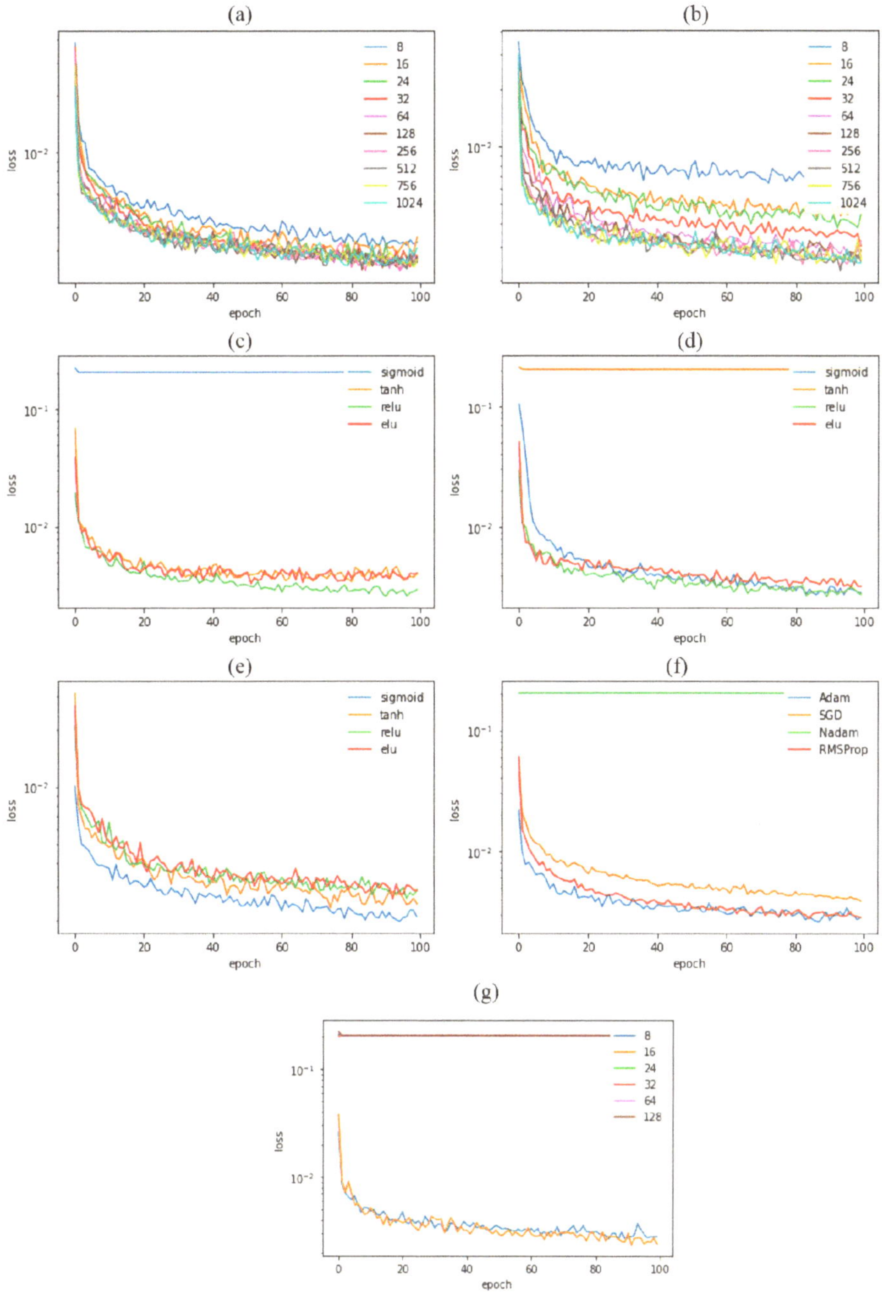

Figure 11.10 Loss function according to MLP hyperparameters. (a) MLP-1 Number Neurons. (b) MLP-2 Number Neurons. (c) MLP-1 hidden activation. (d) MLP-2 hidden activation. (e) Dense activation. (f) Optimizer function. (g) Batches number.

Table 11.1 MLP configuration.

Layer	**Number of neurons**	**Activation function**	**Dropout rate**	**Number of batches**	**Optimizer function**
Dense	256	Relu	–	–	–
Dropout	–	–	20%	–	–
Dense	256	Relu	–	–	–
Dropout	–	–	20%	–	–
Dense	1	Sigmoid	–	–	–
–	–	–	–	8	–
–	–	–	–	–	Adam

Table 11.2 Configuration LSTM.

Layer	**Number of *cells***	**Number of batches**	**Activation function**	**Optimizer function**	**Dropout rate**
LSTM	756	–	Tanh	–	–
Dropout	–	–	–	–	20%
Dense	1	–	Sigmoid	–	–
–	–	8	–	–	–
–	–	–	–	Adam	

cells (see Fig. 11.11(a)). But the computational cost of 1024 is larger than 756 cells. For this reason, 756 cells are used.

For the LSTM layer the nonlinear activation function sigmoid shows larger MSE than Tanh, Elu, and Relu. The best activation function is tanh. On the other hand, for the dense layer, the nonlinear activation function sigmoid shows lower MSE than the linear functions Tanh, Elu, and Relu (see Fig. 11.11(d)). Among the optimizer functions SGD has the worst MSE, because it does not have the updated learning rate like ADAM, RMSProp, and Nadam do. The best optimizers are Adam and RMSProp (see Fig. 11.11(b)), but Adam is faster than RMSProp.

Finally, the number of batches shows the length of the number data before the update weights. A very small number of batches increases the accuracy and the computational cost. The best number of batches is 8, because the MSE is the lowest (see Fig. 11.11(e)). The LSTM hyperparameters are summarized in Table 11.2. The dropout layer is used to reduce overfitting.

Fig. 11.12 displays the MSE, according to the GRU hyperparameters. The GRU includes a GRU layer, a dropout layer, and a dense layer. The best number of cells is 756 (see Fig. 11.12(a)). Only a few cells have poor learning, because they cannot extract all the features. Analyzing the activation function in the GRU layer (see Fig. 11.12(c)) and activation function in the dense layer (see Fig. 11.12(d)), we conclude that sigmoid activation function is better for dense layers than for the GRU layer, because the GRU layer has excessive multiplication. For this reason, the gradient vanishes with sigmoid activation function. Relu was created to solve the issue with the sigmoid function. Therefore the Relu shows the best result on the GRU layer. The best result

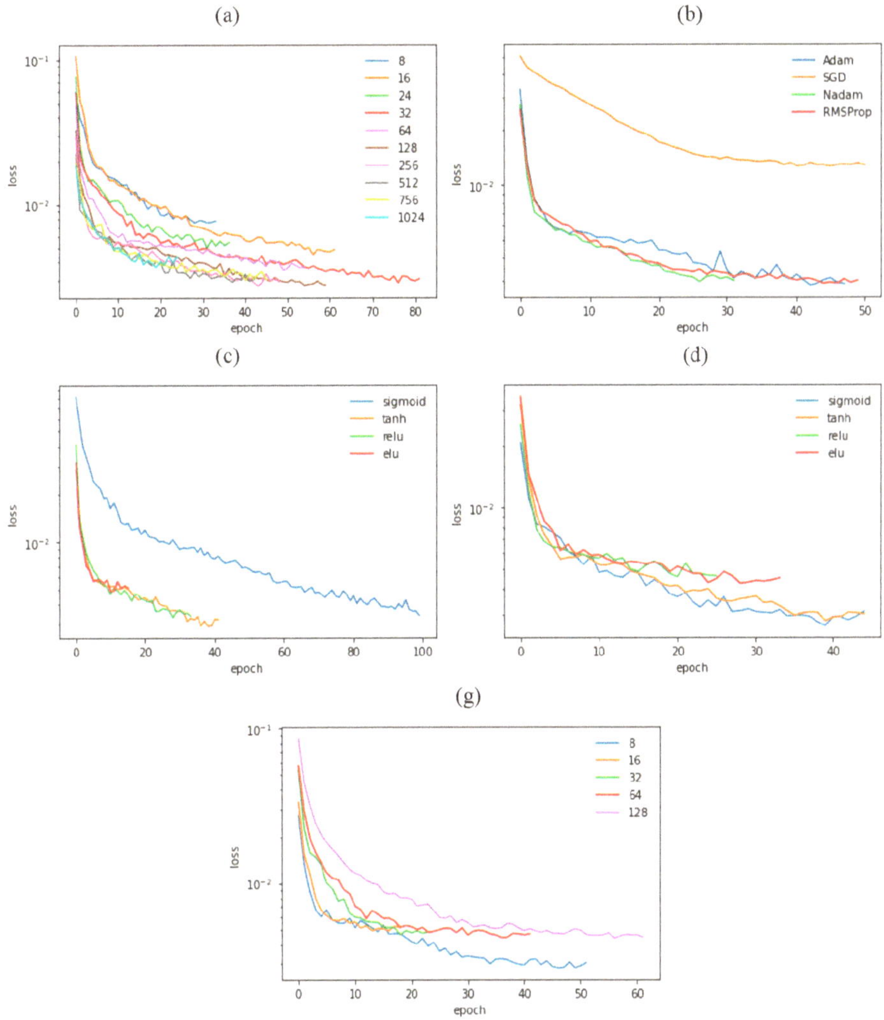

Figure 11.11 Loss function according to LSTM hyperparameters. (a) Number of cell LSTM. (b) Optimizer function. (c) Activation function in LSTM layer. (d) Activation function in Dense layer. (e) Number of batches.

on the dense layer is obtained using Tanh (see Fig. 11.12 (d)). The Nadam optimizer function is unstable, and this optimizer explodes the gradient (see Fig. 11.12(a)). SGD produces poor learning, because it does not update the learning rate. The best optimizers are Adam and RMSProp, but the RMSProp reveals the lowest loss function (MSE). The 64 and 128 batches do not produce good results, because the updating of the weights is slower than 16 and 32 batches. The best number of batches is 32 (see Fig. 11.12(e)). The summary of the GRU configuration is shown in Table 11.3.

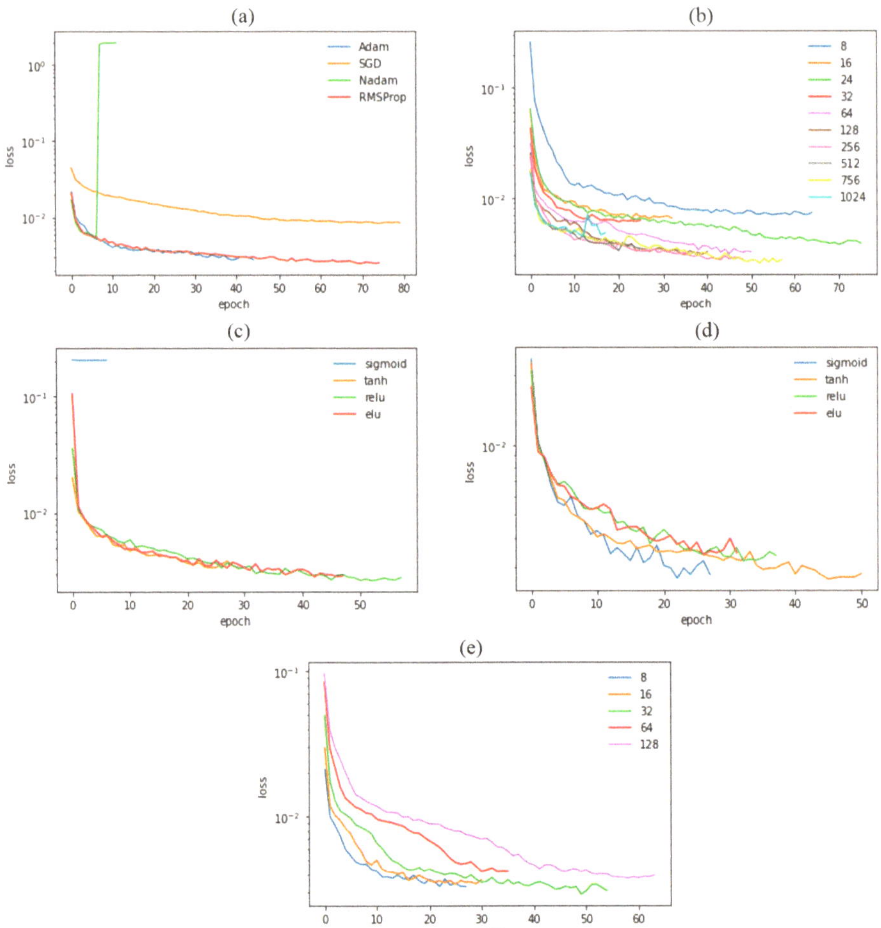

Figure 11.12 Loss function according to GRU hyperparameters. (a) Optimizer function. (b) Number of GRU cell. (c) Activation function in GRU layer. (d) Activation function in dense layer. (e) Number of batches.

Table 11.3 GRU configuration.

Layer	Number of cells	Number of batches	Activation function	Optimizer function	Dropout rate
GRU	756	–	Relu	–	–
Dropout	–	–	–	–	20%
Dense	1	–	Tanh	–	–
–	–	32	–	–	–
				RMSProp	

Table 11.4 MAPE (%) of different neural networks.

Neural networks	LSTM		GRU		MLP	
Layer	**Train (%)**	**Test (%)**	**Train (%)**	**Test (%)**	**Train (%)**	**Test (%)**
One	5.16	7.18	6.93	8.91	5.03	7.29
Two	5.76	8.00	6.40	8.35	5.18	7.60

11.4.2 AI algorithms performance for aggregated load forecast

Table 11.4 shows the MAPE error for MLP, LSTM, and GRU, for each algorithm configuration, as shown in Table 11.1, Table 11.2, and Table 11.3, respectively. The best algorithms for aggregated load forecasting, according to the results, are MLP and LSTM. The difference between MLP and LSTM is less than one percent, because both algorithms use the Adam optimizer. Adam is the main contributor to deep learning, because while the learning rate is both difficult to define and dependent on the dataset, this optimizer has an adaptive learning rate and, for this reason, the difference between the algorithms is quite small. The SGD result (see Fig. 11.10(f) and Fig. 11.11(d)) show a poor result using in the default learning rate (001). This reinforces the fact that the Adam optimizer improves the performance of both classical and deep networks. It should be pointed out that the time series used for this test is not volatile, for this reason the MLP and LSTM have similar results.

Fig. 11.13 shows the load forecasting, according to the three algorithms with real time series for the training and testing processes. GRU and LSTM show similar behavior, because these methodologies have the same basis, whereas MLP is closer to the real curve. The testing process has one day with a different pattern. That day the algorithms showed maximum error. The LSTM and GRU show similar results, and MLP is again closer to the real curve.

11.5 Conclusion

This chapter shows three AI load forecasting algorithms: MLP, LSTM, and GRU. To compare the algorithms, the best hyperparameters were chosen according to several tests. The hyperparameters were number of cells, activation function, optimizer function, and number of batches. The number of epochs was analyzed with early stopping.

The AI algorithms were compared using real dataset with 260 consumers. The consumers were aggregated into a single curve and the three prediction algorithms were compared. The three algorithms presented similar results with a MAPE difference of less than one percent.

Classical MLP neural networks had similar and good result just as deep neural networks, because the improvement of the two deep neural networks algorithms was the Adam optimizer function. Adam calculates the adaptive learning rate, whereas the classical methodologies perform many trials to calculate the best learning rate.

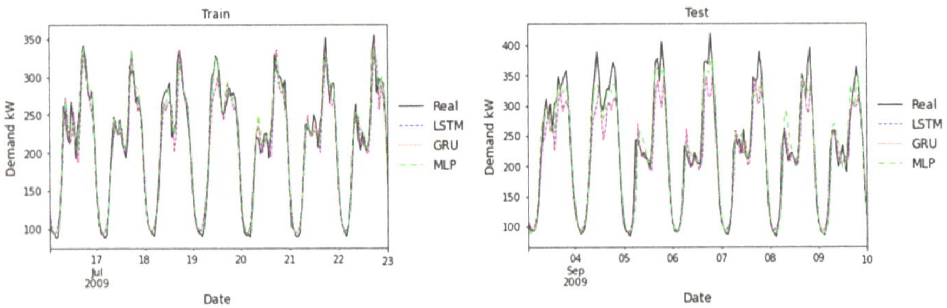

Figure 11.13 Load forecast of the three methodologies and actual data during training and during test.

Classical and deep neural networks with well optimized hyperparameters have similar results for the dataset used. The hyperparameters depend on the dataset, and the selection of the hyperparameters influence the result of neural networks, regardless of whether they are classical or deep neural networks.

Acknowledgments

This study was financed in part by the Coordenação de Aperfeiçoamento de Pessoal de Nível Superior - Brasil (CAPES) – Finance Code 001.

The authors would like to thank the Irish Social Science Data Archive for making the database used in this paper available through "CER Smart Metering Project – Electricity Customer Behaviour Trial, 2009-2010" and also to ISSDA, as follows: "Accessed through the Irish Social Science Data Archive – www.ucd.ie/issda".

References

[1] C.C. Aggarwal, Neural Networks and Deep Learning (Firsty), Springer, 2018.

[2] M. Beccali, M. Cellura, V. Lo Brano, A. Marvuglia, Forecasting daily urban electric load profiles using artificial neural networks, Energy Conversion and Management (2004), https://doi.org/10.1016/j.enconman.2004.01.006.

[3] M. Beccali, M. Cellura, V. Lo Brano, A. Marvuglia, Short-term prediction of household electricity consumption: assessing weather sensitivity in a Mediterranean area, in: Renewable and Sustainable Energy Reviews, 2008.

[4] Y. Bengio, Practical recommendations for gradient-based training of deep architectures, Lecture Notes in Computer Science (Including Subseries Lecture Notes in Artificial Intelligence and Lecture Notes in Bioinformatics) 7700 LECTU (2012) 437–478, https://doi.org/10.1007/978-3-642-35289-8-26.

[5] K. Chapagain, S. Kittipiyakul, Performance analysis of short-term electricity demand with atmospheric variables, Energies 11 (4) (2018) 1–34, https://doi.org/10.3390/en11040818.

[6] CER Smart Metering Project – Electricity Customer Behaviour Trial, 2009–2010 Commission for Energy Regulation, 2012, (CER). Irish Social Science Data Archive. SN: 0012, https://www.ucd.ie/issda/data/commissionforenergyregulationcer.

[7] F. Chollet, Keras, https://keras.io/, 2015.
[8] J. Chung, C. Gulcehre, K. Cho, Y. Bengio, Empirical evaluation of gated recurrent neural networks on sequence modeling, 1–9, http://arxiv.org/abs/1412.3555, 2014.
[9] D.A. Clevert, T. Unterthiner, S. Hochreiter, Fast and accurate deep network learning by exponential linear units (ELUs), in: 4th International Conference on Learning Representations, ICLR 2016 – Conference Track Proceedings, 2016, pp. 1–14.
[10] C. Crowley, F.L. Joutz, Weather effects on electricity loads: modeling and forecasting, Final Report, December 48 (2005).
[11] A.J. da Rocha Reis, A.P. Alves da Silva, Feature extraction via multiresolution analysis for short-term load forecasting, IEEE Transactions on Power Systems 20 (1) (2005) 189–198, https://doi.org/10.1109/TPWRS.2004.840380.
[12] Y.N. Dauphin, H. De Vries, Y. Bengio, Equilibrated adaptive learning rates for non-convex optimization, Advances in Neural Information Processing Systems 2015-Janua (2015) 1504–1512.
[13] T. Dozat, Incorporating Nesterov momentum into Adam, ICLR Workshop 1 (2016) 2013–2016.
[14] J.C. Duchi, E. Hazan, Y. Singer, Adaptive subgradient methods for online learning and stochastic optimization, Journal of Machine Learning Research 12 (2011) 2121–2159, https://doi.org/10.1109/CDC.2012.6426698.
[15] B. Jason, Deep learning for time series forecasting, Ml 1 (1) (2018) 1–50.
[16] T. Exchange, B.S. North, Appendices electricity smart metering customer behaviour trials (CBT) findings report document type: reference: date. trial, https://org/CER/11/080a, May 2011, pp. 1–189.
[17] T. Exchange, B.S. North, Electricity smart metering customer behaviour trials (CBT) findings report document type: reference: date. Trial, https://org/CER/11/080a, May 2011, pp. 1–58.
[18] K. Faceli, A. Lorena, J. Gama, A. Carvalho, Inteligência Artificial. Uma Abordagem de Aprendizagem de Máquina (LTC) (2011).
[19] R. Ge, S.M. Kakade, R. Kidambi, P. Netrapalli, The step decay schedule: a near optimal, geometrically decaying learning rate procedure for least squares, Advances in Neural Information Processing Systems 32 (2019) 1–28.
[20] X. Glorot, Y. Bengio, Understanding the difficulty of training deep feedforward neural networks, Journal of Machine Learning Research 9 (2010) 249–256.
[21] I. Goodfellow, Y. Bengio, A. Courville, Deep Learning, MIT Press, 2016.
[22] S. Haykin, Redes neurais, (Segunda) Bookman. (2008).
[23] V. Hodge, J. Austin, A Survey of Outlier Detection Methodologies, Kluwer Academic Publishers, 2004, pp. 213–241, 1969, https://doi.org/10.4324/9781315744988-22.
[24] T. Hong, S. Fan, Probabilistic electric load forecasting: a tutorial review, International Journal of Forecasting (2016), https://doi.org/10.1016/j.ijforecast.2015.11.011.
[25] T. Hong, P. Wang, L. White, Weather station selection for electric load forecasting, International Journal of Forecasting 31 (2) (2015) 286–295, https://doi.org/10.1016/j.ijforecast.2014.07.001.
[26] T. Hong, P. Wang, L. White, Weather station selection for electric load forecasting, International Journal of Forecasting 31 (2) (2015) 286–295, https://doi.org/10.1016/j.ijforecast.2014.07.001.
[27] T. Hossen, A.S. Nair, R.A. Chinnathambi, P. Ranganathan, Residential load forecasting using deep neural networks (DNN), in: 2018 North American Power Symposium, NAPS 2018 , 2018, pp. 1–5.

[28] T. Hou, R. Fang, J. Tang, G. Ge, D. Yang, J. Liu, W. Zhang, A novel short-term residential electric load forecasting method based on adaptive load aggregation and deep learning algorithms, Energies 14 (7820) (2021) 1–21.
[29] H. Kang, The prevention and handling of the missing data, Korean Journal of Anesthesiology 64 (5) (2013) 402–406, https://doi.org/10.4097/kjae.2013.64.5.402.
[30] Kazil J.K., Data wrangling with Python: tips and tool to make your life easier, in: O'Reilly Media (ed.), 1st ed., vol. 1, 2016.
[31] D.P. Kingma, J.L. Ba, Adam: a method for stochastic optimization, in: 3rd International Conference on Learning Representations, ICLR 2015 – Conference Track Proceedings, 2015, pp. 1–15.
[32] W. Kong, Z.Y. Dong, Y. Jia, D.J. Hill, Y. Xu, Y. Zhang, Short-term residential load forecasting based on LSTM recurrent neural network, IEEE Transactions on Smart Grid 10 (1) (2019) 841–851, https://doi.org/10.1109/TSG.2017.2753802.
[33] W. Kong, Z.Y. Dong, F. Luo, K. Meng, W. Zhang, F. Wang, X. Zhao, Effect of automatic hyperparameter tuning for residential load forecasting via deep learning, in: 2017 Australasian Universities Power Engineering Conference, AUPEC 2017, 2018, pp. 1–6, https://doi.org/10.1109/AUPEC.2017.8282478, 2017.
[34] C. Kuster, Y. Rezgui, M. Mourshed, Electrical load forecasting models: a critical systematic review, Sustainable Cities and Society 35 (August 2017) 257–270, https://doi.org/10.1016/j.scs.2017.08.009.
[35] T. Le, M.T. Vo, T. Kieu, E. Hwang, S. Rho, Multiple electric energy consumption forecasting using a cluster-based strategy for transfer learning in smart building, Sensors (Switzerland) 20 (2668) (2020) 1–17.
[36] X.J. Luo, L.O. Oyedele, Forecasting building energy consumption: adaptive long-short term memory neural networks driven by genetic algorithm, Advanced Engineering Informatics 50 (April 2021) 101357, https://doi.org/10.1016/j.aei.2021.101357.
[37] P. Lusis, K.R. Khalilpour, L. Andrew, A. Liebman, Short-term residential load forecasting: impact of calendar effects and forecast granularity, Applied Energy 205 (March 2017) 654–669, https://doi.org/10.1016/j.apenergy.2017.07.114.
[38] V. Nair, G. Hinton, Rectified linear units improve restricted Boltzmann machines, Haifa 33 (5) (2010) 807–814, https://doi.org/10.1123/jab.2016-0355.
[39] C. Nwankpa, W. Ijomah, A. Gachagan, S. Marshall, Activation functions: comparison of trends in practice and research for deep learning, pp. 1–20, http://arxiv.org/abs/1811.03378, 2018.
[40] F. Pedregosa, G. Varoquaux, A. Gramfort, V. Michel, B. Thirion, O. Grisel, M. Blondel, P. Prettenhofer, R. Weiss, V. Dubourg, J. Vanderplas, A. Passos, D. Cournapeau, M. Brucher, M. Perrot, E. Duchesnay, Scikit-learn: machine learning in python, Journal of Machine Learning Research 12 (2011) 2825–2830.
[41] Y. Peng, Y. Wang, X. Lu, H. Li, D. Shi, Z. Wang, J. Li, Short-term load forecasting at different aggregation levels with predictability analysis, in: 2019 IEEE PES Innovative Smart Grid Technologies Asia, ISGT 2019 , 2019, pp. 3385–3390.
[42] M.Q. Raza, A. Khosravi, A review on artificial intelligence based load demand forecasting techniques for smart grid and buildings, in: Renewable and Sustainable Energy Reviews, 2015.
[43] R. Raza, M.A. Ansari, R.K. Pachauri, Short-term load forecasting using ann technique, in: Communication and Computing Systems – Proceedings of the International Conference on Communication and Computing Systems, ICCCS 2016, vol. 1(2), 2017, pp. 481–485.
[44] S. Ruder, An overview of gradient descent optimization algorithms, arXiv preprint (2017) 1–14, https://doi.org/10.1111/j.0006-341X.1999.00591.x.

[45] H.R. Sadeghi Keyno, F. Ghaderi, A. Azade, J. Razmi, Forecasting electricity consumption by clustering data in order to decline the periodic variable's affects and simplification the pattern, Energy Conversion and Management (2009), https://doi.org/10.1016/j.enconman.2008.09.036.
[46] H. Shi, M. Xu, R. Li, Deep learning for household load forecasting—a novel pooling deep RNN, IEEE Transactions on Smart Grid 9 (5) (2018) 5271–5280, https://doi.org/10.1109/TSG.2017.2686012.
[47] S. Skansi, Introduction to Deep Learning, Springer, 2018.
[48] S.C. Smithson, G. Yang, W.J. Gross, B.H. Meyer, Neural networks designing neural networks: multi-objective hyper-parameter optimization, http://arxiv.org/abs/1611.02120, 2016.
[49] N. Somu, G. Raman M R, K. Ramamritham, A deep learning framework for building energy consumption forecast, Renewable and Sustainable Energy Reviews 137 (December 2020) 110591, https://doi.org/10.1016/j.rser.2020.110591.
[50] M.M. Suarez-Alvarez, D.T. Pham, M.Y. Prostov, Y.I. Prostov, Statistical approach to normalization of feature vectors and clustering of mixed datasets, Proceedings of the Royal Society A: Mathematical, Physical and Engineering Sciences 468 (2145) (2012) 2630–2651, https://doi.org/10.1098/rspa.2011.0704.
[51] S. Sultana, N. Pourebrahim, H. Kim, Household energy expenditures in North Carolina: a geographically weighted regression approach, Sustainability (Switzerland) 10 (5) (2018) 1–22, https://doi.org/10.3390/su10051511.
[52] D. Syed, H. Abu-Rub, A. Ghrayeb, S.S. Refaat, Household-level energy forecasting in smart buildings using a novel hybrid deep learning model, IEEE Access 9 (2021) 33498–33511, https://doi.org/10.1109/ACCESS.2021.3061370.
[53] J. Szoplik, Forecasting of natural gas consumption with artificial neural networks, Energy 85 (2015) 208–220, https://doi.org/10.1016/j.energy.2015.03.084.
[54] J.W. Taylor, R. Buizza, Neural network load forecasting with weather ensemble predictions, IEEE Transactions on Power Systems 17 (3) (2002) 626–632, https://doi.org/10.1109/TPWRS.2002.800906.
[55] G. Upton, I. Cook, A Dictionary of Statistics, 2 ed., Oxford University Press, 2008, 3 ed., Oxford Uni.
[56] J. Xie, Y. Chen, T. Hong, T.D. Laing, Relative humidity for load forecasting models, IEEE Transactions on Smart Grid 9 (1) (2018) 191–198, https://doi.org/10.1109/TSG.2016.2547964.
[57] J. Xie, T. Hong, Wind speed for load forecasting models, Sustainability (Switzerland) 9 (5) (2017) 1–12, https://doi.org/10.3390/su9050795.
[58] Y. Yu, F. Liu, Effective neural network training with a new weighting mechanism-based optimization algorithm, IEEE Access 7 (2019) 72403–72410, https://doi.org/10.1109/ACCESS.2019.2919987.

Study of harmonics in linear, nonlinear nonsinusoidal electrical circuits by geometric algebra

12

Juan M. Ramirez[a], Nitin Sundriyal[a], Mario Roberto Arrieta-Paternina[b], and Alejandro Zamora-Méndez[c]
[a]Center for Research and Advanced Studies of the National Polytechnic Institute, Mexico City, Mexico, [b]National Autonomous University of Mexico, Mexico City, Mexico, [c]Michoacan University of Saint Nicholas of Hidalgo, Morelia, Mexico

12.1 Introduction

Most of the work done to date by researchers has developed power theories for sinusoidal and nonsinusoidal circuits. Sinusoidal power theories are well accepted, but there is still a lengthy debate on methodology and standards for non-sinusoidal, nonlinear circuit theory. The concept of apparent power and active power for these circuits has been a broad research area since the Steinmetz experiment on electric arc [1]. This is because nonlinear loads exhibit distortion in voltage and current waveform, and the power calculation deviates from sinusoidal circuit calculation. Some of the significant research contributions in the above mentioned nonsinusoidal scenario have been listed below.

12.1.1 Sinusoidal theory (classical one): time-domain perspective

Conventionally, in the time domain, power refers to the product of voltage and current. The concept of power is important as it is used for billing purposes, investment cost estimation, energy usage estimation, etc.

The RMS voltage and current for a sinusoidal system with a "*phase shift α*" are expressed by

$$v(t) = \sqrt{2}V cos(\omega t +), \tag{12.1}$$

$$i(t) = \sqrt{2}I cos(\omega t + \alpha \pm \varphi), \tag{12.2}$$

where V and I stand for the amplitude of voltage and current, respectively; ω is the angular frequency; φ is the power factor, and t means time.

Monitoring and Control of Electrical Power Systems Using Machine Learning Techniques
https://doi.org/10.1016/B978-0-32-399904-5.00018-1

The instantaneous power is given by

$$p(t) = v(t)i(t), \tag{12.3}$$

$$p(t) = 2VIcos(\omega t + \alpha)cos(\omega t + \alpha \pm \varphi), \tag{12.4}$$

$$p(t) = VIcos\varphi + VIcos\varphi cos(2\omega t + 2\alpha) + VIsin\varphi sin(2\omega t + 2\alpha). \tag{12.5}$$

Conventionally, the active power is defined as the scalar component of the instantaneous power, first term in (12.5). In the third term's amplitude, i.e., $VIsin\alpha$ represents the reactive power.

The active power is defined as the scalar component of the instantaneous power,

$$P_{avg} = \frac{1}{T}\int_{-\frac{T}{2}}^{\frac{T}{2}} p(t)\,dt = VIcos\varphi. \tag{12.6}$$

Likewise, IEEE standard 270-2006 defines apparent power for a sinusoidal circuit for constant line losses by (12.7) and (12.8)

$$S^2 = P^2 + Q^2, \tag{12.7}$$

$$S = VI. \tag{12.8}$$

In sinusoidal theory, another important concept is the *power factor*. The IEEE standard 1459-2010 and IEEE standard 270-2006 define power factor with constant line losses (12.9).

$$pf = \frac{P}{S} \tag{12.9}$$

According to IEEE 270-2006 standard, this factor gives the degree of line utilization. All the points mentioned above in the mathematical equation are well accepted for sinusoidal circuits with linear loads.

12.1.2 Budeanu power theory

C.I. Budeanu's work on the power theory was presented in detail in his book in 1927, Reactive and Fictive Power [2]. This theory is developed in the frequency domain. He grouped the harmonic current as I_h in response to harmonic excitation voltage V_h. He separated apparent power into average active power (P), reactive power (Q_B), and distortion power (D_B); the subscript B is used to denote Budeanu's reactive power definition.

He first defined active and reactive power and followed it by specifying the distortion power based on the apparent power. He was the first scientist to provide apparent power with a three-dimensional representation (12.10).

$$D_B = \sqrt{S^2 - P^2 - Q_B^2} \tag{12.10}$$

12.1.3 Fryze power theory

Fryze analysis is framed in the time domain [3]. He proposed that the current in the circuit has two components based on the impedance of the circuit. One is the active or in-phase component (due to conductance), and the other one is the reactive or quadrature component (due to susceptance), given by the following:

$$I_a = I\sqrt{2}cos\varphi, \tag{12.11}$$

$$I_r = I\sqrt{2}sin\varphi. \tag{12.12}$$

This theory has a significant effect now. The subsequent work in this field, namely by Depenbrock [4] in FBD theory and Czarnecki in CPC theory [6–8], is heavily influenced by Fryze concepts. FBD is an abbreviation meaning Fryze–Buchholz–Depenbrock, and CPC stands for current's physical component.

12.1.4 IEEE standard 1459-2010

The IEEE standard uses the frequency domain [5]; thus all the equations and quantities belong to the frequency domain. In the standard, voltages and currents are divided into two components: the fundamental frequency components, denoted by the subscript 1, and the nonfundamental elements characterized by the subscript H. The direct voltage and the straightforward current terms (v_0, i_0) are included in the nonfundamental components. Furthermore, this standard defines a new name as nonactive power, and it says nonactive power is not reactive power; only in sinusoidal cases, nonactive power can be known as reactive power.

12.1.5 CPC theory

Czarnecki proposed this theory in the frequency domain, and his timely work in this regard can be seen in [6–9]. This power theory is based on decomposing nonsinusoidal current into a component that can give a physical meaning to them. According to this theory, the supply current (i) can be decomposed into

a. Active current (i_a),
b. Reactive current (i_r),
c. Scattered current (i_s).

Consequently, he defined apparent power (S), active power (P), scattered power (D_s) and reactive power (Q_r) by the following set of equations:

$$S^2 = P^2 + D_s^2 + Q_r^2, \tag{12.13}$$

$$P = \|v\| \, \|i_a\|, \tag{12.14}$$

$$D_s = \|v\| \, \|i_s\|, \tag{12.15}$$

$$Q_r = \|v\| \, \|i_r\|, \tag{12.16}$$

$$S = \|v\| \, \|i\| \,. \tag{12.17}$$

12.1.6 M. Castilla and J.C Bravo's theory

These authors were the first to introduce the idea of a geometric framework [10] to analyze the electric circuit. Their analysis of circuits was dependent on two-port representations. The author compared their result with traditional theories and pointed out the previous method's shortcoming in modeling or reactive power, affecting compensation.

12.1.7 Other contribution

Besides the abovementioned impactful work, numerous contributions to the nonsinusoidal circuit have been made. These works are not discussed in detail in this document. Still, they are worth to be mentioned, as the definition of reactive power, by W. Shepherd and P. Zakikhani [11], the description of apparent power by A.E. Emanuel [12], power based on instantaneous space vector by A. Nabe and T. Tanaka [13], the definition of apparent power based on modified voltage by F. Ghassemi [14]. Peng [15], Dai [16]. Reduced reference frame by Francisco Casado [17], a new interpretation of the Akagi–Nabae power components by Jl Willems [18]; other notable works are given in [19–23].

12.1.8 Geometric algebra power theory

Before the GA power theory work, geometrical algebra was considered better for computer vision and robotic applications [35]. Still, few researchers, such as Menti [25] and Castilla [10] used it to demonstrate power calculation in electrical circuits. Geometrical Algebra power theory is the most recent work of Castro Nunez et al. in non-sinusoidal circuits [26–28]. Castro et al. describe it in a new domain known as "G_n" (*geometric domain*). Also, strong reasons are provided for modeling the linear parameters (R, L, C) in geometric domains. Moreover, a new set of rules of circuit analysis has been defined so that the interaction between different harmonics can be adequately seen, which was not done earlier. Finally, they proposed a compensation scheme for power factors in a nonsinusoidal circuit based on their research work findings.

Although Castro has provided satisfactory results to support geometric algebra as a tool for electric circuits, he has not given a proper answer for nonlinear circuit analysis, as nonlinearity is the root cause of the conflict of interest in the power calculation.

After Castro Nuñez, the pioneer in the geometrical analysis of electrical circuits is Montoya et al., and their continuous effort can be referred to in [24] and [29–32]. They have provided some modifications to the initial proposal of Castro Nunez and have also given a mathematical way for analyzing circuits with subharmonics.

Generally, for the circuits involving disturbances (either voltage or current, or both), there are primarily two approaches of analysis, i.e., time domain and frequency

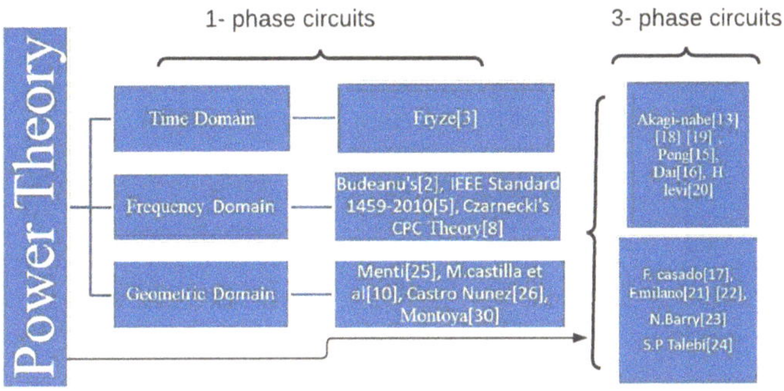

Figure 12.1 Power theory chronology.

domain. The latter has been most widely used for circuit analysis and compensation purposes. In the frequency domain, complex numbers are used for apparent power calculation. Most of the work in this field of research has resulted in various power theories and standards; a very brief summary is shown in Fig. 12.1.

Linear circuit analysis is the primary subject for any electrical engineer. Almost every engineer is well-versed in complex algebra in linear circuit analysis and can easily interpret its result. However, using geometric algebra instead of complex algebra gives a thorough insight into the power flow phenomenon in these circuits. In this chapter, an attempt is made to show the power flow, and side by side, it is validated with another conventional approach (i.e., frequency domain) for a consensus of agreement.

12.1.9 Harmonic domain

Analysis of circuits by the harmonic domain (frequency domain) is a widely accepted technique used by Budenau [2]. It uses Fourier series expansion for representation of the signal in a specific vector arrangement, as shown below,

$$X = \begin{bmatrix} X_{-h} \\ \vdots \\ X_{-1} \\ X_0 \\ X_1 \\ \vdots \\ X_h \end{bmatrix}$$

where X represents the HD signal, "h" represents the highest harmonic number, and "0" represents the DC component. Euler's identity converts the time-domain or excitation signal into an HD arrangement. An example is shown for the HD arrangement:

$$v(t) = \sqrt{2}\left[100\sin(\omega t) + 25\sin(3\omega t)\right]. \tag{12.18}$$

Table 12.1 Power equation in harmonic domain.

PQ index	Mathematical formula
S (Apparent power)	$S = V_{rms} * I_{rms}$
P (Active power)	$P = V_{rms} * I_{rms} cos\varphi$, where φ is the phase difference between voltage and current
Q (Reactive power)	$Q = V_{rms} * I_{rms} cos\varphi$
D (Distortion power)	$D = \sqrt{S^2 - P^2 - Q^2}$

By using Euler's identity, the given signal can be represented as

$$v(t) = 50e^{j\omega t} + 50e^{-j\omega t} - 12.5e^{j3\omega t} + 12.5e^{-j3\omega t}. \tag{12.19}$$

Now, this can be represented in an HD arrangement as

$$V = \begin{bmatrix} 12.5 & 0 & 50 & 0 & 50 & 0 & -j12.5 \end{bmatrix}^T. \tag{12.20}$$

Furthermore, in HD, the impedances are translated as

$$Z_{series} = R + jh\omega_0 L + -\frac{1}{jh\omega_0 C}; \tag{12.21}$$

"h" is the harmonic order, and "ω_0" is the fundamental frequency in the signal.

The admittance can be computed as

$$Y = (Z_{series})^{-1} \tag{12.22}$$

Thereafter current can be calculated using relation in (12.23):

$$I = Y * V, \tag{12.23}$$

where "Y" is an admittance matrix, whose size depends on the number of harmonics.

$$Y = \begin{pmatrix} Y_h^{11} & \cdots & Y_h^{1N} \\ \vdots & \ddots & \vdots \\ Y_h^{N1} & \cdots & Y_h^{NN} \end{pmatrix} \tag{12.24}$$

For power quality indices, the calculation is done by using the relation in Table 12.1.

These power terms form a power cube, whose geometric interpretation is shown in Fig. 12.2 [34].

12.2 Problem statement and research objective

A conclusion is drawn based on the literature survey and investigation results on power flow under sinusoidal and nonsinusoidal linear and nonlinear circuits. However, espe-

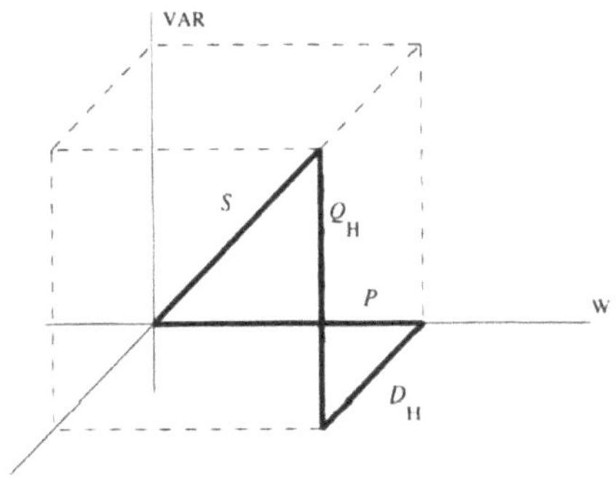

Figure 12.2 Circuit examples used in this study are for mathematical interpretation only, and are extensively solved using geometric algebra. When tackling a problem of great complexity, the need for compact mathematical formulations becomes extremely important, because it allows efficient solutions to be constructed. This is why geometric algebra and differential geometry, based on the mathematical theory of Lie groups and Lie algebras, and Computational Geometry tools for the analysis of evolving interfaces, are consistently used in applications in robotics, artificial vision, etc. These formulations lead to secure and complete solutions applications, numerically stable, and with a precise geometrical interpretation. That is why some of such tools are used in electrical circuit theory and applications to electrical networks.

cially for the nonsinusoidal nonlinear circuits, the debate is still on to identify the standards and definition for power flow (i.e., apparent power), power factor compensation, and a mathematical framework for assessing active and reactive power.

Frequency domain analysis provides a better understanding of power flow in such circuits and can handle the nonlinearity in the circuit. However, most of the work presented in the frequency domain is focused on a single point; i.e., it considers the analysis around the dominant harmonic or analyzing the harmonic one at a time. Yet in [16], [19] researchers demonstrated geometric algebra as a mathematical tool to analyze all the harmonics (disturbances) present in the circuit (voltage and current) to give a unified result on power and power factor and provide the necessary compensation scheme.

The authors use geometrical algebra to explain a power flow in a sinusoidal or nonsinusoidal linear circuit by using theoretical examples to prove its superiority over complex numbers in electrical circuits. Also, GA is a robust mathematical entity with a multidimensional axiom for representing vectors. As complex algebra is the subspace of GA, the choice is appropriate as a mathematical tool, and the computational ease with geometrical algebra has also been demonstrated. Still, these results are yet to be validated for the nonlinear circuit, and modeling the nonlinear element is necessary for the geometric domain to be accepted as a mathematical tool instead of complex numbers.

This study aims to validate the geometric algebra results in sinusoidal, nonsinusoidal linear circuits by frequency domain. Likewise, we revised geometrical circuit analysis techniques involving nonlinear modeling elements, such as a diode, SVC, and saturated core transform.

12.2.1 Contributions

The main contributions of this chapter are

- The use of an old theory, but revived in the light of its possible applications in electrical networks
- We are using geometric algebra to introduce the study of sinusoidal and nonsinusoidal electrical circuits
- The introduction of novel grid devices can be achieved transparently.

12.3 Harmonics in power system

Generally speaking, harmonics are a disturbance in the power system. Mathematically, harmonics are the integer multiple of the system's fundamental frequency (50 Hz or 60 Hz). Harmonics tends to disturb the wave shape of a pure sinusoidal waveform. Harmonics in the system are generally generated due to nonlinear loads, e.g., switching devices and saturated core transformers [33,34].

The presence of harmonics produces several undesirable effects in power systems, such as overheating in lines, motors and transformers, overvoltage, system resonance, and tripping of circuit breakers. These are the power quality issues; they hamper the normal functioning of the power system. In addition, the harmonics in the system lead to a change in the RMS value of current and voltage and, consequently, in total power [33].

The power equipment and appliances are designed to work on a sinusoidal supply. Still, the operating voltage is far from sinusoidal because of the proliferation of RE and power electronics equipment. Then, the appliances have to take the damage of these distorted waveforms. Overall, this deteriorates the power quality of supply, reaching the consumer level. The end-user equipment (e.g. refrigerators, AC, washing machine, computer, LED TV, etc.) is high-risk. The following example plot shows how a voltage waveform source changes from pure sinusoid to non-sinusoidal by the presence of harmonics and inter-harmonics:

$$V = 100\sqrt{2}\,(sinwt + sin3wt + sin3.2wt)\,.$$

In the above waveform, fundamental, the third multiple of fundamental and the inter harmonic between third and fourth and harmonic is present, Fig. 12.3.

12.4 Geometric algebra

Geometric algebra's introduction came in 1878, but it was forgotten for a long time. In 1966 David Hestenes again revived the geometrical interpretation, and, since then, it has been considered a powerful mathematical tool for analysis in various engineering domains.

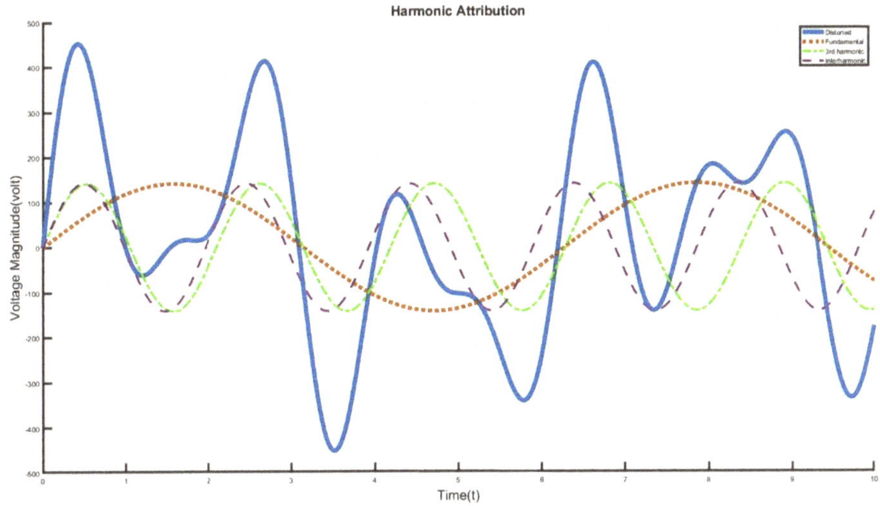

Figure 12.3 Voltage waveform with first, third, and inter harmonic.

The geometric algebra of a vector space is an algebra over a field. It is known for its multiplication operator known as the geometric product on a space of elements called multivector. Geometric algebra can be defined as the Clifford algebra of a vector space. Clifford described a new product known as a geometric product, which is conservative. He combined the work of Hamilton and Grassmann into a single structure and coined it Clifford algebra (CA). Scalars and vector representation have the usual interpretation; detailed theory on the algebraic operator and its properties can be referred to in [35]. The bivectors provide a picture of pseudovector quantities, e.g., area, angular rotation, and torque.

It is essential to realize that complex numbers in physical applications usually have a natural geometric interpretation hidden in conventional formulations. In the following, the authors briefly introduce geometric algebra, emphasizing the geometric meaning of vectors' associative (Clifford) product. It is possible to interpret the unit scalar imaginary number as arising from the geometry of real space.

By geometric representation of two vectors, it is possible to represent vectors in terms of

- Magnitude
- Direction
- Sense
- Grade

For two vectors, the representation in the geometric domain is

$$G_2 = \{1, e_1, e_2, e_1 e_2\}, \tag{12.25}$$

where 1 is a scalar, e_1 and e_2 are vectors, and e_1, e_2 represent a pseudoscalar. e_i (here $i = 1$ to 2) is the basis vector that generates the multivector G_2. In general, for n-

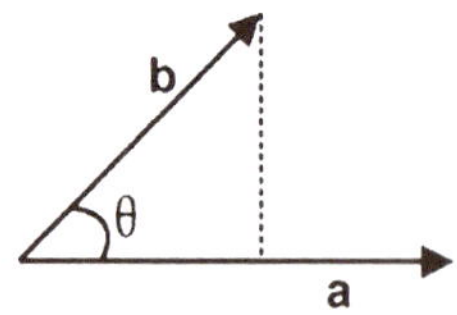

Figure 12.4 Scaler or dot product or inner product.

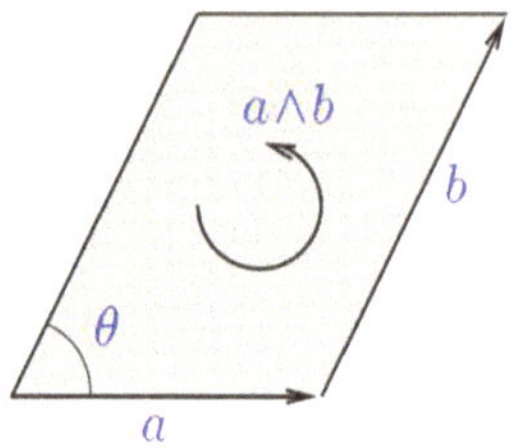

Figure 12.5 Wedge or exterior product (grade 2 element).

vectors, the geometric subspace becomes

$$G_n = gen\{1, \{e_1, e_2, ...e_n\}, \{e_{23}, e_{31}, ...e_{n(n-1)}\}, \{e_{123}, e_{124}, ...e_{n(n-1)(n-2)}, ...\{I\}\}. \quad (12.26)$$

12.4.1 Geometric product

A way to imagine the outer product is to interpret **a** ∧ **b** as the area "swept out" by displacing **a** along **b**, with the orientation given by traversing the parallelogram so formed first along a vector, then along a **b** vector. This notion provides a generalization of objects with higher dimensionality or *grade* products. Thus if the bivector **a** ∧ **b** (*grade* 2) is swept out along another vector **c** (*grade* 1), we obtain the directed volume element (**a** ∧ **b**)∧**c**, which is a trivector (*grade* 3).

The geometric product between two vectors **a** and **b** is defined as Figs. 12.4–12.5,

$$M = ab = a.b + a \wedge b \quad (12.27)$$

Similarly, when we move to a higher subspace of GA, i.e., grade 4, grade 5, different shapes are observed for the exterior product of vectors.

Adding a scalar and a bivector doesn't seem right, because they are different types of quantities. But it is precisely what you want an addition to do. Adding a scalar to a bivector is an object with both scalar and bivector parts. The addition of real and imaginary numbers yields an object with both real and imaginary parts. We call such entity a "*complex number*," and we shall refer to a (scalar + bivector) as a "*multivector*," accepting that we combine objects of different types. It has been demonstrated in [26–28] that the use of the complex number as a fundamental tool in the frequency domain can be replaced by geometric algebra (GA), as complex numbers are subspaces

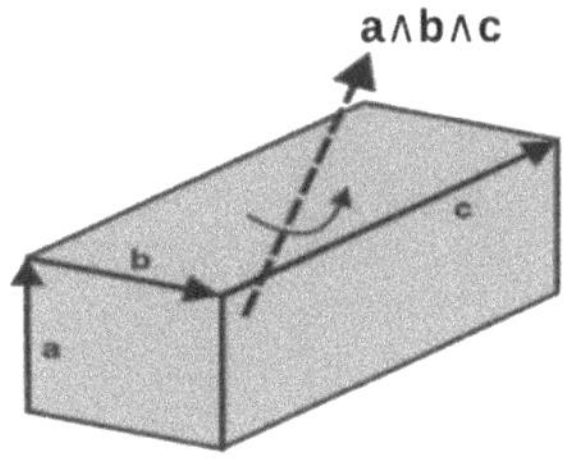

Figure 12.6 Trivector (grade 3 element).

of GA. Also, they do not impart any physical significance to the calculation. It is a mere way of representing magnitude and direction; however, with GA representation, we get information about the sense or shape of an interacting vector, i.e., a geometric shape. Also, GA gives information about the frequency component interaction, which is not possible with complex numbers only. The following equation shows the transformation from the time domain to the geometric domain for circuit analysis:

Transformation equation

$$\left.\begin{aligned} x_{c1}(t) &= X_{rms}\cos(\omega t) \leftrightarrow e_1 \\ x_{s1}(t) &= X_{rms}\sin(\omega t) \leftrightarrow -e_2 \\ &\vdots \\ x_{cn}(t) &= X_{rms}\cos(\omega t) \leftrightarrow \wedge_{i=2}^{n+1} e_i \\ x_{sn}(t) &= X_{rms}\sin(\omega t) \leftrightarrow \wedge_{\substack{i=1 \\ i\neq 2}}^{n+1} \end{aligned}\right\} \tag{12.28}$$

Multivector addition and multiplication obey the associative and distributive laws; then, we have the geometric algebra of the plane. Next is a summary of basic GA [26]:

- The geometric product of two parallel vectors is a scalar number: the product of their lengths.
- The geometric product of two perpendicular vectors is a bivector: the directed area formed by the vectors.
- Parallel vectors commute under the geometric product: perpendicular vectors anticommuting.
- The bivector $e_1 \wedge e_2$ has the geometric effect of rotating the vectors $\{e_1$ and $e_2\}$ in its plane by 90° clockwise when multiplying them on their left. It turns vectors by 90° anticlockwise when multiplying on their right. This can be used to define the orientation of e_1 and e_2.
- The square of the bivector area e_1 and e_2 is a scalar: bivector $(e_1 \wedge e_2)^2 = -1$. By the last two properties, the bivector $(e_1 \wedge e_2)$ becomes our first candidate for the role of the unit imaginary i, and in 2-dimensional applications; it fulfills this role admirably. Indeed, we see that the even-grade elements $z = x + y(e_1 \wedge e_2)$ form a natural subalgebra, equivalent to complex numbers.

Table 12.2 Geometric circuit equations.

Geometric circuit equation	**Interpretation**
$M = VI = P + CN_{r(ps)} + CN_{r(hi)} + CN_d$	Net multivectorial power (commonly referred to as apparent power in an AC circuit)
$CN_{r(ps)} + CN_{r(hi)} = Q$	Reactive power because of power source and harmonic interaction
$\|M\| = \sqrt{\left\langle \tilde{M}M \right\rangle_0}$	Norm or absolute value of a multivector or RMS value
$Zh = R + \left(\frac{1}{hwC} - hwL\right) e_1 e_2$	Impedance evaluated at harmonic level $h = 0...n$
$I = Y.V$	I= current, Y= admittance, V=voltage
$Y = G + j\,B$	G=conductance, B=susceptance
$ig = \sum_{i=1}^{n} V * G_i$	ig = current component due to conductance
$ib = \sum_{i=1}^{n} V * B_i$	ib= current component due to conductance
$pf = cos^{-1}\left(\frac{P}{\|M\|}\right)$	pf= power factor

The transformation converts the time domain signal into a geometric domain for analysis and adapts to Fourier series expansion. The power quality parameters of the circuit differ from their meaning in complex notation. By using GA, they give multivectorial representation. For example, the total power calculated by GA is known as apparent net power, which is different from apparent power obtained by complex algebra calculation. For power quality indices calculation of any linear circuit by GA, the relations used are given below in Table 12.2 for the reader's convenience.

12.4.2 Circuit expressions in the geometric domain

Despite the above table, we'd like to discuss a few of the already highlighted relationships for the reader's benefit.

1. **Power multivector** produces voltage and current vector in the geometric domain. Therefore, geometrically, it is decomposed as [27].

$$M = VI = P + CN_{r(ps)} + CN_{r(hi)} + CN_d \tag{12.29}$$

The above interpretation of apparent power in the geometric domain is different from the traditional definition of apparent power; in (12.22), the interaction of various frequency components is observed more precisely, where

- M is the complete multivector power involving scalers, vector, bivector, etc.
- P is the scalar part of the active power
- $CNr(ps)$ is a reactive power term that involves the product of voltage and current ib (due to susceptance)

- $CNr(hi)$ is the reactive part of power due to harmonics
- CNd is the reactive part of power corresponding to the cross-frequency term between voltage and ig (due to circuit conductance) or degraded capacity. (ig and ib are defined separately)
- $CNr(ps) + CNr(hi) = Q$ is the total reactive power

In the case of pure sinusoidal excitation in (12.22) the terms $CNr(hi)$= 0, and CNd= 0 and $Q = CNr(hi)$, i.e., fundamental component of frequency in voltage and current contribute to the reactive power.

2. **Admittance**

It is the reciprocal of impedance $Y = Z^{-1}$; it is used for current calculation in geometric algebra as

$$I = Y.V \tag{12.30}$$

Admittance can be written in the form of conductance (G) and susceptance (B):

$$Y = G + jB. \tag{12.31}$$

Thus in geometric domain,

$$Y = G + Be_1e_2. \tag{12.32}$$

3. **Current calculation**

$$I = Y.V$$

Thus current can be separated into two parts based on G and B:

$$ig = \sum_{i=1}^{n} V * G_i, \tag{12.33}$$

$$ib = \sum_{i=1}^{n} V * B_i, \tag{12.34}$$

where "n" is the harmonic order, ig and ib are, correspondingly, the parallel and quadrature (perpendicular) component of current. In geometrical algebra, the current component "ig" is defined as projection, and "ib" is defined as the rejection of a space defined by 'I'.

12.5 Case studies and results

The case studies presented here in the form of circuits have been adopted from the previous work on power theories and validated by the harmonic domain, which the

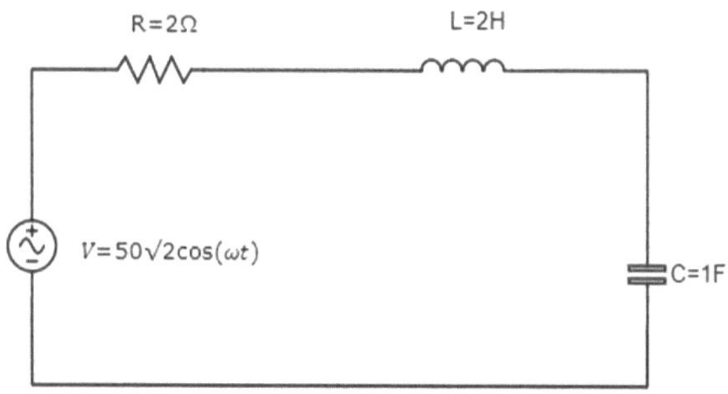

Figure 12.7 RLC circuit.

Table 12.3 Result for Sinusoidal case 1.

Geometric domain	Harmonic domain
Vrms= 100 V	Vrms=100 V
Irms=44.72 A	Irms=44.72 A
M=4.472 kVA	S=4.472 kVA
P= 4000 W	P=4000 W
Q=2 kVA	Q=2 kVA
D=0	D=0
Pf=0.8944	Pf=0.8944

authors above have not done. GA concept has been extensively applied to the problems and compared with harmonic domain results. The case studies presented here are meant for a theoretical understanding of power flow and power factor only. More practical applications of the proposed methods will be formally achieved in the systems involving measurements in the electrical power system.

Case 1: Sinusoidal

The circuit shown above is analyzed in different conditions (sinusoidal and nonsinusoidal) by changing R, L, C parameters, and results are presented in both harmonic and geometric domain:

For sinusoidal case taking R=2Ω, L=2H, and C= 1F, w=1 rad/sec and the supply voltage as $V = 50\sqrt{2}\ coswt$.

The results in the geometric and harmonic domains are summarized in Table 12.3.

A regular visual inspection of the results above implies that all the parameters calculated by geometrical algebra are the same as those calculated by the harmonic domain. So we can consider for the pure sinusoidal excitation that there is an agreement in circuit analysis, as it is hard to find any difference in calculated values. Also, the principle of energy conservation can be verified by using the above results.

Case 2: Nonsinusoidal:

For the sake of simplicity, the same circuit is taken, but the source is made nonsinusoidal:

$$V\_s = \sqrt{}2[(100\sin(wt) + 100(sin3wt)] \tag{12.35}$$

and R=1Ω, L=1/2 H, C= 2/3F (adopted from [27]).

For calculating current, harmonic impedance is calculated at each harmonic and voltage is transformed from the time domain to the geometric domain by using the transformation in (12.19).

$$V = -100e_2 + 100e_{134}, \tag{12.36}$$

and the impedance in the geometric domain by (12.32) as

- $Z1$ represents the impedance at fundamental harmonic

$$Z1 = \in + e_{12} \tag{12.37}$$

- $Z3$, represents the impedance at third multiple of fundamental or 3rd harmonic

$$Z3 = \in - e_{12} \tag{12.38}$$

Using the reciprocal, the admittance is calculated as

$$Y1 = 0.5000 \in -0.5000e_{12}, \tag{12.39}$$

$$Y3 = 0.5000 \in +0.5000e_{12}. \tag{12.40}$$

Since

$$i = V * Y, \tag{12.41}$$

$$i = 50.000e_1 - 50.000e_2 + 50.000e_{134} - 50.000e_{234}. \tag{12.42}$$

The current can be separated into two parts using Eq. (12.33) and (7)34

$$ig = -50.000e_2 + 50.000e_{134}, \tag{12.43}$$

$$ib = 50.000e_1 - 50.000e_{234}. \tag{12.44}$$

The power multivector and its component are calculated using the relation in Table 12.1:

$$M = 10000 + 1000e_{12} + 10000e_{34}, \tag{12.45}$$

and the power factor according to relation in Table 12.1

$$Pf = 0.577. \tag{12.46}$$

The power flow solution for case study 2 has been tabulated in Table 12.4. The nonsinusoidal behavior of current and voltage is shown in Fig. 12.6, which is done by geometric algebra, and the Maple GA toolbox is used for it. Also, in Fig. 12.7, waveforms obtained by harmonic analysis in Matlab® are presented.

There is a difference in the harmonic domain and CPC result of reactive power in the above. However, the CPC theory author does not use the Fourier coefficient for reactive power. (See Figs. 12.8 and 12.9.)

Table 12.4 Result for nonsinusoidal case 2.

Geometric domain	Harmonic domain	CPC theory
Vrms=141.4 V	Vrms=141.4 V	Vrms=141.4
Irms=100 A	Irms = 100 A	Irms= 100 A
M=17.32 kVA	S=14 kVA	S=14 kVA
P= 10 kW	P=10 kW	P=10 kW
Q=10 kVA	Q=118 VA	Q=10 kVA
D=0	D= 10 kVA	D=0
Pf=0.577	Pf=0.71	Pf=0.71

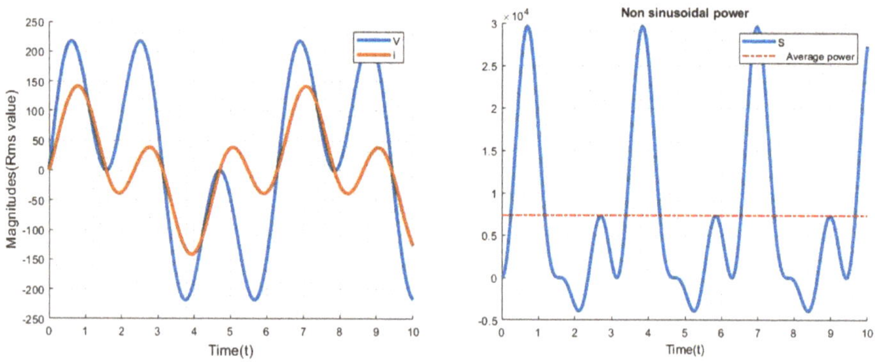

(a) (b)

$V := [100 \sin(t) + 100 \sin(3\,t)]$ $i := 50 \sin(t) + 50 \sin(3\,t) + 50 \cos(t) - 50 \cos(3\,t)$

Figure 12.8 (a) Voltage and current waveform in the nonsinusoidal circuit for case 2 (GD); (b) nonsinusoidal power for case 2 (GD).

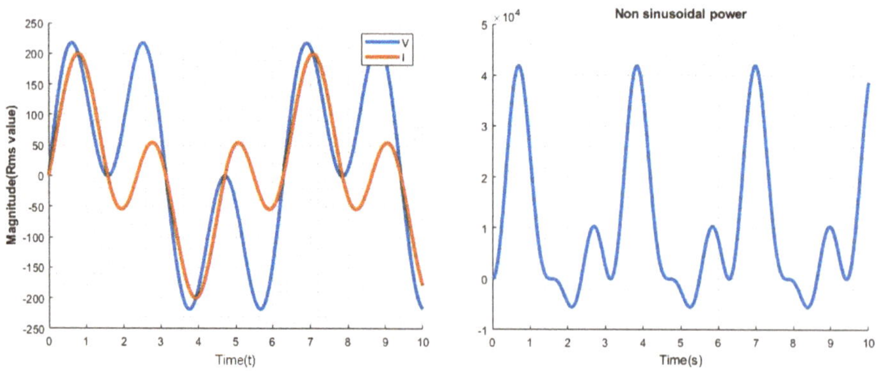

(a) (b)

Figure 12.9 (a) Current and voltage waveform for case 2 (HD); (b) nonsinusoidal power for case 2 (HD).

Table 12.5 Result for odd harmonics of higher-order case 3.

Geometric domain	Harmonic domain
Vrms=109.18 V	Vrms=109.18 V
Irms=75.16 A	Irms = 75.16 A
M=8.28 kVA	S=8.278 kVA
P= 4.5 kW	P=5.65 kW
Q=6.8 kVA	Q=1.39 kVA
D=1.25 kVA	D= 5.95 kVA
pf=0.54	Pf=0.6885

In the above figures, all the waveforms presented are instantaneous; in the waveform by GD domain, the results are converted from the geometric domain to the time domain using the transformation provided in (12.21).

Case 3: Nonsinusoidal with higher-order harmonics

In this case, the same circuit as in Fig. 12.6 is taken. However, the voltage harmonics are increased, such that the voltage is given by an odd multiple of the fundamental frequency. $v(t) = 100\sqrt{2\left(sinwt + \frac{sin3wt}{3} + \frac{sin5wt}{5} + \frac{sin7wt}{7} + \frac{sin9wt}{9} + \frac{sin11wt}{11}\right)}$ and R=1Ω, L=1/2 H, C= 2/3F. These values are also unaltered. The results are shown in Table 12.5.

The results are almost the same in both domains for apparent power, voltage and current, but there is a change in outcomes for P, Q, D, and pf. This is because, in the harmonic domain, it captures only a partial part of reactive power. Although reactive power is not a physical entity, it impacts the overall power quality of the system and has a significant impact on the compensation of circuits. In addition, it affects the design of filters if it is evaluated in the wrong manner.

12.6 Conclusions

The inclusion of power electronics-based devices in electrical networks means they handle signals that are not purely sinusoidal, but are transformed into nonsinusoidal signals with harmonic content. This chapter has summarized basic concepts of geometric algebra (GA), which represent a viable and theoretically well-founded alternative for studying nonsine electrical signals. The algebraic operations defined in this context make it possible to correctly calculate the currents, voltages, and powers required to identify an electrical circuit. For example, it is also possible to handle unconventional definitions of reactive power. This is thus of transcendent use in new electrical technologies.

In this chapter, the authors have adopted GA to analyze the power quality parameters for linear circuits and presented a comparative study and another analysis tool, the harmonic domain. Together, we have analyzed and reviewed previous work also.

In addition, the authors support GA circuit analysis through harmonic domain comparison.

The examples presented in previous theories have been rigorously examined and compared with the harmonic domain in the proposed work. It is observed that there is computational ease in operating in the geometric domain. For sinusoidal case (1), both tools give the almost identical result. However, when we proceed into nonsinusoidal analysis, the reactive power computed in the geometric domain is more significant. Other power characteristics are similarly different from those calculated in the harmonic domain, implying that capturing the harmonic interaction is ambiguous. In the presence of harmonics and nonsinusoidal excitation, all preceding experiments indicate some type of divergence in power computation. GA is used as a technique for analysis in this work, because it provides accurate power outputs in the sinusoidal situation and is mathematically simple.

References

[1] C.P. Steinmetz, Theory and calculation of alternating current phenomena, vol. 4, McGraw-Hill Book Company Incorporated, 1916.

[2] C.I. Budeanu, Puissances Réctives et Fictives, Institut National Roumain, 1927.

[3] S. Fryze, Active, reactive and apparent power in circuits with nonsinusoidal voltages and currents, Przeglad Elektrotechniczny 7 (1931) 193–203, 8 (1931) 225–234, 22 (1932) 673–676.

[4] V. Staudt, Fryze-Buchholz-depenbrock: a time-domain power theory, in: International School on Nonsinusoidal Currents and Compensation, ISNCC 2008, IEEE, 2008, pp. 1–12.

[5] EEE Standard Definition for the Measurement of Electric Power Quantities under Sinusoidal, Nonsinusoidal, Balanced, or Unbalanced Conditions, IEEE Std. 1459-2010, March 2010.

[6] L.S. Czarnecki, What is wrong with the Budeanu concept of reactive and distortion power and why it should be abandoned, IEEE Transactions on Instrumentation and Measurement 1001 (3) (1987) 834–837.

[7] L.S. Czarnecki, Considerations on the reactive power in nonsinusoidal situations, IEEE Transactions on Instrumentation and Measurement IM-34 (3) (Sep. 1985) 399–404.

[8] L.S. Czarnecki, Dynamic, power quality oriented approach to theory and compensation of asymmetrical systems under nonsinusoidal conditions, European Transactions on Electrical Power 5 (1994) 347–358.

[9] L.S. Czarnecki, Currents' Physical Components (CPC) Concept: a fundamental of power theory, in: Nonsinusoidal Currents and Compensation, 2008. ISNCC 2008 International School on Łagów, 10–13 June 2008, pp. 1–11.

[10] M. Castilla, J.C. Bravo, M. Ordóñez, J.C. Montaño, A. Borrás, A. López, J. Gutierrez, The geometric algebra as a power theory analysis tool, in: IEEE IX Conference-Seminar on Nonsinusoidal Currents and Compensation, Łagów, Poland, June 10-13 2008.

[11] W. Shepherd, P. Zand, Energy Flow and Power Factor in Nonsinusoidal Circuits, Cambridge Univ. Press, Cambridge, U.K., 1979.

[12] A.E. Emanuel, Powers in nonsinusoidal situations – a review of definitions and physical meaning, IEEE Transactions on Power Delivery 5 (3) (Jul 1990) 1377–1389.

[13] H. Akagi, Y. Kanazawa, A. Nabae, Instantaneous reactive power compensators comprising switching devices without energy storage components, IEEE Transactions on Industry Applications IA-20 (3) (1984) 625–631.
[14] F. Ghassemi, What is wrong with electric power theory and how it should be modified, in: IEE Ninth International Conference on Metering and Tariffs for Energy Supply, Birmingham, May 25-28, 1999, pp. 109–114.
[15] Peng Fang Zheng, Lai Jih-Sheng, Generalised instantaneous reactive power theory for three-phase power systems, IEEE Transactions on Instrumentation and Measurement 45 (1) (1996) 293–297.
[16] Dai Xian zhong, Liu Guohai, Gretsch Ralf, Generalised theory of instantaneous reactive quantity for multiphase power system, IEEE Transactions on Power Delivery 19 (3) (2004) 965–972.
[17] Francisco Casado-Machado, Jose L. Martinez-Ramos, Manuel Barragan-Villarejo, Jose A. Rosendo-Macias, Reduced reference frame transform: deconstructing three-phase four-wire systems, IEEE Access 8 (2020) 143021–143032, https://doi.org/10.1109/ACCESS.2020.3012510.
[18] J.L. Willems, A new interpretation of the akagi-nabae power components for nonsinusoidal three-phase situations, IEEE Transactions on Instrumentation and Measurement 41 (4) (1992) 523–527.
[19] H. Akagi, S. Ogasawara, H. Kim, The theory of instantaneous power in three-phase four-wire systems: a comprehensive approach, in: Proc. Conf. Rec. IEEE Ind. Appl. Conf. 34th IAS Annu. Meeting, vol. 1, Oct. 1999, pp. 431–439.
[20] H. Lev-Ari, A.M. Stankovic, Instantaneous power quantities in polyphase systems—a geometric algebra approach, in: Proc. IEEE Energy Convers. Congr. Expo., Sep. 2009, pp. 592–596.
[21] F. Milano, A geometrical interpretation of frequency, arXiv:2105.07762, 2021.
[22] F. Milano, G. Tzounas, I. Dassios, T. Kërçi, Applications of the Frenet frame to electric circuits, IEEE Transactions on Circuits and Systems I: Regular Papers 69 (4) (April 2022) 1668–1680.
[23] N. Barry, The application of quaternions in electrical circuits, in: Proc. 27th Irish Signals Syst. Conf. (ISSC), Jun. 2016, pp. 1–9.
[24] S.P. Talebi, S. Werner, D.P. Mandic, Quaternion-valued distributed filtering and control, IEEE Transactions on Automatic Control 65 (10) (Oct. 2020) 4246–4257.
[25] A. Menti, T. Zacharias, J. Milias-Argitis, Geometric algebra: a powerful tool for representing power under nonsinusoidal conditions, IEEE Transactions on Circuits and Systems. I, Regular Papers 54 (3) (2007) 601–609.
[26] M. Castro-Núñez, R. Castro-Puche, E. Nowicki, The use of geometric algebra in circuit analysis and its impact on the definition of power, in: 2010 International School on Nonsinusoidal Currents and Compensation (ISNCC), IEEE, 2010, pp. 89–95.
[27] M. Castro-Núñez, The use of geometric algebra in the analysis of nonsinusoidal networks and the construction of a unified power theory for single phase systems-a paradigm shift, Ph.D thesis, University of Calgary, 2013.
[28] M. Castro-Núñez, D. Londoño-Monsalve, R. Castro-Puche, M the conservative power quantity based on the flow of energy, The Journal of Engineering 2016 (7) (2016) 269–276.
[29] Francisco G. Montoya, Raúl Baño, Alfredo Alcayde, Francisco M. Arrabal-Campos, Analysis of power flow under non-sinusoidal conditions in the presence of harmonics and interharmonics using geometric algebra, International Journal of Electrical Power & Energy Systems 111 (2019) 486–492, 52.

[30] Francisco G. Montoya, Raúl Baños, Alfredo Alcayde, Francisco M. Arrabal-Campos, A new approach to single-phase systems under sinusoidal and non-sinusoidal supply using geometric algebra, Electric Power Systems Research 189 (2020) 106605.
[31] Francisco G. Montoya, Raúl Baños, Alfredo Alcayde, Francisco M. Arrabal-Campos, Geometric algebra forteaching AC circuit theory, International Journal of Circuit Theory and Applications (2021) 1–15, https://doi.org/10.1002/cta.3132.
[32] F.G. Montoya, R. Baño, A. Alcayde, F.M. Arrabal-Campos, J. Roldán-Pérez, Vector geometric algebra in power systems: an updated formulation of apparent power under non-sinusoidal conditions, Mathematics 9 (11) (Jun. 2021) 1295.
[33] F. de la Rosa, Harmonic & Power System, Ed Taylor & Francis, 2005.
[34] E. Acha, M. Madrigal, Power System Harmonics, John Wiley: Cambridge University Press, 2001.
[35] Eduardo Bayro Corrochano, Geometric Algebra Applications, vol. 1, Springer, ISBN 978-3-319-74828-3, 2019.

Harmonic sources estimation in distribution systems

13

Sara Sulis[a], Paolo Attilio Pegoraro[a], Antonio Vincenzo Solinas[a], and Daniele Carta[b]
[a]University of Cagliari, Cagliari, Italy, [b]Forschungszentrum Jülich GmbH, Julich, Germany

13.1 Distribution grids and power quality issues

The distribution systems are the systems that traditionally have been transporting electrical energy, generated by large plants and passed through transmission systems to the final users. They are operated by distribution system operators (DSOs), which maintain and develop the system, and also manage interconnections with other systems.

A typical distribution system is composed of the high-to-medium voltage (HV/MV) transformation centers, the MV system, the medium-to-low voltage (MV/LV) transformation centers, and the LV grid. Large industrial facilities are commonly connected to the MV, whereas residential and commercial customers are mainly connected to the LV. The LV grid can present three-phase, but also two-phase and single-phase loads. This characteristic, in addition to the spread of nonlinear loads and events in the grid management and operating conditions, creates the so-called power quality (PQ) issues [1]. This scenario changed due to several factors.

First, it is possible to mention the energy market deregulation that has led to a constant growth in the small-scale renewable energy sources (which are commonly called distributed generation, DG) located in different points of the grid and more and more often accompanied by energy storage. DG is installed in an unscheduled manner and commonly works in a substantially unpredictable and intermittent way.

Then, recently, there has been a change of pace in climate policy: in July 2021, the European Union declared that a cut more of 50% of its carbon emissions would be done by 2030 [2]. Such an ambitious goal will require fundamental changes in many sectors, most notably the automotive industry. This decision will push the required transition to electric vehicles (EVs) in an unexpectedly fast way.

DG and EVs are generally united under the more general name of distributed energy resources (DERs).

DERs bring both economic and environmental benefits, however, they also lead to technical issues in the management and safe operation of the grid. For example, DG can introduce an input of unused local energy that can lead to bidirectional power flows. For this reason, the traditional passive distribution grids have been transformed into complex and potentially active networks. As for EVs, new management methodologies must be studied, because a contemporary request of charge by many vehicles could determine a power demand that has not been foreseen, creating overloads and

Monitoring and Control of Electrical Power Systems Using Machine Learning Techniques
https://doi.org/10.1016/B978-0-32-399904-5.00019-3

significant voltage drops in the distribution grids. DERs are connected to the grid through power electronics interfaces that make the PQ issues increasingly present and relevant.

For economical, legal, and technical reasons, the complexity of this changing scenario brings the need of new management tools and protection schemes to coordinate and manage suitably such energy resources, avoiding unintended disconnection for the end-users or unintended supply of disconnected parts of the grid, while monitoring of the quality of the supplied energy.

DSOs are interested in identifying the causes of the poor quality to improve the effectiveness of the supply, so that failures of sensitive devices and increase in power losses and in management costs could be addressed, and the quality of the service provided to the customers, particularly crucial in the liberalized market where the customers can introduce clauses for the quality of the service in their supply contracts. The placement and the usage of PQ mitigation equipment are up to the system operators, but the costs associated with their operations and their maintenance did not help the spread of such devices.

As far as the harmonic distortion is concerned, the number and size of polluting loads continuously increase, so that it is increasingly important to define suitable methodologies to attribute the responsibilities for the deterioration of the service. It is indeed recognized that harmonic pollution could lead to a reduction in the lifetime of the components of both large networks and small systems, incrementing the management costs [3].

For these reasons, new distributed monitoring systems providing accurate and reliable information on the actual operating conditions of the grids are needed, so that effective management and control of modern power networks is addressed also in the presence of harmonic pollution.

13.2 Background context of harmonics monitoring

Monitoring techniques based on distributed and/or hierarchical measurement systems are common solutions to obtain a reasonable level of knowledge of the actual operating condition of the grids.

Obviously, measurement devices able to provide information about harmonics must be available on the field. In [4], for example, the problem of a fair allocation of the costs for PQ mitigation equipment, by means of an appropriate metering infrastructure, was discussed. Further analysis on the architecture and capability requirements of a measurement device for sensing the harmonic distortion have been discussed in [5], with the focus not only on the quantification of the harmonic distortion but also on its time variations.

In literature, several studies on PQ issues and indexes focused in particular on harmonics have been presented in the last decades. In [6], for example, PQ indexes for the evaluation of harmonic distortion have been compared, and their performances have been evaluated in presence of capacitors, see also [7] discussing the detrimental effects of capacitors in presence of harmonics. In [8], the relation between three different non-active power quantities, determined at the same measurement point, has been used to detect the presence of polluting loads in a system. In [9], neural networks are used to determine the variation of the source impedance, representing the equivalent upstream network from the point of common coupling (PCC), and thus predict the harmonic distortion on the customer side. In [10], a discrete wavelet transform-based architecture for harmonic estimation, aimed at improving the estimation performance of the fast Fourier transform, has been implemented on FPGA for real-time applications.

Several methodologies were designed to reach a given level of knowledge of the "quality" at harmonic level on the whole networks. These can generally be grouped in two main categories: harmonic state estimation and harmonic source estimation methods.

The harmonic state estimation (HSE) is the name assigned to a technique evaluating the harmonic spread on the entire grids based on the mathematical relations between system state variables and measurements gathered from the network (see, e.g., [11–13]). Therefore, similarly to traditional state estimation techniques for power grids, the aim of HSE is the evaluation of the voltage profile for the harmonic orders of interest.

The HSE techniques have been designed firstly to be applied for transmission systems, where the availability of real measurements is commonly redundant. Even though not directly measured, with redundant measurement systems more and more often based on accurate measurement chains, the harmonics can be estimated with an adequate accuracy level.

In [12], a dedicated harmonic measurement system for HSE was presented, introducing the concepts of observable, semi-observable and unobservable islands, to approach the lack of observability. The proposed infrastructure relied on the use of local computers, installed in the substations to generate GPS-synchronized measurements, and a master workstation, working as a data concentrator and performing the actual HSE. In [14], the optimal configuration of a monitoring system for HSE has been obtained by means of an evolutionary algorithm. The algorithm aims at minimizing the monitoring cost, while preserving the full observability of the system.

In any case, the network topology and the knowledge of the network parameters are key elements of estimation solutions. Nevertheless, as underlined in [15], the knowledge of the network parameter is not trivial. Nominal parameter values are typically used in HSE approaches, but physical aspects, such as the cable geometry, placement, and transposition, as well as external factors (such as the operating temperature), can lead to significant deviations from the nominal values. In [16], a statistical approach based on Monte Carlo simulations has been used to evaluate the impact of measurement accuracy on HSE.

In distribution systems, the small number of measurement devices implies the requirement of differently designed HSE techniques and, commonly, to obtain the observability of the system, large amount of a priori information is required (i.e., knowledge available before the measurement data are taken) resulting from historical and available data. This kind of derived information also permits the so-called pseudo measurements to be derived, but is characterized by low accuracy.

In [17], a solution based on honey-bee mating optimization was presented, where assumptions such as a limited number of nonlinear loads and knowledge on their installation node, known values of contracted load and generation at each node, and others, have been made to guarantee the observability of the system. In [18], an offline estimation algorithm has been proposed, built on harmonic currents and voltages collected at the PCC. The proposed solution is based on the WLS method and uses a nonlinear aggregated model for the representation of polluting loads.

A perspective, different from traditional HSEs, is represented by the harmonic source estimation techniques (HSoEs), generally aimed at providing a localization of the harmonic producing loads [19]. To this aim, under given operating conditions, the nonlinear loads can be modeled as if composed of a combination of a linear load, representing the linear characteristic of the load, in parallel with ideal current sources, representing the injection of harmonic currents. In the HSoE problem, the vector composed by the harmonic current sources to be estimated is usually considered as the forcing vector.

Due to the limited real time information coming from the field, an effective approach for distribution grids is mainly designed to identify and estimate the main harmonic sources, i.e., the prevailing harmonic pollutants.

As presented in [20], information on the possible presence of polluting loads could be provided to the network operators so that they can act locally, where the suspicious loads are, instead of focusing on the whole network. The full observability of the system considered in [21] is based on the combined use of real-time measurements and a priori information, and the proposed estimation solution is also validated on experimental measurements. Due to the high uncertainties related to prior information, in [22] the authors proposed a Bayesian approach, with the a posteriori validation of the obtained results, to verify their coherence with the initial assumption. In particular, the posterior distributions of Bayesian estimation processes were evaluated by means of an adaptive Metropolis–Hastings (M–H) approach, which can indeed produce a given number of samples driven from the target distribution, allowing the desired moments and statistics to be calculated. According to this approach, the harmonic sources are estimated along with their complete a posteriori distribution, providing a full uncertainty description of the results (for instance, in terms of standard deviation, quantiles, or even histograms).

Following the idea of identifying and estimating the prevailing harmonics in a network, other procedures have been recently proposed in the literature.

In this chapter, a paradigm to find the prevailing harmonic sources in the network and based on the estimation of a sparse set of injected harmonic currents will be presented and discussed, first introducing the framework, and then discussing the problem definition, solution techniques, and strategies to deal with measurement uncertainty in a metrologically correct way. The approach is specifically conceived to cope with the typical measurement scarcity of distribution systems.

Based on the results of the preliminary estimation given by HSoE, it is possible to perform a more accurate and expensive analysis only on limited areas, where the suspicious loads are located, rather than on the entire network.

13.3 Instruments and architectures

New generation distribution networks will require an upgraded monitoring system to have a better picture of the network status and of the ongoing conditions. Modern distribution grids are expected to be equipped with new smart devices able to generate vast amounts of data about the network status, which will be key components for the efficient, dynamic, and prompt management of such complex grids.

In this regard, it is worth noting that a sort of time tag, stating the reference time of the measurements considered in the estimation process, would be necessary as for the coordination and comparison of the measured data.

Smart meters are expected to be "intelligent" instruments providing different measurement types and able to communicate the obtained data from the locations spread across the network to the control centers. However, the evaluation of these measurements is commonly based on different time window lengths, but information on the time observation windows is rarely provided in the datasheet of these devices and a clear standardization does not exist.

PQ meters, when complying with Class A synchronization level of standard IEC-61000-4-30 [23], must have a time-clock uncertainty that does not exceed ±20 ms (considering a rated frequency of 50 Hz). This synchronization accuracy is not enough for an accurate monitoring of harmonics, in terms of both amplitude and phase angle, particularly for higher-order harmonics since time errors directly affect phase-angle errors.

For this reason, in the scenario of a coordinated monitoring on a wide area of harmonic pollutions, synchronized harmonic phasor measurements appear promising.

Currently, phasor measurement units (PMUs) are the most advanced synchronized instrument for power system monitoring. PMUs are indeed able to measure synchronized voltage and current phasors (the so-called synchrophasors, dealing with fundamental frequency phasors only) on a common universal coordinate time (UTC) reference, with unprecedented accuracy and reporting rate (rates that can vary from 10 to 100 measurements per second, when the rated frequency is 50 Hz). The IEEE Standards C37.118.1-2 for PMUs were released in 2011 by means of two documents, [24] and [25], with measurement and data communication specifications, respectively. The main metric to evaluate the accuracy of PMUs is the total vector error (TVE), which represents the relative absolute vector error between measured and reference phasor. The standardization is in progress and, in particular, it is worth noticing that IEC technical committee 95: Measuring relays and protection equipment, in cooperation with the Power system relaying committee of the IEEE Power and Energy Society, under the IEC/IEEE Dual Logo Agreement, published a joint synchrophasor standard, the IEEE/IEC 60255-118-1, in 2018 [26]. The standard defined these limits for TVE: below 1% in case of steady-state conditions (1.3% in case of strong interharmonic pollution), below 3% in case of dynamic conditions.

The signals presented in [24] and [25] were thought for transmission systems, but the current trend is to extend the benefits of the synchrophasor technology also to distribution networks. In the SRA2035 [27], it is highlighted that "Usage of measurement devices as PMU ... opens new system awareness and even an early warning approach for the control systems at all levels in the grid".

As mentioned, PMUs are the most accurate instruments for power system monitoring, but PMUs employment in distribution networks requires specific design processes, because they are asked to provide trustworthy information also in the presence of signals that can occur in typical distribution grid operating conditions. Considering the current standards, signals unforeseen but realistic for distribution systems could bring to PMU measurements strongly affected in their accuracy by PQ issues [28]. In the last decade, research activity has been devoted to these studies, and PMUs for distribution systems applications have been developed [29], [30].

Moreover, the evaluation of quantities such as harmonic synchrophasors is being studied worldwide. For this type of quantities, the impact of the uncertainty, due to the employed voltage transformers (VTs) [31] and to the other devices that compose the measurement chain, must be analyzed in detail [32], [31].

To measure harmonics, an upgraded monitoring infrastructure is thus necessary, relying on a new generation of harmonic-devoted PMU-like devices or PQ meters that can include also harmonic synchrophasor functionalities [30]. For highly accurate harmonic measurements, the impact of instrument transformers can be critical [31], and new compensation techniques should also be introduced to support the novel instrumentation [32].

Such a monitoring infrastructure requires a widespread time propagation that can be based on Global satellite navigation systems' (GPS, Galileo, etc.) receivers integrated with each instrument or on a packet-based time synchronization (e.g., through precision time protocol, PTP, [33]). These instruments are connected through standard network protocol (typically, via UDP or TCP) and transmit the packets, including measurements to the control center or to local processing units. This new measurement architecture can represent, particularly for MV distribution grids, a disruptive factor for new PQ assessment tools. Clearly, considering the costs and the need of an incremental upgrade of the monitoring system, a redundant measurement system that allows tracking harmonics in each node of the network is unlikely to be installed in the next future.

HSoE techniques will probably rely on synchronized harmonic measurements available at strategical points of the network, and these measurements—as will be described in what follows—are perfectly suited to define linear relationships between measured values and sought harmonic sources and help in discriminating the influence of different polluting sources.

13.4 Estimation methods for harmonic source estimation

Every estimation method for HSoE is based on a measurement model that allows linking the collected measurements to the unknown harmonics sources to seek. In the following, the method proposed in [34] is mainly described and discussed with references to other related approaches. In the next section, some examples of its application are reported to show how it can help in identifying prevailing sources of harmonic pollution and in estimating them.

The main idea is to find the harmonic currents injected into the network by each load or generator, which can act as a potential source of harmonic injection. This approach can be applied to every considered harmonic order h and, in particular, to the most relevant ones, which are commonly the low odd orders. For this reason, the measurement mode can be written as a relationship between the considered measurements and the injected harmonic currents.

Considering the novel monitoring systems based on synchronized measurement, it is foreseeable that synchronized harmonic phasor measurements will be available to monitor the network at relevant locations. PQ meters equipped with appropriate levels of synchronization (at least below 1 μs) or PMU-like devices extended to harmonic measurements can build the coordinated monitoring infrastructure for HSoE tasks. Focusing thus on measured phasors, which can be reported on the same timescale and referred to the same time instants, i.e., the reporting instants of the instruments used to tag the measured values, it is possible to write the measurement model with a linear representation:

$$\overline{\mathbf{y}}_h = \overline{\mathbf{A}}_h \overline{\mathbf{u}}_h + \overline{\mathbf{e}}_h, \tag{13.1}$$

where the measurement vector $\overline{\mathbf{y}}_h$ is expressed as a linear function of the current injections included in vector $\overline{\mathbf{u}}_h$, and $\overline{\mathbf{e}}_h$ represents the vector of measurement errors that can be considered as a random quantity and has typically zero mean. All the quantities are complex, since both measurements and injected currents are phasors that correspond to the same measurement instant (thanks to the synchronization). Eq. (13.1) is referred to a given time tag, which is not reported here for the sake of brevity. If M is the number of measured harmonic phasors, which can be both voltages and currents (node or branch currents), and N is the number of potentially polluting nodes, i.e., the number of potential harmonic sources, $\overline{\mathbf{y}}_h$ is an $M \times 1$ vector, and $\overline{\mathbf{A}}_h$ is a $M \times N$ complex matrix (the measurement matrix) that defines the relationships. $\overline{\mathbf{A}}_h$ depends on the network topology and on the considered load models since it includes the line and load impedances. Its entries change with the harmonic order, because the model of the network changes accordingly.

Vector $\overline{\mathbf{u}}_h$ (indicated as forcing vector from here on) is the vector of unknowns that needs to be estimated to find the points of the network where pollution is higher. The forcing terms can be very different, depending on the actual contribution of each potential source, and thus it is likely to think of the forcing vector as a sparse vector or, at least, a quasi-sparse (sparse from a practical viewpoint) one. Each contribution can be classified as polluting or as weakly-/nonpolluting, depending on the practical rules a system operator can define based on its experience and needs. For instance, standard limits for polluting appliances can be used to set thresholds to discriminate a prevailing polluting term from a negligible one.

With these considerations, the problem in (13.1) becomes an inverse problem with a sparse solution, and is thus possible to address it, even though a limited number of monitoring stations and of measurements is available, as typical in distribution systems. It is extremely important to consider the uncertainty associated with the measured quantities, and thus it is possible to reformulate (13.1) in rectangular coordinates so that the uncertainty can be explicitly associated in a suitable way. The formulation in real-valued vectors can be expressed as

$$\mathbf{y}_h = \begin{bmatrix} \mathbf{y}_h^r \\ \mathbf{y}_h^i \end{bmatrix} = \mathbf{A}_h \mathbf{u}_h + \mathbf{e}_h = \begin{bmatrix} \mathbf{A}_h^r & -\mathbf{A}_h^i \\ \mathbf{A}_h^i & \mathbf{A}_h^r \end{bmatrix} \begin{bmatrix} \mathbf{u}_h^r \\ \mathbf{u}_h^i \end{bmatrix} + \begin{bmatrix} \mathbf{e}_h^r \\ \mathbf{e}_h^i \end{bmatrix}, \tag{13.2}$$

where the real and imaginary parts of each vector and matrix in (13.1) is indicated by the corresponding symbol with superscript ^{r} and ^{i}, respectively. Therefore the measurement matrix $\mathbf{A}_h$ is a $2M \times 2N$ real matrix, and $\mathbf{y}_h$ and $\mathbf{u}_h$ are a $2M \times 1$ and $2N \times 1$ real vectors, respectively.

The solution of the estimation problem in (13.2) requires two important steps: the definition of the uncertainty associated with the measurements in $\mathbf{y}_h$, and the definition of the optimization problem to solve. In the following, subscript h will be dropped for the sake of simplicity, since the focus is on a single and generic harmonic.

Each measurement has its own error and the associated accuracy. To weight each measured value according to the associated uncertainty, it is useful to perform a transformation of the inputs, so that they become uncorrelated and with unitary standard uncertainty (i.e., standard deviation). This operation is extremely powerful to include uncertainty information in the HSoE problem, as will be described in what follows. In practice, applying the so-called whitening transformation, the variance-covariance matrix of the measured quantities becomes an identity matrix. The new inputs are given by

$$\mathbf{y}_w = \mathbf{W}\mathbf{y} \Rightarrow \mathbf{\Sigma}_{\mathbf{y}_w} = \mathrm{I}_{2M}, \tag{13.3}$$

where $\mathbf{y}_w$ is the whitened measurement vector in rectangular coordinates, $\mathbf{\Sigma}_{\mathbf{y}_w}$ denotes its covariance matrix, and I_{2M} is the $2M \times 2N$ identity matrix. To obtain this effect, any matrix $\mathbf{W}$, so that $\mathbf{W}^T\mathbf{W} = \mathbf{\Sigma}_{\mathbf{y}}^{-1}$, serves the purpose.

The usual representation of measurement uncertainty for phasor quantities is given by amplitude and phase-angle uncertainties. The definition of the estimation problem relies thus also on the information on instrument metrological characteristics that can be extracted from available specifications or from characterization campaigns. For this reason, the covariance matrix of $\mathbf{y}$ can be obtained through the application of the law of propagation of uncertainty (see [35]) as

$$\mathbf{\Sigma}_{\mathbf{y}} = \mathbf{J}\mathbf{\Sigma}_{\rho\phi}\mathbf{J}^T = \mathbf{J} \begin{bmatrix} \sigma_{\rho_1}^2 & & \\ & \ddots & \\ & & \sigma_{\phi_M}^2 \end{bmatrix} \mathbf{J}^T, \tag{13.4}$$

where $\mathbf{\Sigma}_{\rho\phi}$ is the diagonal matrix, including the variances $\sigma_{\rho_i}^2$ $(i = 1, ..., M)$ of all the measured synchrophasor amplitudes and phase angles, and $\mathbf{J}$ is the Jacobian matrix of the polar to rectangular coordinates transformation for all phasors. From this covariance matrix, the whitening transformation can be found and used to define the estimation problem

$$\mathbf{y}_w = \mathbf{WAu} + \mathbf{We} = \mathbf{A}_w\mathbf{u} + \mathbf{e}_w. \tag{13.5}$$

To find the sparse solution of (13.5), signal processing techniques, such as compressive sensing, also known as compressed sensing (CS), can be used. This is a mathematical approach to recover sparse signals with a limited number of measurements. In the HSoE problem the number of available synchrophasors can be less than the number of potential polluters, but is much higher than the sought prevailing sources in **u**. The number of nonzero entries in the forcing vector is the sparsity level S and is much smaller than the vector size. In practice, the forcing terms are approximately sparse, and thus **u** can be defined as "compressible," approximately sparse or relatively sparse, and compressing sensing theory still applies when looking for the prevailing harmonic sources.

Finding a sparse solution of (13.5) means solving the following optimization problem:

$$\arg\min_{\mathbf{u}} \|\mathbf{u}\|_0 \text{ s.t. } \mathbf{A}_w\mathbf{u} = \mathbf{y}_w, \tag{13.6}$$

where $\|\cdot\|_0$ denotes the pseudonorm ℓ_0, i.e., the number of nonzero entries or the sparsity. The problem is NP-hard, specific algorithm for approximated solutions exist. Clearly this problem is equivalent to the corresponding one defined without using the whitening approach, but in its approximate solutions and modifications the normalized version can be handy and advantageous, as discussed in what follows.

Two of the most widespread solution algorithms for CS problems as (13.6) are the orthogonal matching pursuit (OMP) [36], and the ℓ_1-minimization approach [37], [38]. OMP for HSoE has been proposed in [39] and extended to multiple harmonic orders in [40]. This is a greedy approach that iteratively finds the most promising candidate from the possibly polluting sources looking at the column in $\mathbf{A}_w$ that best "correlates" with the residuals vector at each iteration (maximum inner product is chosen) and computes the best solution (in the least squares sense) at each step before proceeding. The algorithm is fast and terminates when either the assumed sparsity level is obtained or the energy of the residuals is below a given threshold.

The second approach looks for an optimization in norm ℓ_1 as in

$$\arg\min_{\mathbf{u}} \|\mathbf{u}\|_1 \text{ s.t. } \mathbf{A}_w\mathbf{u} = \mathbf{y}_w. \tag{13.7}$$

When no measurement error is considered and $\mathbf{A}_w$ has given properties (restricted isometry property), perfect recovery can be obtained for a sparse **u**. This method was introduced for HSoE in [41] and is often referred to as basis pursuit.

In a real application, as mentioned above, measurement error is unavoidable, and thus the ℓ_1-optimization problem can be reformulated as (basis pursuit denoising):

$$\arg\min_{\mathbf{u}} \|\mathbf{u}\|_1 \; s.t. \; \|\mathbf{A}_w\mathbf{u} - \mathbf{y}_w\|_2 < \epsilon, \tag{13.8}$$

where the equality constraints of the measurements have been relaxed to take measurement errors into account. In particular, in (13.8), is considered defining an upper bound for the considered mismatch [42]. This formulation allows better dealing with measurement uncertainty, but the limit ϵ needs to be carefully defined, since it summarizes the actual measurement scenario. For this reason, in [34], a specific approach has been proposed to cope with error energy computation. In particular, ϵ needs to be a valid limit to the energy of the measurement residuals, and thus a good choice is to set it as a meaningful bound to the energy of the measurement error vector. The normalized vector $\mathbf{e}_w$ is the results of the whitening transformation, and can be thus used to evaluate the energy and compute ϵ.

Considering that the weighing matrix $\mathbf{W}$ is chosen so that $\mathbf{W}^T\mathbf{W}= \boldsymbol{\Sigma}_{\mathbf{y}}^{-1}$ and is invertible (the covariance matrix is positive definite), it can be decomposed through its QR factorization as $\mathbf{W} = \mathbf{Q}\ \mathbf{R}$ in a unique way, where $\mathbf{Q}$ is orthonormal, and $\mathbf{R}$ is upper triangular. Then it follows that

$$\boldsymbol{\Sigma}_{\mathbf{y}}^{-1} = \mathbf{W}^T\mathbf{W} = \mathbf{R}^T\mathbf{Q}^T\mathbf{Q}\mathbf{R} = \mathbf{R}^T\mathbf{R}, \tag{13.9}$$

and thus $\mathbf{R}$ gives the Cholesky factorization of the inverse of the covariance matrix of the measurements. This simple result has an important consequence on the evaluation of the error energy. In fact, whatever weighing matrix $\mathbf{W}$ is chosen to normalize the measurements, it is always obtained as an orthonormal transformation of the matrix $\mathbf{R}$ given by Cholesky factorization. Another interesting aspect is that orthonormal transformations do not alter the energy of a vector, and thus, regardless the chosen $\mathbf{W}$, $\mathbf{e}_w$ has the same energy. This consideration allows choosing a peculiar weighing that helps in finding a limit for $\|\mathbf{e}_w\|$, and thus ϵ.

In the case at hand, the distribution of measurement errors can be assumed as uniform both for amplitude and phase-angle errors when information other than the accuracy of the instrument or the maximum deviations is available. Passing from polar to rectangular coordinates and whitening can change the distribution of the entry of $\mathbf{e}_w$, but when $\mathbf{W}$ is chosen so that it is equal to $\boldsymbol{\Sigma}_{\mathbf{y}}^{-\frac{1}{2}}\ \mathbf{J}^{-1}$ a scaled version of the original errors in polar coordinates (under the assumption of small measurement errors, as it is in relevant situations) is obtained, thus preserving also the distributions. This apparently elaborate procedure leads to a very simple result: it is possible to consider zero-mean uniform distributions with unitary standard deviations to define a reasonable limit to measurement errors. In particular by defining the following ($E_{\mathbf{e}_w}$ is the energy of $\mathbf{e}_w$):

$$\epsilon^2 = \mu_{E_{\mathbf{e}_w}} + k\sigma_{\mathbf{e}_w} \tag{13.10}$$

i.e., defining the squared limit as the average energy of the error vector $\mu_{E_{\mathbf{e}_w}}$, plus k-times its standard deviation, where k is chosen so that a desired confidence level for the coverage of the energy is guaranteed, it is possible to correctly frame the problem (13.8) and couple the knowledge about measurement uncertainty with the HSoE target. With M harmonic phasor measurements, $\mu_{E_{\mathbf{e}_w}} = 2M$ and $\sigma_{\mathbf{e}_w} = 2\sqrt{\frac{2}{5}M}$, while k

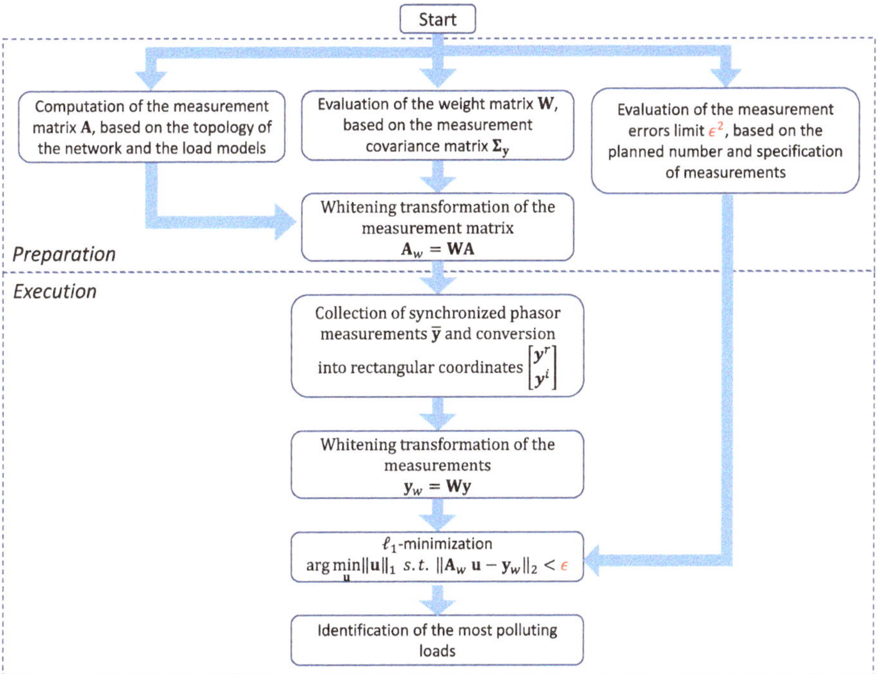

Figure 13.1 Flow chart of the HSoE.

can be chosen, for instance, to guarantee that the limit is valid at 95% without overestimating the misfit.

The solution of (13.7) and (13.8) is a vector, whose entries are usually nonzero, so the found forcing vector is not sparse, but shows prevailing elements. For the solution of the HSoE problem, a given number of sources can be retained (e.g., fixing S or limiting the magnitude of the polluting currents). The most polluting terms can be thus identified, and their contribution evaluated.

The flow chart summarizing the described approach can be found in Fig. 13.1.

13.5 Examples of estimation scenarios

The HSoE methods described in the previous section can be applied to different networks, and its performance depends on the available monitoring system and on the network characteristics. For this reason, in what follows, some examples of application and performance will be reported using two simulated distribution grids designed for testing purposes. In the following, the results are reported focusing on two methods: the basis pursuit problem defined by (13.7), which is indicated as Method A, and the basis pursuit denoising in (13.8), which will be referred to as Method B.

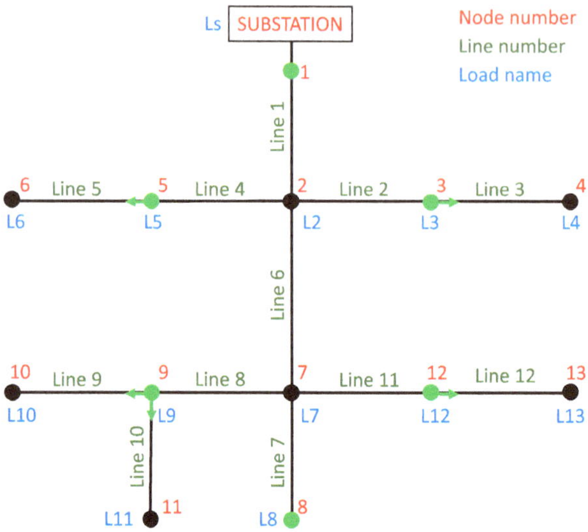

Figure 13.2 13-node MV test network.

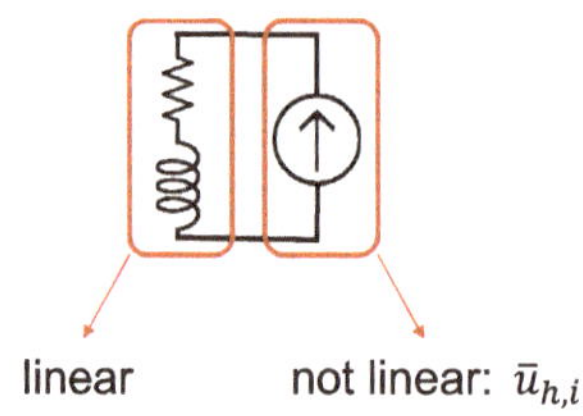

Figure 13.3 Considered load model.

The first test grid is a single-phase distribution network, operating at a rated voltage of 4.16 kV and at a rated frequency of 60 Hz, whose topology is inspired from the IEEE 13 bus feeder [43]. Fig. 13.2 shows the network scheme and reports the node numbers and the branch indexes. In this network every node is connected to a potential polluting source, save node 1, i.e., the substation. The load model is composed of an ohmic–inductive impedance in parallel with an ideal current generator, which is the harmonic current generator for each harmonic order, and represents the harmonic source possibly present (depending on the actual injected current).

Fig. 13.3 shows the considered load model, where the generic ith element of the forcing vector is associated with the injected harmonic current.

The harmonic current generators thus represent the forcing term corresponding to the nonlinearity of the loads. Every generator injects a current that is the result of all the harmonic currents at all the considered harmonic orders. In particular, in the following, harmonic orders up to $h = 15$ are used. The default harmonic pattern, in terms of percent amplitudes with respect to each load nominal current, is reported in Table 13.1 and is chosen so that it complies with the harmonic current emission limits

Table 13.1 Default magnitude for each harmonic order.

Harmonic order h	3	5	7	9	11	13	15
Magnitude [%]	5	5	3	3	3	1	1

indicated by the IEC Technical Report 61000-3-6 [44] for the connection of distorting installations to MV and HV power systems.

It is extremely important to understand which is the uncertainty of the considered synchronized harmonic phasor measurements. Generally speaking, this is a challenging task since the uncertainty of the overall measurement chain depends on the harmonic order, on the harmonic level, on the instrument transformers, and on the adopted instrument.

In the reported tests, measurement errors for both amplitude and phase angle are extracted randomly from uniform distributions, which represent the typical case in real applications, where no additional information is available beside the maximum errors. If any further indications on the measurement uncertainties can be derived from instrument datasheets or from models and characterization procedures of the different contributions in the chain, they need to be suitably considered. As an example, standard IEC 61000-4-7 [45] gives the accuracy requirements for both harmonic voltage and current measurements and defines two accuracy classes for the instruments: Class I and Class II. Class I is intended for accurate measurements, as those required by compliance verification and emission assessment, whereas instruments belonging to Class II have general purpose. Clearly, in the HSoE context, Class I instruments would be an appropriate choice. Phase angle accuracy is not discussed in [45], but in the upcoming new edition of the standard IEC 61869-1 for instrument transformers [46], accuracy classes for wider bandwidth will be included, which defines also phase error limits recommended for harmonic measurements. This is not the only source of uncertainty, since synchronized instruments also introduce errors. Just as an example, a synchronization error directly translates into a phase-angle error linearly with the harmonic order. The HSoE method should include the overall uncertainty in the problem description.

In what follows, results are reported with different measurement accuracies, with the aim to illustrate the role that measurement uncertainty plays in the pursuit of polluting sources. In particular, three scenarios are considered: low (0.5% and 0.6 crad for maximum amplitude and phase-angle deviations, respectively), medium (2.5% and 3%), and high (5% and 6 crad) uncertainty. In a real application, the appropriate infrastructure should be designed and implemented depending on the identification and estimation targets the utility wants to achieve.

As a first example, synchronized harmonic measurement devices are assumed to be available in the network on six nodes (node 1, 3, 5, 8, 9, 12, see Fig. 13.2). Each device measures the node voltage at the installation node and the currents on the connected branches 3, 5, 9, 10, and 12 are also monitored. To assess the performance of the HSoE algorithm, repeated tests are performed using the same configuration and changing the polluting sources. Two possible prevailing sources are set simultaneously, with a sparsity level of two (considering the injected current phasors), modifying them at each trial. In each test, a polluting source includes all the harmonics of Table 13.1.

A first index of the HSoE capabilities of the approach is given by the identification accuracy in repeated trials. Considering twelve possible sources and two simultaneous polluters, sixty-six different pairs of sources can be introduced. When no measurement uncertainty is present (unrealistic but useful benchmark condition) correct identification is always achieved, thus confirming the capability of the method to retrieve the sparse solution. It is important to highlight that, in this case, problem (13.7) perfectly fits the underlying condition and gives the same results as (13.8).

When measurement errors are introduced, the identification starts decreasing. In what follows, the so-called "recall" is used, which is typical of pattern recognition studies, i.e., the ratio of correct identifications on the number of polluting sources. This index gives the sensitivity or the correct detection rate of the algorithm. Since each configuration of polluting sources is estimated in a thousand different trials, where the measurement errors are extracted randomly from a given error range so that the uncertainty in the measurement chain can be simulated, it is possible to give the average recall value, the average recall value for each configuration of the polluters, and other statistics (e.g., the standard deviation among different configurations). This approach gives an idea of the overall performance and highlights the most critical conditions.

Fig. 13.4 reports the recall variability bars for different harmonic orders (third and seventh), different uncertainty scenarios and consider Method A and Method B. Recall values are clearly between 0 and 1 (0% and 100% of correct detection), and thus the distribution of their values is highly asymmetric when good identification performance can be found. As expected, for both methods, the average recall decreases, whereas its variability increases when the uncertainty in the measurement process increases. In the figure, the average detection ratio is 99.3% and 99.1% ($h = 3$ or 7, respectively) for Method A, whereas Method B achieves 99.8% and 99.9%, thus proving that a custom integration of uncertainty-related information, since from the algorithm, can surely help the HSoE. The graphs also show that the dispersion of the values is quite large, and Method A always suffers a greater variability among different configurations.

These results show that the impact of uncertainty is different, depending on the considered harmonic order, because the different injections are considered and, in addition, a different frequency of interest changes the impedances in the network, and thus the measurement model.

It is also interesting to investigate the pollutants that make the identification more sensitive to measurement errors and more prone to errors. In this regard, it is interesting to notice that there are nodes and pairs or nodes where the identification is particularly effective and others where it is more challenging, and that the two methods behave differently. In Fig. 13.5, the percentage of correct detection of load L11 is reported considering $h = 3$ when, in all the repeated trials, L11 is imposed as always polluting, and the second injecting load is changed to verify its role in the identification of L11. Load L11 is always found as a polluter through Method B, regardless of the considered uncertainty level. Method A for low and medium uncertainty levels shows the same performance, since the effect of the uncertainty, which is the main difference among them, appears negligible. When the higher uncertainty level is con-

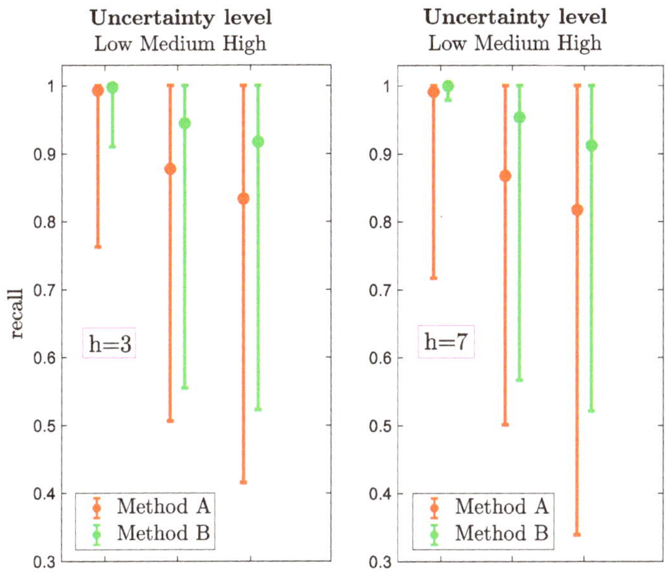

Figure 13.4 Detection rate results for third and seventh harmonics and different uncertainty levels.

sidered, Method A starts degrading its detection capabilities, in particular for specific choices of the second polluter (L10, for instance). This effect, which was slight in the previous case, appears determinant when focusing on load L2 (always on the third harmonic order). It is possible to show that if measurement uncertainties are correctly modeled in the problem definition, exploiting the metrological characteristics of the measurement chain, Method B can significantly outperform Method A, depending on actual conditions. The correct detection rate of the source in L2 (always polluting) can be computed when it is coupled with a second injecting load that is added and changed across different tests.

Fig. 13.6 shows the rate as a function of the other polluting load. It is possible to see that with low uncertainty there are no detection errors with both methods, whereas when the measurement uncertainty starts affecting the performance, the two methods respond quite differently, and Method A is less robust. Furthermore, the detection accuracy is highly variable with respect to the simultaneous configuration of pollutants and only when L7 is also considered a correct identification of L2 is possible in more than 50% of cases also with Method A, which becomes almost unsuitable in these conditions. This underlines once more the importance of a correct representation of uncertainty within the model to achieve good detection performance.

Tests performed on larger networks with a great number of possible combinations of pollutants confirm the presented outcomes, better highlighting also the differences among the methods. Indeed, Method B always guarantees the best detection results at different harmonic orders and in the presence of different measurement configurations. The advantages of uncertainty modeling are more significant as the measurement uncertainty increases, as can easily occur in realistic conditions.

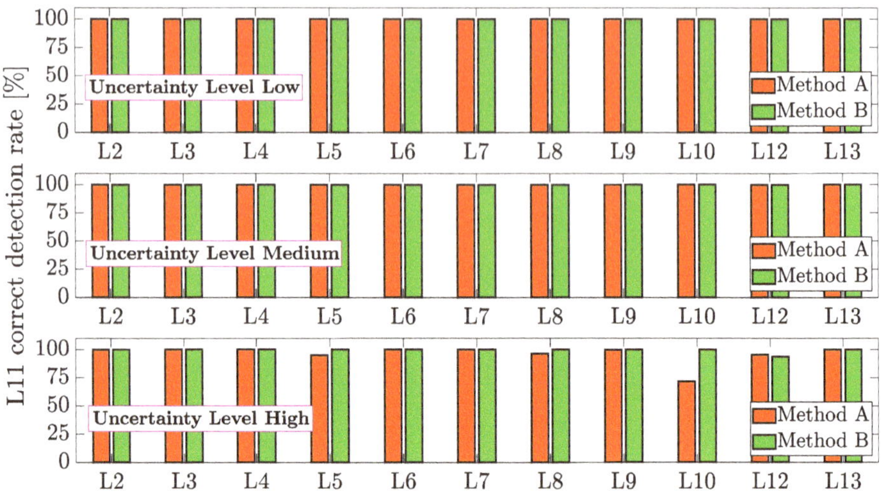

Figure 13.5 Detection rate results of polluting load L11 for third harmonic and different uncertainty levels.

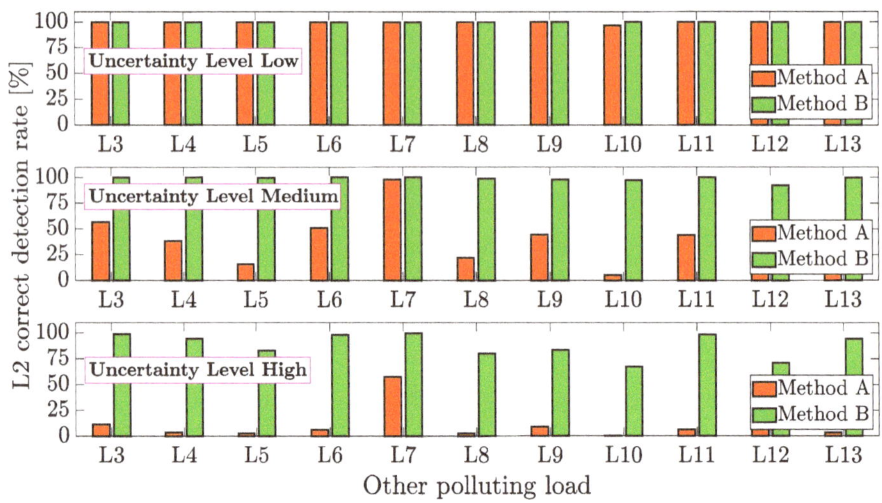

Figure 13.6 Detection rate results of polluting load L2 for third harmonic and different uncertainty levels.

From the above examples, the potentialities of CS-based methods for HSoE emerge rather clearly. Even though the reported examples refer to single harmonics or, at least, to harmonics treated independently, it is interesting to add that it can be also used in a CS-based approach where a load is considered as polluting if its whole spectrum of harmonic injections is significant. Very often, a polluting source involves more than a single harmonic, and an approach as those described in [40] can be very helpful. Method A and B can be also extended to consider multiple harmonics altogether.

It is also possible to observe that HSoE and machine learning (ML) can be considered linked in several ways.

ML is now used intensively to deal with estimation methods in different fields. As illustrative examples, it is possible to mention ML studies for automotive, [47] and [48], industrial equipment [49], and power systems applications, [50] and [51]. In this context, the performance of the HSoE can also be improved by ML. In particular, ML-based prediction can help improving the measurement matrix definition. To define the measurement matrix properly, it is indeed essential to have an accurate model that depends also on the knowledge of the loads. In this regard, ML can be useful to perform load forecasting, as discussed in [52], where an artificial neural network load forecasting is used to improve the accuracy of a distribution system state estimation technique.

Moreover, it is also possible to interpret the ℓ_1-optimization problem in (13.8) as a least absolute shrinkage operator (LASSO) regression [53], which was introduced by Tibishirani in 1996, and is often used in the ML applications to regularize linear estimation\regression problems by obtaining an automatic feature selection [54]. Both (13.8) and LASSO are based also on the definition of a regularization parameter. In the LASSO methodology, designed and applied to overdetermined problems, the choice of the parameter can be made through statistical techniques, such as cross-validation and using a large dataset, whereas in the HSoE methodology, as presented previously, the choice of the parameter is achieved, thanks to the estimation of the energy of the measurement error and exploiting the few measurements available on the field and any possible knowledge about the measurement error.

13.6 Conclusions

In this chapter, the topic of the identification and estimation of harmonic sources has been discussed and, in particular, a recent paradigm based on compressive sensing has been presented. The main idea is to find the prevailing sources of harmonic pollution, which are much fewer with respect to all the possible points of injection in a distribution network. This approach allows dealing with synchronized PQ monitoring devices, which, in distribution grids, are and will be, even in the foreseeable future, not widespread and able to define a redundant measurement system as it is instead for transmission networks. Nodes where aggregated loads are considered can be seen as the connection point of potential injectors of significant harmonic currents, representing the nonlinear behavior of modern loads and generators.

A methodology as the one presented is intended for the distribution system operator to perform a preliminary screening on the network looking for the source of harmonic disturbances to limit their negative effects.

Identification and, as a consequence, estimation performance depends on the available monitoring infrastructure, on the uncertainty associated with the measurements, on the network characteristics and conditions, but it is interesting to note that HSoE tools appear promising to raise alarms also in real-time, and allow further inspection on suspect portions of the network.

Refined measurements or ad hoc campaigns can thus be triggered by the presented method's outcomes, thus reducing the need for invasive, complicated, and long interventions.

References

[1] M.H.J. Bollen, What is power quality?, Electr. Power Syst. Res. 66 (1) (Jul. 2003) 5–14, https://doi.org/10.1016/S0378-7796(03)00067-1.

[2] https://ec.europa.eu/clima/eu-action/european-green-deal/2030-climate-target-plan_en.

[3] https://leonardo-energy.pl/wp-content/uploads/2017/08/Poor-power-quality-costs-european-business-more-than-150-billion-a-year.pdf.

[4] E.J. Davis, A.E. Emanuel, D.J. Pileggi, Harmonic pollution metering: theoretical considerations, IEEE Trans. Power Deliv. 15 (1) (Jan. 2000) 19–23, https://doi.org/10.1109/61.847223.

[5] J. Barros, R.I. Diego, M. de Apraíz, A discussion of new requirements for measurement of harmonic distortion in modern power supply systems, IEEE Trans. Instrum. Meas. 62 (8) (Aug. 2013) 2129–2139, https://doi.org/10.1109/TIM.2013.2267451.

[6] N. Locci, C. Muscas, S. Sulis, On the measurement of power-quality indexes for harmonic distortion in the presence of capacitors, IEEE Trans. Instrum. Meas. 56 (5) (Oct. 2007) 1871–1876, https://doi.org/10.1109/TIM.2007.903597.

[7] N. Locci, C. Muscas, S. Sulis, Detrimental effects of capacitors in distribution networks in the presence of harmonic pollution, IEEE Trans. Power Deliv. 22 (1) (Jan. 2007) 311–315, https://doi.org/10.1109/TPWRD.2006.877088.

[8] A. Cataliotti, V. Cosentino, S. Nuccio, Comparison of nonactive powers for the detection of dominant harmonic sources in power systems, IEEE Trans. Instrum. Meas. 57 (8) (Aug. 2008) 1554–1561, https://doi.org/10.1109/TIM.2008.925338.

[9] J. Mazumdar, R.G. Harley, F.C. Lambert, G.K. Venayagamoorthy, M.L. Page, Intelligent tool for determining the true harmonic current contribution of a customer in a power distribution network, IEEE Trans. Ind. Appl. 44 (5) (Sep. 2008) 1477–1485, https://doi.org/10.1109/TIA.2008.2002213.

[10] V.K. Tiwari, S.K. Jain, Hardware implementation of polyphase-decomposition-based wavelet filters for power system harmonics estimation, IEEE Trans. Instrum. Meas. 65 (7) (Jul. 2016) 1585–1595, https://doi.org/10.1109/TIM.2016.2540861.

[11] G.T. Heydt, Identification of harmonic sources by a state estimation technique, IEEE Trans. Power Deliv. 4 (1) (Jan. 1989) 569–576.

[12] A.P.S. Meliopoulos, F. Zhang, S. Zelingher, Power system harmonic state estimation, IEEE Trans. Power Deliv. 9 (3) (Jul. 1994) 1701–1709, https://doi.org/10.1109/61.311191.

[13] H. Liao, Power system harmonic state estimation and observability analysis via sparsity maximization, IEEE Trans. Power Syst. 22 (1) (Feb. 2007) 15–23, https://doi.org/10.1109/TPWRS.2006.887957.

[14] C.F.M. Almeida, N. Kagan, Harmonic state estimation through optimal monitoring systems, IEEE Trans. Smart Grid 4 (1) (Mar. 2013) 467–478, https://doi.org/10.1109/TSG.2012.2235472.

[15] C. Rakpenthai, S. Uatrongjit, N.R. Watson, S. Premrudeepreechacharn, On harmonic state estimation of power system with uncertain network parameters, IEEE Trans. Power Syst. 28 (4) (Nov. 2013) 4829–4838, https://doi.org/10.1109/TPWRS.2013.2273943.

[16] K.K.C. Yu, N.R. Watson, J. Arrillaga, Error analysis in static harmonic state estimation: a statistical approach, IEEE Trans. Power Deliv. 20 (2) (Apr. 2005) 1045–1050, https://doi.org/10.1109/TPWRD.2004.833895.
[17] A. Arefi, M.R. Haghifam, S.H. Fathi, T. Niknam, J. Olamaei, A novel algorithm based on honey bee mating optimization for distribution harmonic state estimation including distributed generators, in: 2009 IEEE Bucharest PowerTech, Jun. 2009, pp. 1–7.
[18] J.M.M. Ortega, A.G. Exposito, A.L.T. Garcia, M.B. Payan, A state estimation approach to harmonic polluting load characterization in distribution systems, IEEE Trans. Power Syst. 20 (2) (May 2005) 765–772, https://doi.org/10.1109/TPWRS.2005.846201.
[19] R. Sinvula, K.M. Abo-Al-Ez, M.T. Kahn, Harmonic source detection methods: a systematic literature review, IEEE Access 7 (2019) 74283–74299, https://doi.org/10.1109/ACCESS.2019.2921149.
[20] G. D'Antona, C. Muscas, S. Sulis, State estimation for the localization of harmonic sources in electric distribution systems, IEEE Trans. Instrum. Meas. 58 (5) (May 2009) 1462–1470, https://doi.org/10.1109/TIM.2009.2014504.
[21] G. D'Antona, C. Muscas, S. Sulis, Localization of nonlinear loads in electric systems through harmonic source estimation, IEEE Trans. Instrum. Meas. 60 (10) (Oct. 2011) 3423–3430, https://doi.org/10.1109/TIM.2011.2126830.
[22] G. D'Antona, C. Muscas, P.A. Pegoraro, S. Sulis, Harmonic source estimation in distribution systems, IEEE Trans. Instrum. Meas. 60 (10) (Oct. 2011) 3351–3359, https://doi.org/10.1109/TIM.2011.2126910.
[23] IEC, Electromagnetic compatibility (EMC) – Part 4-30: Testing and measurement techniques – Power quality measurement methods, 2015.
[24] IEEE Standard for Synchrophasor Measurements for Power Systems, IEEE Standard C37.118.1-2011, Dec. 2011.
[25] IEEE Standard for Synchrophasor Data Transfer for Power Systems, IEEE Standard C37.118.2-2011, 2-2011, Dec. 2011.
[26] IEEE/IEC International Standard – Measuring relays and protection equipment – Part 118-1: Synchrophasor for power systems – Measurements, in IEC/IEEE 60255-118-1:2018, pp.1-78, 19 Dec. 2018, https://doi.org/10.1109/IEEESTD.2018.8577045.
[27] https://www.etip-snet.eu/wp-content/uploads/2017/04/sra2035.pdf.
[28] P. Castello, C. Muscas, P.A. Pegoraro, S. Sulis, PMU's behavior with flicker-generating voltage fluctuations: an experimental analysis, Energies 12 (17) (2019) 3355, https://doi.org/10.3390/en12173355.
[29] P. Castello, J. Liu, C. Muscas, P.A. Pegoraro, F. Ponci, A. Monti, A fast and accurate PMU algorithm for P+M class measurement of synchrophasor and frequency, IEEE Trans. Instrum. Meas. 63 (12) (Dec. 2014) 2837–2845.
[30] G. Frigo, A. Derviskadic, P.A. Pegoraro, C. Muscas, M. Paolone, Harmonic phasor measurements in real-world PMU-based acquisitions, in: Proc. IEEE Int. Instrum. Meas. Technol. Conf. (I2MTC), Auckland, New Zealand, May 2019, pp. 1–6.
[31] M. Faifer, C. Laurano, R. Ottoboni, S. Toscani, M. Zanoni, Harmonic distortion compensation in voltage transformers for improved power quality measurements, IEEE Trans. Instrum. Meas. 68 (10) (Oct. 2019) 3823–3830.
[32] P. Castello, C. Laurano, C. Muscas, P.A. Pegoraro, S. Toscani, M. Zanoni, Harmonic synchrophasors measurement algorithms with embedded compensation of voltage transformer frequency response, IEEE Trans. Instrum. Meas. 70 (2021) 1–10, https://doi.org/10.1109/TIM.2020.3042317.
[33] IEEE Standard for a Precision Clock Synchronization Protocol for Networked Measurement and Control Systems, in: IEEE Std. 1588-2019 (Revision of IEEE Std 1588-2008), 16 June 2020, pp. 1–499, https://doi.org/10.1109/IEEESTD.2020.9120376.

[34] D. Carta, C. Muscas, P.A. Pegoraro, A.V. Solinas, S. Sulis, Compressive sensing-based harmonic sources identification in smart grids, IEEE Trans. Instrum. Meas. 70 (2021) 1–10, https://doi.org/10.1109/TIM.2020.3036753.
[35] JCGM, Evaluation of data – guide to the expression of uncertainty in measurement, JCGM 100 (Sep. 2008) 2008.
[36] J.A. Tropp, Greed is good: algorithmic results for sparse approximation, IEEE Trans. Inf. Theory 50 (10) (2004) 2231–2242.
[37] E.J. Candès, T. Tao, Decoding by linear programming, IEEE Trans. Inf. Theory 51 (12) (Dec. 2005) 4203–4215.
[38] E.J. Candès, J. Romberg, T. Tao, Robust uncertainty principles: exact signal reconstruction from highly incomplete frequency information, IEEE Trans. Inf. Theory 52 (2) (2006) 489–509.
[39] D. Carta, C. Muscas, P.A. Pegoraro, S. Sulis, Harmonics detector in distribution systems based on compressive sensing, in: 2017 IEEE International Workshop on Applied Measurements for Power Systems (AMPS), Sep. 2017, pp. 1–5.
[40] D. Carta, C. Muscas, P.A. Pegoraro, S. Sulis, Identification and estimation of harmonic sources based on compressive sensing, IEEE Trans. Instrum. Meas. 68 (1) (Jan. 2019) 95–104, https://doi.org/10.1109/TIM.2018.2838738.
[41] D. Carta, C. Muscas, P.A. Pegoraro, A. Vincenzo Solinas, S. Sulis, Impact of measurement uncertainties on compressive sensing-based harmonic source estimation algorithms, in: 2020 IEEE International Instrumentation and Measurement Technology Conference (I2MTC), May 2020, pp. 1–6.
[42] E.J. Candès, J.K. Romberg, T. Tao, Stable signal recovery from incomplete and inaccurate measurements, Commun. Pure Appl. Math. 59 (8) (2006) 1207–1223.
[43] IEEE PES Distribution Test Feeders, Online, Available, https://cmte.ieee.org/pes-testfeeders/resources/.
[44] IEC, Electromagnetic Compatibility (EMC) Part 3–6: Limits Assessment of Emission Limits for the Connection of Distorting Installations to MV, HV and EHV Power Systems, document IEC TR 61000-3-6:2008, Feb. 2008.
[45] IEC. Electromagnetic Compatibility (EMC)—Part 4–7: Testing and Measurement Techniques—General Guide on Harmonics and Interharmonics Measurements and Instrumentation, for Power Supply Systems and Equipment Connected Thereto, document IEC 61000-4-7:2002+AMD1:2008 CSV, Oct. 2009.
[46] IEC. Instrument Transformers—Part 1: General Requirements, IEC Standard 61869-1 eD 2, Draft for approval, forecast publication date 2023.
[47] W. Zine, Z. Makni, E. Monmasson, L. Idkhajine, B. Condamin, Interests and limits of machine learning-based neural networks for rotor position estimation in EV traction drives, IEEE Trans. Ind. Inform. 14 (5) (May 2018) 1942–1951.
[48] G.O. Sahinoglu, M. Pajovic, Z. Sahinoglu, Y. Wang, P.V. Orlik, T. Wada, Battery state-of-charge estimation based on regular/recurrent Gaussian process regression, IEEE Trans. Ind. Electron. 65 (5) (May 2018) 4311–4321.
[49] X. Zhang, Y. Qin, C. Yuen, L. Jayasinghe, X. Liu, Time-series regeneration with convolutional recurrent generative adversarial network for remaining useful life estimation, IEEE Trans. Ind. Inform. 17 (10) (Oct. 2021) 6820–6831.
[50] T. Skrjanc, R. Mihalic, U. Rudez, A non-intrusive approach for enhancing power-system frequency stability, IEEE Trans. Power Syst. (2021), https://doi.org/10.1109/TPWRS.2021.3131098.
[51] A.S. Zamzam, N.D. Sidiropoulos, Physics-aware neural networks for distribution system state estimation, IEEE Trans. Power Syst. 35 (6) (Nov. 2020) 4347–4356.

[52] S. Carcangiu, A. Fanni, P.A. Pegoraro, G. Sias, S. Sulis, Forecasting-aided monitoring for the distribution system state estimation, in: Complexity, 2020.
[53] R. Tibshirani, Regression shrinkage and selection via the lasso, J. R. Stat. Soc. B 58 (1) (1996) 267–288, [Online]. Available http://www.jstor.org/stable/2346178.
[54] T. Hastie, R. Tibshirani, J. Friedman, The Elements of Statistical Learning: Data Mining, Inference and Prediction, 2nd ed., Springer, 2009, [Online]. Available, http://www-stat.stanford.edu/~tibs/ElemStatLearn/.

Index

E

F

FSC
www.fsc.org
MIX
Papier | Fördert gute Waldnutzung
FSC® C083411

Kolibri 360 GmbH
Ferdinand-Jühlke-Straße 7
99095 Erfurt, Deutschland
produktsicherheit@kolibri360.de